57

Topics in Heterocyclic Chemistry

Aims and Scope

The series Topics in Heterocyclic Chemistry presents critical reviews on present and future trends in the research of heterocyclic compounds. Overall the scope is to cover topics dealing with all areas within heterocyclic chemistry, both experimental and theoretical, of interest to the general heterocyclic chemistry community.

The series consists of topic related volumes edited by renowned editors with contributions of experts in the field. All chapters from Topics in Heterocyclic Chemistry are published OnlineFirst with an individual DOI. In references, Topics in Heterocyclic Chemistry is abbreviated as Top Heterocycl Chem and cited as a journal.

More information about this series at http://www.springer.com/series/7081

László Somsák
Editor

Carbohydrate-spiro-heterocycles

With contributions by

J.-B. Behr · P. Compain · D. Hazelard · R. Hensienne · S. Josse · A. Martín · N. Pellegrini-Moïse · M. Pommier · D. Postel · M. Richard · H. Rodríguez-Solla · R. Soengas · M. Soto · P. R. Sridhar · E. Suárez · S. Vidal · Y. Yamamoto

Springer

Editor
László Somsák
Department of Organic Chemistry
University of Debrecen
Debrecen, Hungary

ISSN 1861-9282 ISSN 1861-9290 (electronic)
Topics in Heterocyclic Chemistry
ISBN 978-3-030-31941-0 ISBN 978-3-030-31942-7 (eBook)
https://doi.org/10.1007/978-3-030-31942-7

This Springer imprint is published by the registered company Springer Nature Switzerland AG.
The registered company address is: Gewerbestrasse 11, 6330 Cham, Switzerland

Preface

Spiro compounds that contain a bicyclic or tricyclic system to share a single atom between two rings represent an increasing class of molecules of both natural and synthetic origin. These compounds exhibit various biological activities and can be used in materials science applications due to, e.g. photochromism and hole transporting abilities. The unique and well-defined 3D structure of spiro derivatives makes them attractive targets in drug discovery, offering a deviation from the traditional aromatic/heterocyclic "flatland" chemistry. The conformational constraints resulting from the spiro structure lend rigidity to these molecules, and this may end up with a diminishing entropic penalty during binding to the biological targets. The spirocyclic constructs can also meet the needs of discovering new regions of chemical space and thereby facilitate uncovering new properties, biological effects as well as medical and other applications. In addition, these scaffolds may be advantageous in view of patentability as well. On the other hand, the construction of spirocycles especially in terms of regio- and stereoselectivities poses a highly provoking task to the preparative chemist. Due to the above important and interesting properties and the demanding synthetic challenges, the spiro compounds attract more and more attention both in academia and industry that is reflected in the appearance of five to six thousands of primary publications and patents per annum during the last decades. Furthermore, the utility of spirocyclic derivatives as commodities is demonstrated by marketed drugs, ophthalmic lenses and sunglasses, auxiliary compounds in stereoselective syntheses, electronic displays, optical data storage devices, etc. Several other utilizations towards, e.g. new medications, chemical biosensing, controlled release drug delivery, molecular switches and solar cells are in a developmental phase.

Carbohydrates are ubiquitous molecules in nature and participate in a vast number of biological interactions. Their conjugates, including all kinds of primary and secondary metabolic small molecules and also biomacromolecules, represent valuable tools for glycobiology research and also lead compounds for drug discovery. While monosaccharides per se appear as heterocycles, their natural conjugates frequently exhibit spiro(hetero)cyclic derivatives, in many cases of high therapeutic relevance. Well-known carbohydrate-spiro-heterocycles are, e.g. the antifungal papulacandins, the antibiotic orthosomycins, the herbicidal hydantocidin, each of natural origin, and the synthetic tofogliflozin, the active ingredient of approved antidiabetic medications.

Monosaccharides with their multiple stereogenic centres and various intramolecular interactions involving the substituent groups on the sugar ring make the formation of spirocycles on such a skeleton even more challenging in controlling selectivities. Thus, the outcomes of a particular spirocyclization may well depend on the sugar moieties' stereochemical and conformational peculiarities resulting in different products or product ratios when the sugar is changed.

This book as a whole as well as its individual chapters intends to give an insight into the world of carbohydrate-spiro-heterocycles from various perspectives. In the introductory chapter, the cyclization methodologies to form a spiro-fused ring at the anomeric carbon of pyranoid sugars are categorized and a selection is presented by *Pommier* and *Vidal*, who also highlight some important biomedical applications of such compounds.

Specific methods of spiro ring formation are emphasized in the next chapters. *Pellegrini-Moïse* and *Richard* highlight 1,3-dipolar cycloadditions to form carbohydrate-derived spiro-isoxazolines and spiro-isoxazolidines and their transformations into other interesting compounds. *Martín* and *Suárez* have compiled a plethora of radical reactions to demonstrate their unique potential and versatility to achieve a wide range of spirocycles on both pyranoid and furanoid sugar units. The formation of spiroketals and related lactones based on unsaturated monosaccharide derivatives such as *endo*- and *exo*-glycals is surveyed by *Sridhar*. *Josse* and *Postel* summarize the uses of sugar-derived cyanohydrins and α-aminonitriles for the formation of spirocycles at ring positions of carbohydrates also involving biomedically outstanding derivatives.

Some important compound types are reviewed in the remaining chapters. The multifaceted chemistry of spirocyclic nucleosides is overviewed by *Soto*, *Rodríguez-Solla* and *Soengas* also pointing out their biological utility. *Yamamoto* presents the syntheses and uses of phthalane spiro-*C*-glycosidic compounds including papulacandins and tofogliflozin. Finally, a special and emerging type of glycomimetics, the spiro-iminosugars, is surveyed by *Hazelard*, *Hensienne*, *Behr* and *Compain*.

I greatly appreciate the meticulous work of the contributors, and I am also indebted to those colleagues from all over the world who voluntarily reviewed the manuscripts, thereby providing invaluable help in the editorial work.

Debrecen, Hungary
March 2019

László Somsák

Contents

Top Heterocycl Chem (2019) 57: 1–26
DOI: 10.1007/7081_2019_33

Published online: 22 August 2019

Anomeric Spiro-Annulated Glycopyranosides: An Overview of Synthetic Methodologies and Biological Applications

Maxime Pommier and Sébastien Vidal

Contents

Abstract Organic chemistry developed a series of synthetic strategies toward spiro-annulated carbohydrates as potential pharmaceutical drugs or developed new organic synthetic methodologies. The present chapter gives a general overview of the spiro-annulation of carbohydrates at the anomeric position. The main synthetic strategies can be summarized in five paths. Intramolecular cyclizations can be performed through two short tethers with their reactive ends generating the

M. Pommier and S. Vidal (✉)
Institut de Chimie et Biochimie Moléculaires et Supramoléculaires (UMR 5246),
Université Claude Bernard Lyon 1 and CNRS, Bâtiment Lederer, Villeurbanne, France
e-mail: sebastien.vidal@univ-lyon1.fr

spirocycle or through a single tether reacting at the anomeric position for cyclization. The three other strategies rely on intermolecular reactions with a portion of the spirocycle only in the external substrate or also on the carbohydrate. Radical-mediated cyclization and cycloaddition reactions are the main strategies toward spiro-annulated carbohydrates. A special attention is paid to discussion of the stereocontrol of the anomeric configuration and also to yields in industrial syntheses or biological activities of the molecules. A specific attention is devoted to tofogliflozin and glycogen phosphorylase inhibitors both used as antihyperglycemic drugs and drug candidates, respectively.

Keywords 1,3-Dipolar cycloaddition · Cycloaddition · Glycogen phosphorylase · Hydrogen atom transfer (HAT) · Medicinal chemistry · Radical cyclization · Ring-closing metathesis · SGLT2 · Spiroketal · Spiro-lactam · Type 2 diabetes

1 Introduction

Carbohydrates are a major class of natural products with biological implications in bacterial or viral infections, cancer signaling or metastasis, and inflammation. Chemists have intensively investigated the synthesis of not only oligosaccharides for applications in vaccines but also mimetics of monosaccharides for the design of potential drugs or in a medicinal chemistry approach. These efforts have generated literature on the protecting group strategies [1] and also glycosylation methodologies [2, 3] for the optimal syntheses of such natural oligosaccharides or glycomimetics. Spiro-annulated carbohydrates are the focus of investigations to design pharmaceutical drugs or to develop new synthetic approaches for their preparation. Natural products such as papulacandin or the recently approved synthetic tofogliflozin as an antihyperglycemic drug used for the treatment of type 2 diabetes are some leading examples in this series (Fig. 1).

Fig. 1 Structure of papulacandin D and tofogliflozin

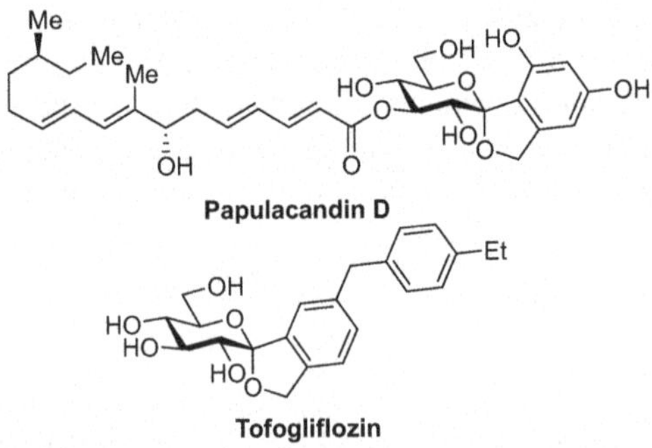

The present chapter is prepared as a general overview of the synthetic strategies reported for the spiro-annulation of carbohydrates at the anomeric position. Typical examples have been selected, and the discussion is not intended to be an exhaustive description of the topic. A specific focus on spiro-annulation of hexopyranoses at the anomeric position will also explain the selected examples described herein, and several additional strategies might be possible toward the same spiro-glycoside discussed. For instance, synthetic strategies based on pyranoside ring closure will not be discussed herein.

The main synthetic strategies toward spiro-annulated glycosides can be summarized in five paths (Fig. 2). The cyclization is the key step reaction and can be performed by pre-installing two reactive moieties (X and Y) with a short tether at the anomeric position (Path A) or a single arm of the spirocycle can be incorporated at the anomeric carbon atom with a reactive end (Y) that will cyclize at the anomeric position (X) (Path B). Paths A and B are intramolecular strategies which are typically more efficient in terms of yields. While Path A will have the stereochemistry of the anomeric center pre-defined in the starting material, the stereoselectivity will need to be controlled under Path B. The other strategies rely on intermolecular reactions with a portion of the spirocycle only in the external substrate (Path C) or also on the carbohydrate (Path D). In both Path C and D, the stereochemistry at the anomeric position will have to be controlled. Finally, another strategy creating two bonds (Path E) will not involve the creation of the stereogenic anomeric center with the functionalities already present at the anomeric carbon atom but with addition of a new molecular entity to generate the spirocyclic system. Paths A and E are hence very similar but differ only in the number of chemical bonds created during the cyclization process.

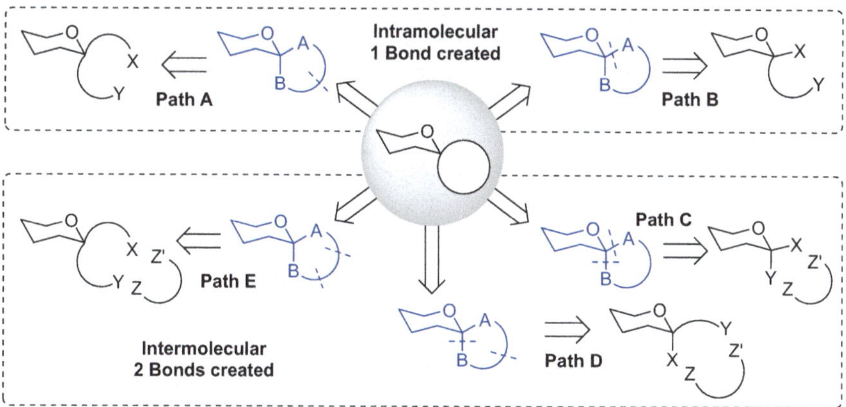

Fig. 2 Spiro-annulation strategies at the anomeric position

2 Intramolecular Strategies Toward Spiro-Annulated Carbohydrates

2.1 Through Two Pre-installed Tethers and Without Stereoselective Outcome (Path A)

2.1.1 Ring-Closing Metathesis

Ring-closing metathesis is one of the most common and applied cyclization strategies wherein polymerization can be controlled over cyclization by using diluted solutions of substrates and low catalyst loadings. Anomeric 3,3′-(glycopyranosylidene)bis(1-propene) (e.g. **1**) could be readily converted into the spiro-cyclopentene derivatives **2** via ring-closing metathesis using Grubbs I catalyst from acetylated D-*gluco*-, D-*manno*-, or D-*galacto*-configured precursors (e.g., **1**) [4] or Grubbs II catalyst from the benzyl-protected D-*gluco*- or D-*galacto*-configured bis-propenes (e.g., **1a**) toward the corresponding spiro compounds (e.g., **2a**) [5] (Scheme 1).

In a similar approach, the synthesis of spiro-annulated isofagomine analogue **5** could be performed through RCM cyclization of the precursor *C*-vinyl glucoside **3** followed by dihydroxylation of the double bond of intermediate **4** and unmasking the protecting groups (Scheme 2) [6]. Spiro-azepanes could also be obtained from the corresponding *C*-allyl glucoside. Some of these compounds were identified as selective yet moderate α-mannosidase inhibitors.

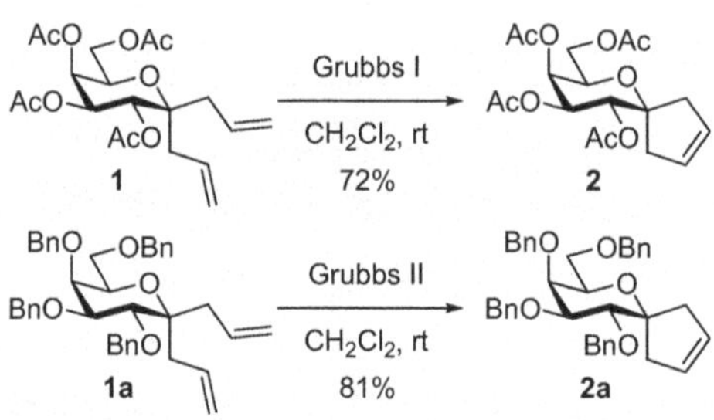

Scheme 1 Spiro-cyclization from bis-*C,C*-allyl glycoside via ring-closing metathesis

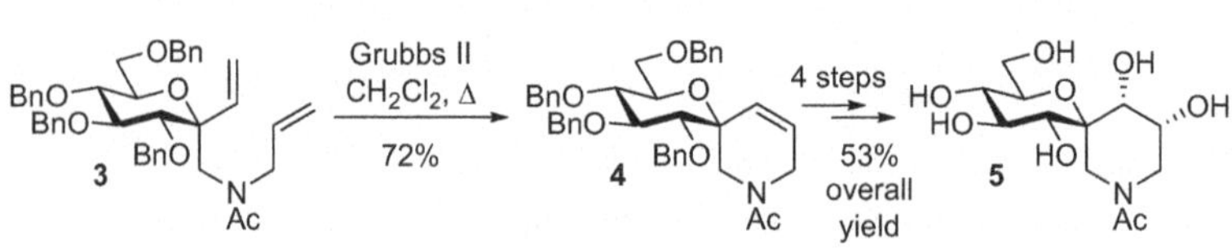

Scheme 2 RCM-mediated spiro-cyclization toward amino sugars

2.1.2 Radical-Mediated Ring Closure

Intramolecular addition of a radical to an allyl glycoside was reported from an (iodo-deoxy) *O*-allyl ketoside **6** (Scheme 3) [7]. Tributyltin hydride was used as the hydrogen donor, while the radical reaction was triggered thermally using AIBN. The stereochemistry of the anomeric carbon was unchanged from that of the starting material **6**, while the configuration of the carbon in the tetrahydrofuran ring bearing the methyl group was clearly identified as depicted from further chemical derivatizations and NMR characterizations. This single isomer **7** was rationalized through a chair-transition-state model and based on the anomeric effects.

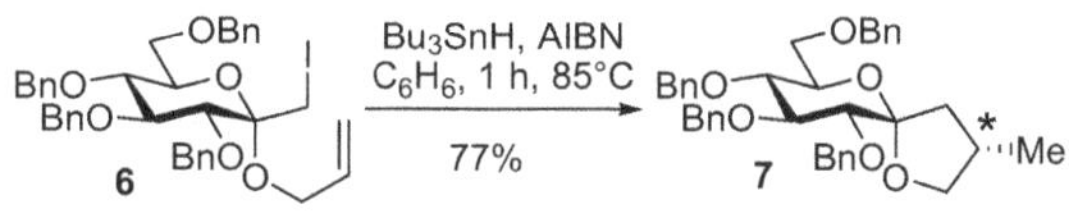

Scheme 3 Spiro-cyclization by intramolecular radical-mediated addition

2.1.3 Spiro-Annulation Toward Crown Ethers

A 15-crown-5 ether incorporating a spiro-glucoside was synthesized through a lactonization of the carboxylic acid **8** to the lactone **9** (Scheme 4) [8] with potential phase transfer catalytic properties and providing access to chiral crown ether systems. Larger crown ethers could be obtained from glycosylation cyclization strategies [9]. The 15-crown-5 derivative **11** incorporating two spiro-glycosylidene units was prepared from the corresponding disaccharide **10** through Fraser-Reid glycosylation. The corresponding deprotected chiral crown ethers and larger macrocycles (incorporating more carbohydrate moieties up to four) were then evaluated as asymmetric ligands in Cram model phenyl acetate-acrylate addition with only moderate stereocontrol.

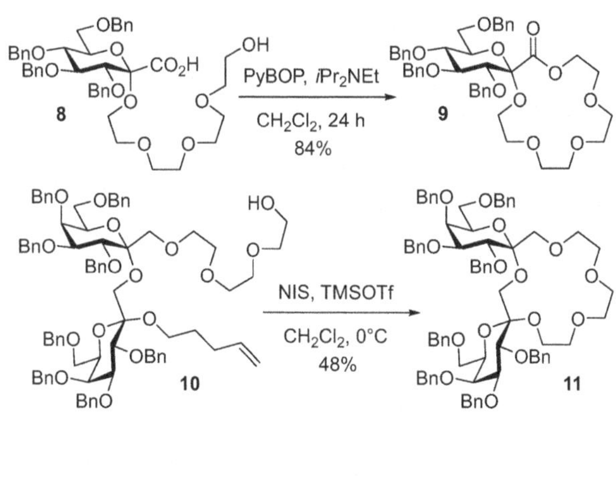

Scheme 4 Cyclization toward spirketal crown ethers

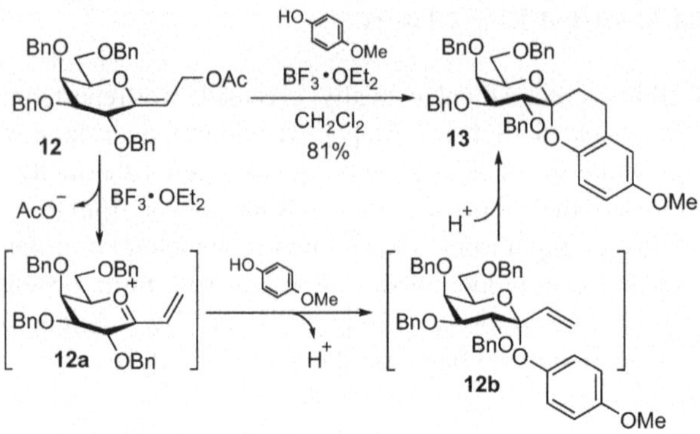

Scheme 5 Spiro-cyclization through a Ferrier-type rearrangement and Friedel-Crafts reaction toward a spiro-pyran

2.1.4 Ferrier-Type Rearrangement and Friedel-Crafts Strategy

A two-step process starting from the *exo*-galactal derivative **12** was designed to access six-membered spiro-annulated galactosides [10] (Scheme 5). The activation of the allyl acetate with boron trifluoride etherate generated a *C*-vinyl oxonium intermediate **12a** under a Ferrier-type rearrangement. Nucleophilic substitution with 4-methoxy-phenol led to the *C*-vinyl *O*-galactoside intermediate **12b** which upon acid activation underwent an intramolecular Friedel-Crafts reaction to lead to the benzo-fused spiro-galactoside **13**.

Access to benzo-fused spiro-cyclopentane *C*-glycosides **16** was also reported starting from the *exo*-glycal **14** through Heck-type *C*-arylation to obtain the aldehyde **15** (Scheme 6) [11]. Subsequent Friedel-Crafts spiro-cyclization occurred in the presence of a thiol under $InCl_3$ catalysis. Activation of the aldehyde led to intermediate **15a**, and addition of the butanethiol provided the thioacetal **15b**. Elimination of the catalyst generated the sulfonium cation **15c** which underwent intramolecular Friedel-Crafts-type cyclization to the oxonium ion **15d**. Elimination of a proton afforded the desired spiro-glycoside **16**. The presence of an electron donating group in the *meta* position (*para* to the created bond) on the aromatic ring was required to allow the cyclization. While the Heck-type *C*-arylation was highly α-stereoselective, a 3:1 diastereoisomeric ratio was observed for the thioether stereogenic center generated in the reaction.

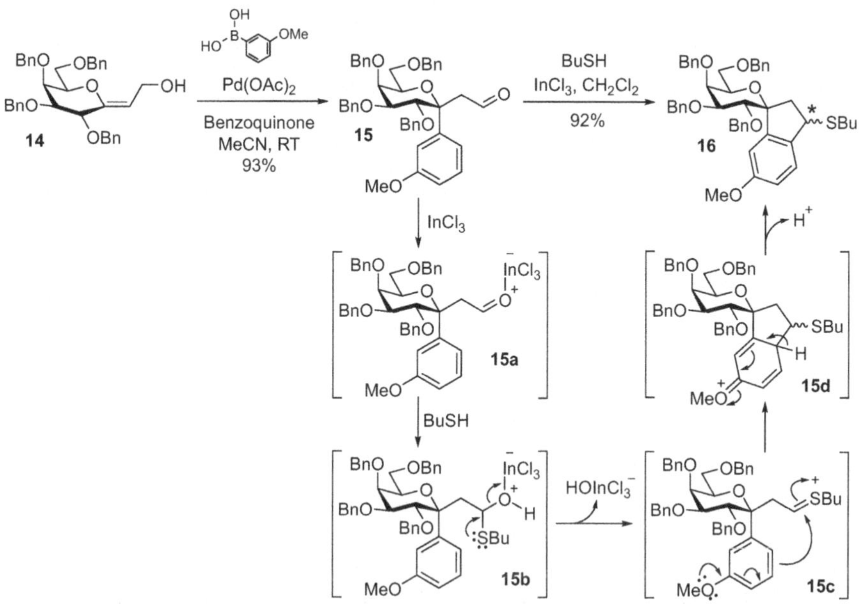

Scheme 6 Spiro-cyclization through a Ferrier-type rearrangement and Friedel-Crafts reaction toward a benzo-fused spiro-pentane

2.1.5 Spiro-Lactamization

The ketosyl azide **17** was reacted under reductive conditions to generate in situ a hemiaminal whose amine moiety led to the spiro-lactams **18α** and **18β** in equimolar ratio (Scheme 7) [12, 13]. Formation of the spiro-epimers can be explained by the anomerization of the hemiaminal intermediate similar to mutarotation for hemi-acetals. Six-membered rings could be obtained using the same strategy.

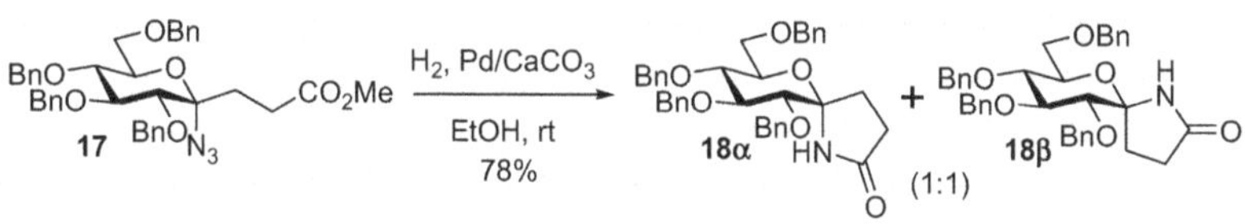

Scheme 7 Spiro-lactamization from a ketosyl azide

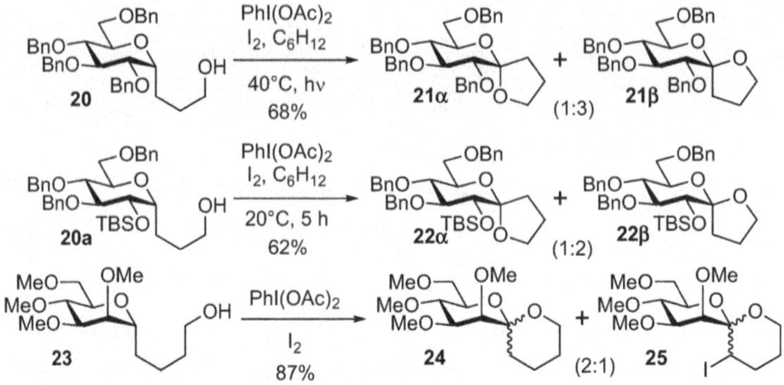

Scheme 8 Spiro-cyclization through hydrogen atom transfer (HAT) to alkoxyl radicals

2.2 *Through One Pre-installed Tether and with Stereoselective Outcome (Path B)*

2.2.1 Radical-Mediated Spiro-Annulation

Intramolecular hydrogen atom transfer (HAT) to alkoxyl radicals from the anomeric position of a *C*-glucoside **20** appeared as a reliable access to spiro-furans **21α** and **21β** with moderate stereocontrol (1:3 ratio) in a short synthetic sequence and in good yield (Scheme 8) [14]. The protecting group pattern around the pyranose ring was found to influence the outcome of the reaction since a 3,4,6-tri-*O*-benzyl-2-*O*-*t*-butyldimethylsilyl-*C*-glucopyranoside **20a** gave an α/β 1:2 ratio of spiro-furans **22α** and **22β** with 62% yield [15]. On the other hand, the *O*-permethylated *C*-mannoside **23** gave four compounds as epimeric mixtures of spiro-pyran **24** and 3-iodinated spiro-pyran **25** [16].

The same HAT strategy could be applied to phosphoramidate **26** through a nitrogen-centered radical (Scheme 9) [17]. The hydrogen abstraction involves the β-hydrogen atom at the anomeric carbon atom, while the nucleophilic cyclization of the nitrogen moiety occurs on the less hindered α-face leading to the spiro-aminal **27**. The *C*-mannosyl amide **28** could also react under HAT conditions to lead to the spiro-lactam **29** in good yield and with high stereocontrol [18]. It is worth pointing out that in the case of such amide used as the nucleophile in the HAT cyclization, the *O*- or *N*-cyclization product can be observed, but the *O*-cyclization would occur only with carba-sugars (carbon atom replacing the endocyclic oxygen), and the pyranoside would selectively provide the spiro-lactams [18].

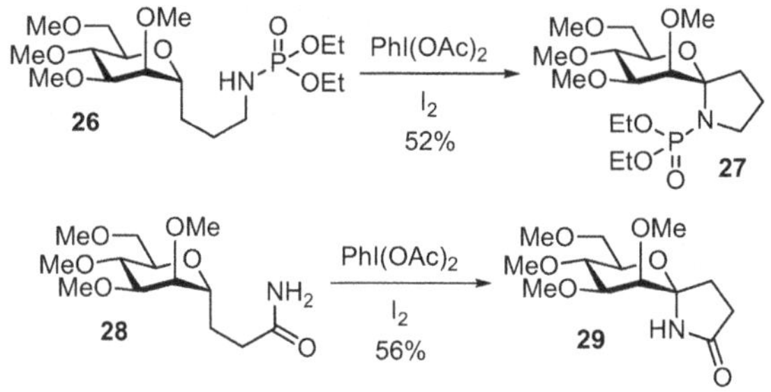

Scheme 9 Spiro-cyclization through hydrogen atom transfer (HAT) to nitrogen-centered radicals

2.2.2 Spiro-Annulation Through C-H Activation

An anomeric C-H activation of a *C*-(2-deoxy-glycosyl) carboxamide **30** afforded a spiro-oxindole derivative **31** with a moderate diastereoselectivity (dr = 3:2) (Scheme 10) [19].

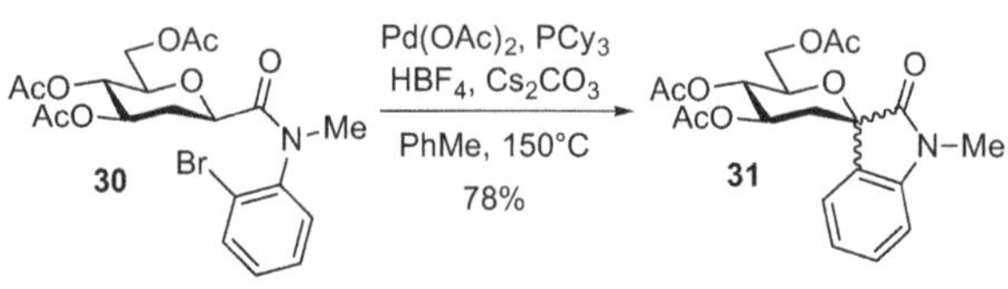

Scheme 10 Spiro-cyclization through anomeric C-H activation

2.2.3 Spiro-Lactonization on a Sialic Acid Scaffold

SmI_2-Mediated intramolecular spiro-cyclization of the thiosialoside **32** provided a robust access to the sialic acid-derived spiro-δ-lactone **33** as a 1:1 mixture of diastereoisomers on the newly created stereogenic carbon atom (Scheme 11) [20]. Five- and seven-membered ring lactones could not be obtained through this strategy probably due to steric constraints of the samarium enolate intermediate.

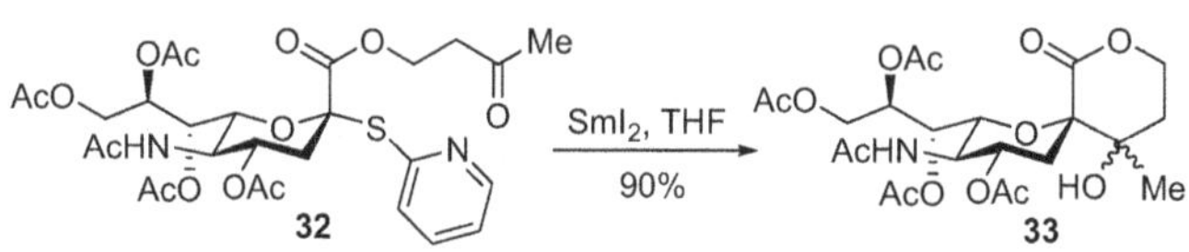

Scheme 11 Spiro-cyclization from sialic acid scaffolds

3 Intermolecular Synthetic Strategies Toward Spiro-Annulated Carbohydrates

3.1 Strategies Toward Spiro-Annulated Derivatives with Three-Membered Rings (Path C)

Glycosylidene-spiro-diazirines **36** are precursors of glycosylidene carbenes which can then be reacted with various electrophiles. The synthesis of such diazirines started from [(glycosylidene)amino]methanesulfonates **34** to the diaziridine **35** (actually a Path D-type reaction) which were then oxidized to the diazirines **36** (Scheme 12) [21]. Both compounds **35** and **36** could be crystallized from the reaction mixtures. The glycosylidene diaziridines **35** could be also characterized through its crystallographic structure and displayed slightly shorter C-N bond lengths and slightly longer N-N bond length when compared to noncyclic C-N and N-N bond lengths [22].

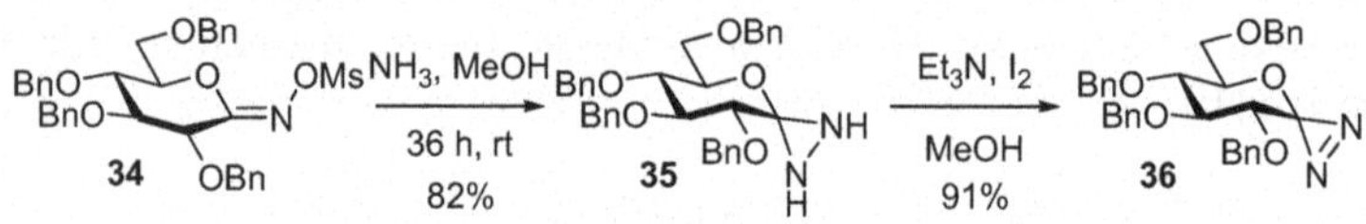

Scheme 12 Synthesis of glycopyranosylidene-spiro-diazirines

Such glycosylidene carbene precursors could then be applied to the synthesis of glycosylidene cyclopropane derivatives (Scheme 13). The reaction of diazirine **36** under photolysis conditions provided a mixture of all four possible regio- and stereoisomers **37a** and **37b** [23, 24]. The same cyclopropane derivatives **37a–b** could be obtained by the photolysis of tosylhydrazone salt **38** [23, 25]. The acetylated spiro-cyclopropanes **40a** and **40b** were prepared from the diazido-glucoside **39** as a mixture of four diastereoisomers [26]. Such glycosylidene cyclopropane scaffolds were identified as glycosidase inhibitors [27].

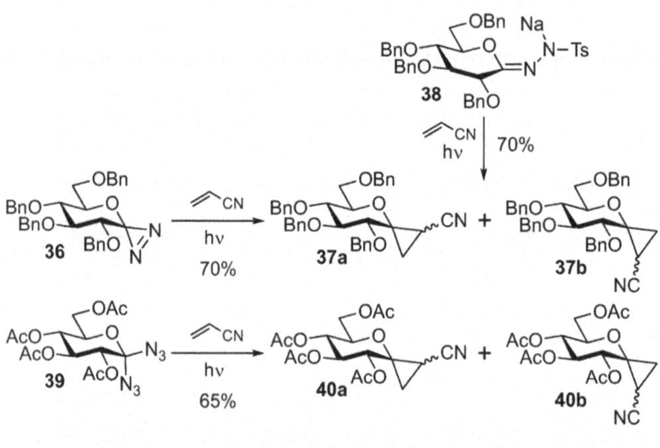

Scheme 13 Synthesis of glycopyranosylidene-spiro-cyclopropanes under photochemical conditions

Glycosylidene cyclopropanes could also be accessed by [2 + 1] cycloaddition of ethyl diazoacetate with *exo*-glycal **41** (although through Path D) to afford the corresponding cyclopropane derivative **42** (Scheme 14) [28]. Although the stereochemistry of the cycloaddition could not be controlled, the reduction and then rearrangement under 2-iodoxybenzoic acid (IBX) activation led to the spiro-furan derivative **43** again without stereocontrol.

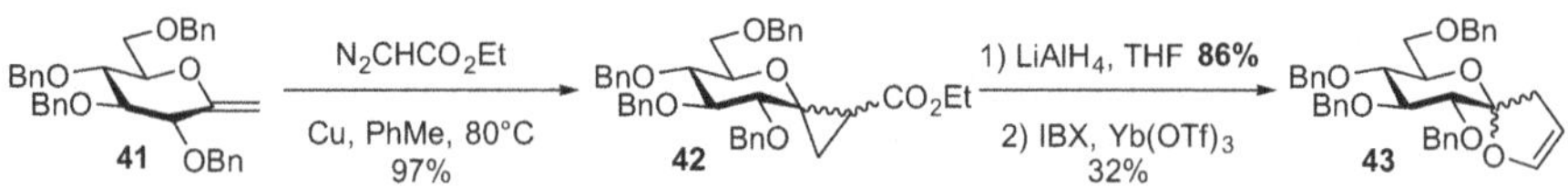

Scheme 14 Synthesis of glycopyranosylidene-spiro-cyclopropanes from *exo*-glycal and conversion into spiro-furan derivatives

The reaction of diazirine **36** under either thermolysis [29] or photolysis conditions in the presence of acetone led to the spiro-epoxide **44** without stereocontrol (Scheme 15) [30]. Similar epoxides could be obtained from the oxidation of *exo*-glycal **41** with DMDO (reaction type of Path D) to afford epoxide **45** as a mixture of diastereomers [31]. Reaction of the anion, obtained by lithiation of methyl bromoacetate, with the benzylated gluconolactone **46** afforded yet another pathway to the epoxide **47** with good stereocontrol (path D) [32–36].

Scheme 15 Synthetic strategies toward glycopyranosylidene-spiro-epoxides

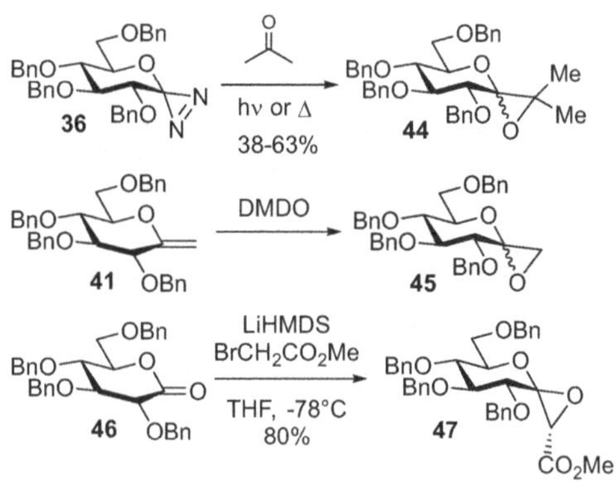

3.2 *Strategies from C-1* gem-*Di-Activated Pyranosides*

gem-Bromo-chloro glucopyranose **48** was reacted with ethylene glycol in the presence of silver triflate and *sym*-collidine used as a base to generate a first "glycosylation" with an intermolecular substitution of alcohol at the anomeric carbon atom, followed by an intramolecular second "glycosylation" to obtain the orthoester **49a** (Scheme 16) [37, 38]. The same reaction when performed with catechol required the absence of *sym*-collidine to afford the benzo-fused spiro-orthoester **50**. *sym*-Collidine was suspected to oxidize catechol to the corresponding *o*-quinone. Inter- and intramolecular reactions are competing in this cyclization process as verified by the 65% yield of compound **49a** obtained with ten equivalents of ethylene glycol while a larger excess (90 equivalents) provided a much lower yield (9%) of the same compound, while the bis-ethylene glycol product represented 43% yield. A similar approach from the bromo-cyano-galactose **51** exemplified the use of cyanide as a leaving group to obtain the spiro-galactoside **49b** [39]. The access to the 2-deoxy series **49c** was reported from the corresponding 2-deoxy-1-bromo-glucosyl cyanide [39] and also from tri-*O*-acetyl-glucal **52** under Pd/Cu catalysis via a Wacker-type mechanism [40].

Scheme 16 Access to spiro-orthoesters (or ortholactones) from *gem*-di-activated glycosides glycals (ligand = *N*,*N*,*N*′, *N*′-tetramethyl-1,2-*trans*-cyclohexanediamine)

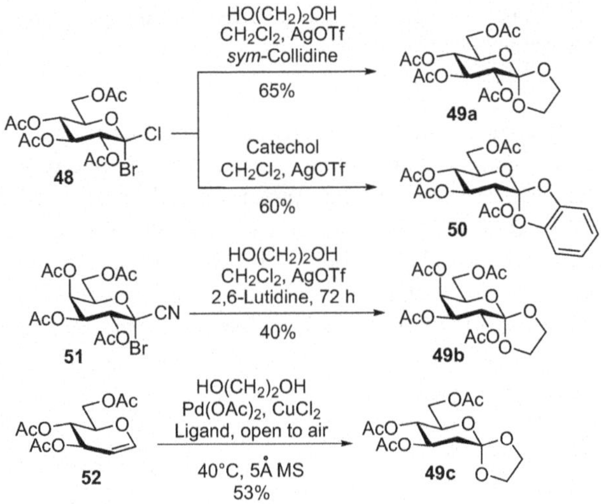

3.3 Through a Tether and Activation at the Anomeric Position (Path D)

The peracetylated sialic acid **53** was reacted with carbodiimides to afford spiro-hydantoins **54** in high yields (Scheme 17) [41]. The reaction proceeded through a one-pot sequential process with carbodiimides adding to the carboxylic acid and then boron trifluoride etherate induced intramolecular *N*-sialylation with high α-selectivity.

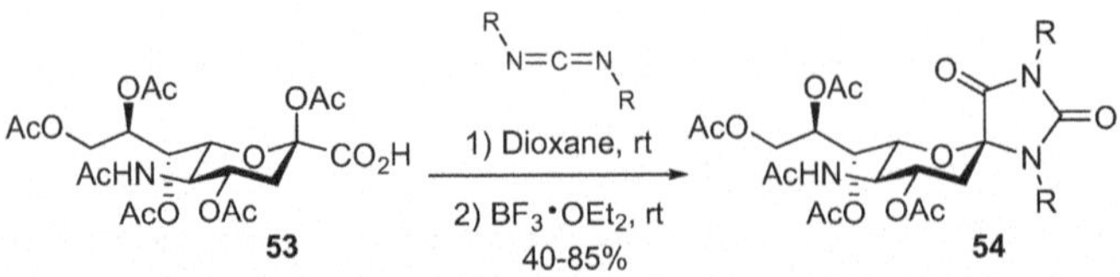

Scheme 17 Spiro-cyclization from sialic acid scaffolds

3.4 *Through Two Tethers (Path E)*

3.4.1 Cycloaddition Reactions

The [2 + 2 + 2] cycloaddition strategy of alkynes could be readily applied for the synthesis of spiro-annulated glucosides under either rhodium [42] or ruthenium catalysis [43]. The cycloaddition of the propargyl ketoside with a 1,2-triple bond **55** with acetylene afforded the desired benzo-fused spiro-glucoside **56** in high yields (Scheme 18). Technically, this reaction scheme could also be classified under Path A when considering the creation of only one carbon-carbon bond in the benzene ring.

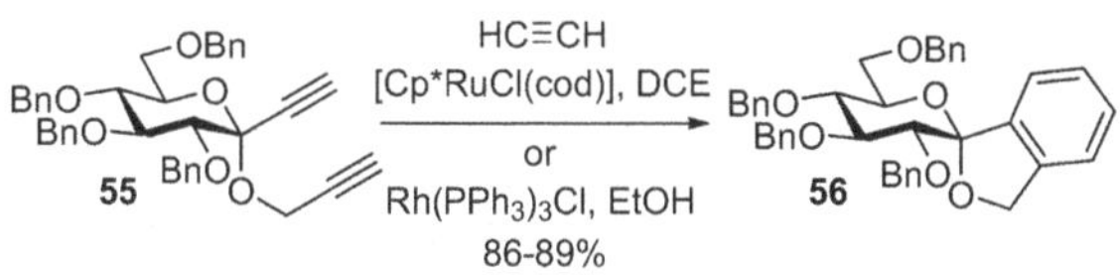

Scheme 18 [2 + 2 + 2] spiro-annulation

A Diels-Alder [4 + 2] cycloaddition strategy was described for the synthesis of a six-membered ring spiro-glucoside (Scheme 19) [44]. The *exo*-glucal **41** was used as a dienophile with the diene **58** generated in situ by simple treatment of the diketone **57** with a weak base (pyridine) to trigger the elimination of phthalimide. The mixture of epimers **59α** and **59β** was obtained in 91% yield although with 2.5:1 ratio of separable isomers.

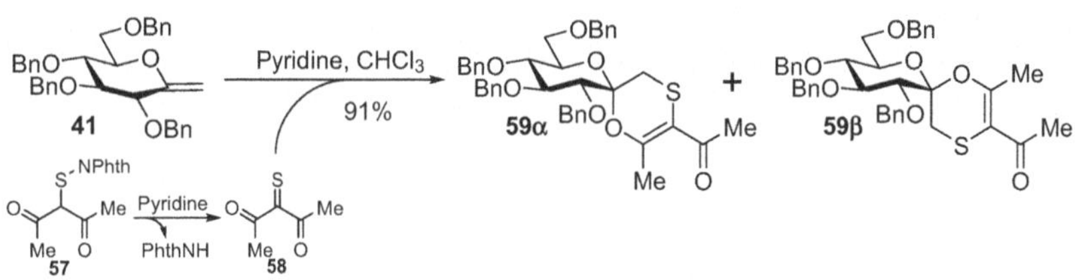

Scheme 19 [4 + 2] spiro-annulation from an *exo*-glycal

3.4.2 Triphosgene Addition

Five-membered heterocyclic spiro-bicyclic glycosides can be readily obtained by ring closure on the amino-deoxy ketose **60** using a phosgene equivalent (triphosgene) (Scheme 20) [45, 46]. The conjugation of triphosgene with both the hemiacetal alcohol and the amine moieties provided the spiro-bicyclic carbamate **61** in high yields with a series of substituents on the amine group. It is worth pointing out that this cyclization occurred on the unprotected carbohydrate.

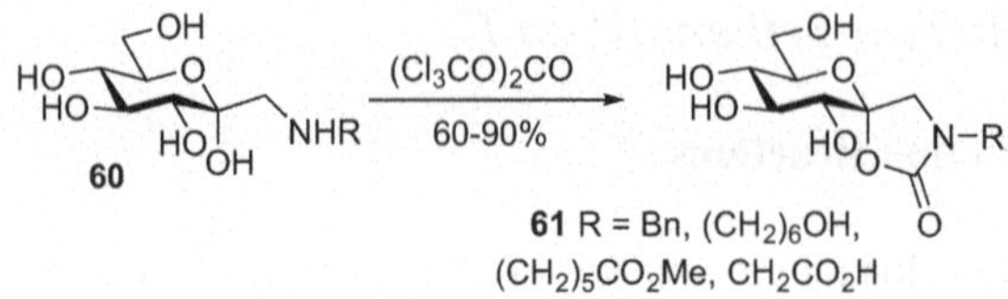

Scheme 20 Spiro-cyclization from a heptose-based Amadori rearrangement product

4 Strategies Toward Biologically Active Spiro-Bicyclic Glycopyranosides

4.1 Papulacandin (Path B)

Papulacandin is a glycolipid with potent antifungal activities with a benzo-fused spiroketal moiety on a glucopyranose scaffold (Scheme 21). The total synthesis reported by Denmark will be discussed briefly herein as this was one of the first reports to describe an interesting 1,2-epoxide on the pyranose ring. This synthesis was based on an esterification at position 3 of the glucopyranose ring, and hence the other key step was the creation of the spiro-annulated glucopyranose (Scheme 21) [47, 48]. The *C*-aryl glucal **62** was converted into the corresponding epoxide **63** with *m*-chloroperbenzoic acid in the presence of a base ($NaHCO_3$) to avoid the spiro-ketalization with the benzylic alcohol on the glucal, prior to the epoxidation. Epoxidation was highly stereoselective for the α-face, and the resulting configuration of C-2 could be ascertained from NMR experiments [48]. This epoxide could then undergo spiro-ketalization in situ leading to compounds **64α** and **64β** in a 5:1 diastereoisomeric ratio. The β-anomer **64β** could be readily converted into the desired α-anomer **64α** in the presence of hydrochloric acid under thermodynamic control.

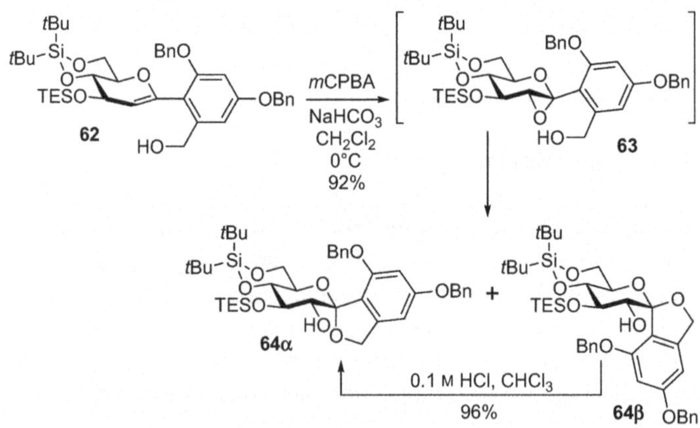

Scheme 21 Exemplary synthesis of papulacandin D

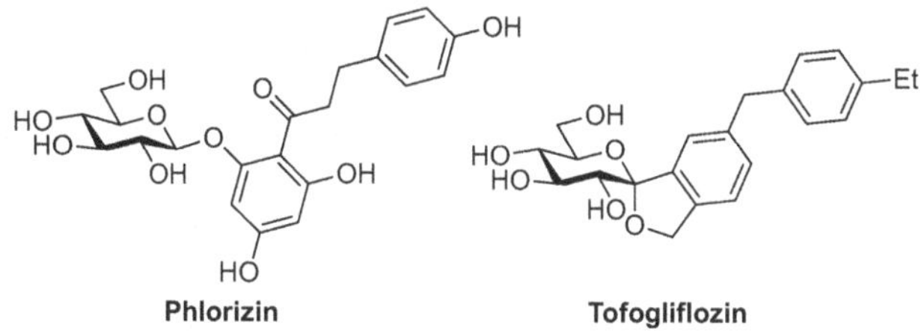

Fig. 3 Phlorizin and gliflozins as SGLT2 inhibitors

4.2 Tofogliflozin (Path B)

Glucose transporters are targets to control glycemia as they allow glucose to cross cell membranes. Among those proteins, sodium-dependent glucose transporters (SGLTs) play an important role for glucose absorption and renal reabsorption. They are located in the small intestine (SGLT1) and in the proximal tubule (SGLT1 and SGLT2) of the nephron. SGLTs use the energy of the sodium gradient to transport glucose across the apical membrane, against the uphill glucose gradient, via a symport mechanism [49–51].

Under hyperglycemia, SGLTs are saturated and excess of glucose is eliminated in the urine. In patients with type 2 *diabetes mellitus* (T2DM), SGLT2 is overexpressed and have higher activity. Therefore, less glucose is excreted in the urine leading to a constant hyperglycemia in the blood stream. Hence, a control of hyperglycemia through selective SGLT2 versus SGLT1 inhibition has been investigated to control glycemia.

A class of inhibitors, known as gliflozins marketed since 2013, was inspired by phlorizin known as a natural compound inhibiting glucose transport mediated by SGLTs (Fig. 3) [52–54]. Among those, tofogliflozin was selected for the purpose of this chapter due to its spiro-annulated carbohydrate moiety.

Tofogliflozin was developed by Chugai Pharmaceuticals and is used as a drug for the treatment of T2DM since 2014 [55]. Tofogliflozin is based on a *C*-aryl β-D-glucoside attached to an aglycon with a spiro-bicycle moiety at the anomeric carbon. The aglycon itself is composed of a diphenylmethane skeleton. Many syntheses of tofogliflozin have already been described in the literature along with challenges going from lab scale (~100 mg) to industrial scale (~4 kg). The silylated lactone **67** and the aryl moiety **66** were condensed followed by the construction of the aglycon (Scheme 22) [56]. The spiro-bond was then created in the last steps of the synthesis leading to the crude tofogliflozin after reduction. The silylated lactone **67** avoided the hydrogenolysis of benzyl ethers in the original procedure disclosed [57, 58] and thus circumvented the possible traces of heavy metals and hazardous use of hydrogen. The final preparation of the tetracarbonate **69** as a crystalline solid allowed a simple filtration for the purification of tofogliflozin.

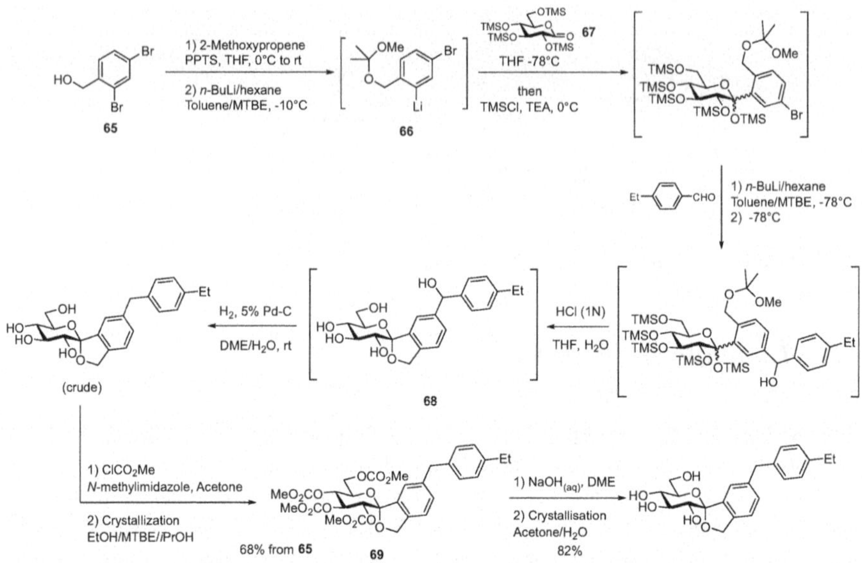

Scheme 22 Lab-scale synthesis of tofogliflozin

Although purification issues were improved from chromatography [57, 58] to crystallization, the use of hydrogen or a highly volatile reagent such as 2-methoxypropene required cautious handling and therefore needed an improvement to conduct an industrial synthesis. Furthermore, the use of palladium catalyst for the hydrogenolysis of **68** may contaminate the final active pharmaceutical ingredient with traces of the heavy metal. Therefore, a 7-step industrial route toward tofogliflozin has been developed (Scheme 23) [59]. The key steps of this synthetic

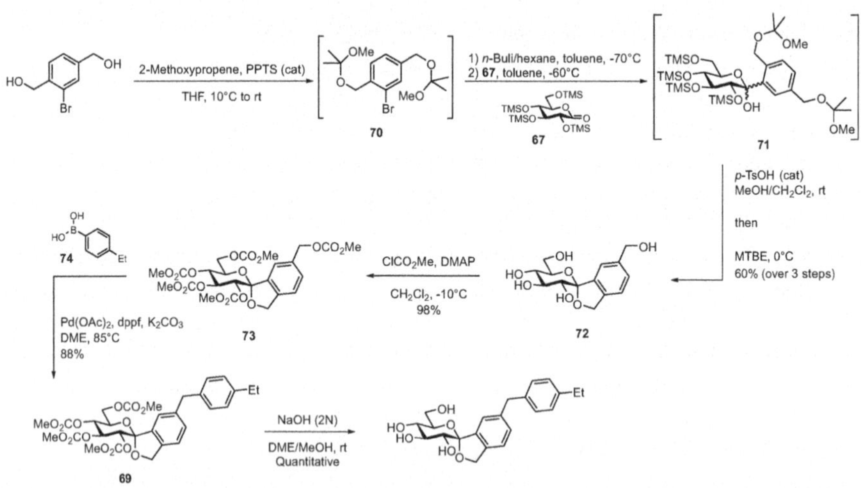

Scheme 23 Scalable (~50 g) route with safety and purification improvements

strategy are the formation of the spiroketal intermediate **72** and then the construction of the methylene bridge to build the aromatic moiety. The protected diol **70** was condensed with the silylated lactone **67** to give the lactol **71** which was then converted to the desired spiroketal intermediate **72**. After a step of protection with carbonate, the formation of the aglycon was performed with a Suzuki coupling of benzyl carbonate **73** with the corresponding boronic acid **74** leading to the tetracarbonate of tofogliflozin **69** which was purified by recrystallization. A final step of deprotection afforded the target tofogliflozin in a 52% overall yield.

A second industrial synthesis was reported through a 12-step sequence including purifications without column chromatography (Scheme 24) [60]. Unlike the previous methodologies, the aryl aglycon was built earlier with a regioselective Friedel-Crafts reaction between the acyl chloride **75** and ethylbenzene. After reduction and protection, the aryl intermediate **76** was reacted with the silylated lactone **67** in the presence of PrMgCl·LiCl in THF to form the corresponding lactol **77**. Subsequent treatment with MsOH allowed the spiro-cyclization through the concomitant acid-catalyzed desilylation to afford the crude tofogliflozin. Preparation of the carbonated tofogliflozin **69** allowed the purification by crystallization. Even though there was no further improvement in terms of yield, this methodology required the use of less expensive reagents.

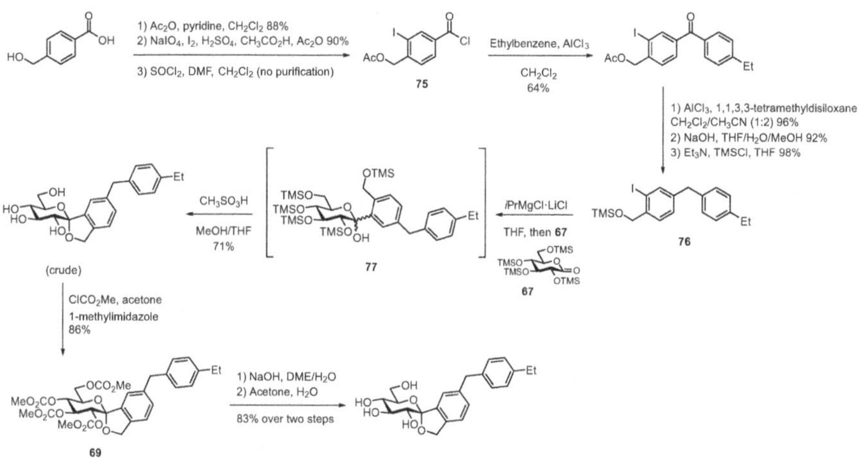

Scheme 24 Scalable (~100 g) route toward tofogliflozin

These syntheses pointed out the different challenges going from lab-scale synthesis to multi-kilogram synthesis. Benzyl protecting groups were changed to silyl ether, avoiding a deprotection with hazardous hydrogen and heavy metal palladium. The introduction of carbonates as protecting groups allowed the crude tofogliflozin to be purified by crystallization. A difference in strategy was also discussed as the aglycon was formed after or before the *C*-glycosylation and the construction of the spiro-junction.

These synthetic strategies bring out stereo- and/or regio-selectivity concerns for the construction of the spiro-junction [59, 61]. In the reactions leading to **68** and **71** → **72**, the spiro-cyclization to the five-membered ring was always preferred over the six-membered ring since the latter would be a highly strained 1,3-cyclophane. Regarding the stereoselectivity, the axial position of the oxygen atom is thermodynamically favored due to the anomeric effect.

4.3 Synthesis of Spiro-Bicyclic Systems as Potential Glycogen Phosphorylase Inhibitors

The control of hyperglycemia is of prime importance in the context of T2DM to avoid severe side effects such as retinopathy, nephropathy, or neuropathy [62]. Glucose is a source of energy for cells and is stocked in the liver as glycogen, a glucose-α-(1,4)-glucose polymer with α-(1,6)-glucose ramifications. Glycogen phosphorylase is the enzyme responsible for the depolymerization of glycogen by releasing a glucose-1-phosphate unit from the nonreducing end of the polymer. Thus, design and synthesis of glycogen phosphorylase inhibitors appear as a therapeutically relevant strategy for the treatment of hyperglycemia in type 2 diabetes [63–70]. Typical inhibitors are based on glucopyranoside scaffolds, and the spiro-bicyclic compounds are among the most potent inhibitors of this enzyme [65, 67, 71, 72].

4.3.1 Radical-Mediated Cyclization (Path B)

Glucosylidene-spiro-oxathiazoles **79** can be obtained by the radical-mediated cyclization of glucopyranosyl-thiohydroximates **78** (Scheme 25) with a 4:1 stereocontrol of the reaction in favor of the epimer **79** [73, 74]. Both epimers can be separated by chromatography. The deprotected derivatives of **79** were identified as sub-micromolar glycogen phosphorylase inhibitors [75, 76].

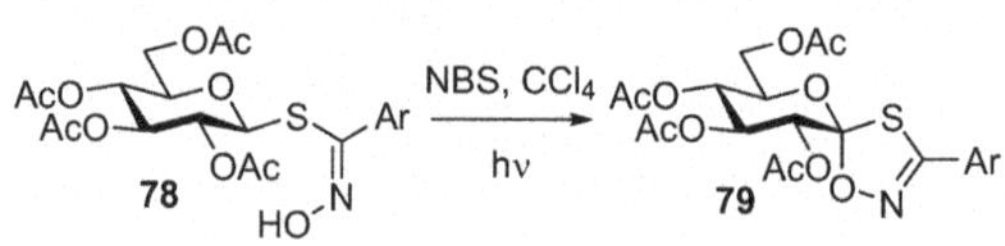

Scheme 25 Spiro-cyclization from a thiohydroximate to the corresponding glycosylidene-spiro-oxathiazole

4.3.2 1,3-Dipolar Cycloadditions (Path D)

Stereoselective 1,3-dipolar cycloaddition of nitrile oxides to *exo*-glucals **41** could afford the corresponding spiro-isoxazolines **80** in high yields (Scheme 26) [77–80]. The stereoselectivity was in favor of the exclusive addition from the α-face of the pyranoid ring. This spiro-isoxazoline bicyclic scaffold was later identified as a major pharmacophore motif for the submicromolar inhibition of glycogen phosphorylase as well as in vivo antihyperglycemic effects in the context of type 2 diabetes [81, 82].

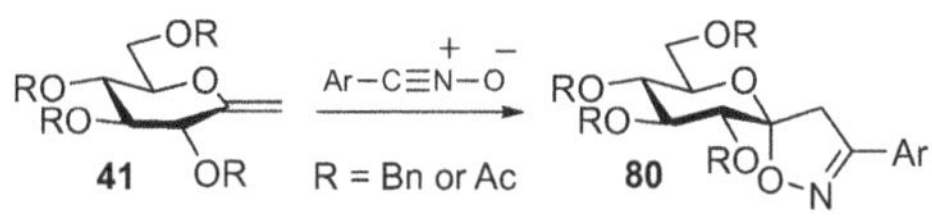

Scheme 26 Spiro-cyclization from an *exo*-glycal and nitrile oxide through 1,3-dipolar cycloaddition

4.3.3 Ionic Ring Closures (Path D)

Spiro-cyclization of the 1,2-aminoalcohol **81** using a Burgess-type reagent (*t*-butyl *N*-(triethylammoniumsulfonyl)-carbamate) was accomplished in good yields leading to the corresponding spiro-sulfamide **82** (Scheme 27). After removal of the protecting groups (Boc/TFA and Bn/H_2), the spiro-glycosides were evaluated as glycogen phosphorylase inhibitors with encouraging micromolar activity [83]. The same strategy when applied to the 1,2-diol **83** afforded the corresponding spiro-sulfamidates **84** (Scheme 27) [84]. The stability of the Burgess reagent is poor, and reflux temperature was required to obtain optimized yields by shortening the reaction time to 2 h from 24 h at room temperature. The reaction was highly regio- and stereoselective for the α-anomer. The deprotected spiro-glycosides obtained after acid-catalyzed removal of the carbamate and hydrogenolysis of the benzyl ethers provided weak but selective α-glucosidase and amyloglucosidase inhibitors. Opposite β-configured spiro-sulfamidate **86** could be obtained in the 2-deoxy-glycoside series (Path B) from the β-*C*-glycoside **85** by Rh-mediated C-H activation of the anomeric carbon-hydrogen bond followed by Boc-protection of the crude mixture to facilitate purifications (Scheme 27) [85]. Surprisingly, the α-anomer of compound **85** decomposed under similar conditions.

Scheme 27 Spiro-cyclization from 1,2-aminoalcohol and 1,2-diol or through anomeric C-H activation

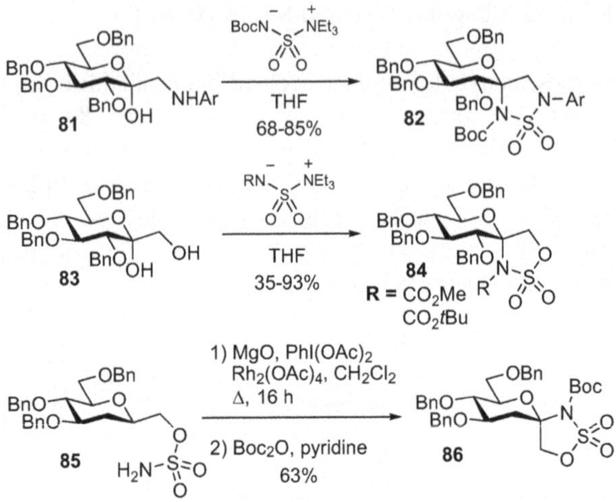

The *C*-(1-bromo-1-deoxy-β-D-glucopyranosyl) formamides were applied to numerous transformations leading to a series of five-membered spiro-bicyclic systems (Schemes 28 and 29). The *C*-(glucosyl)formamide **87a** upon reaction with silver carbonate in acetone, used as a reagent and solvent, afforded the spiro-bicyclic compound **88a** (Scheme 28) [86]. While silver ions were responsible of abstracting the halogen to generate the oxocarbenium intermediate, acetone acted as a nucleophile and added to the cation and intramolecular cyclization from the amide led to the spiro-bicyclic structure. A similar sequence could be performed from the benzoylated hemiacetal **89** through activation with triflic acid to generate the oxocarbenium intermediate in acetone leading to compound **88b** as the benzoylated analogue of compound **88a** (Scheme 28) [87]. A series of ketones has been used in this approach, and further debenzoylation could be performed affording inhibitor candidates for glycogen phosphorylase which unfortunately turned out to have no inhibitory effect.

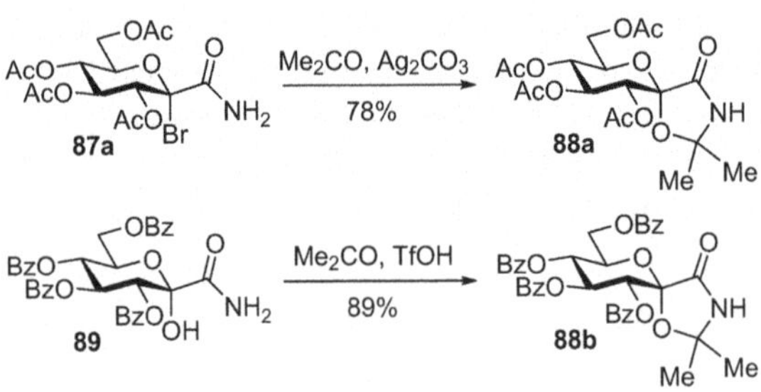

Scheme 28 Spiro-cyclization from *C*-(glucopyranosyl)formamides

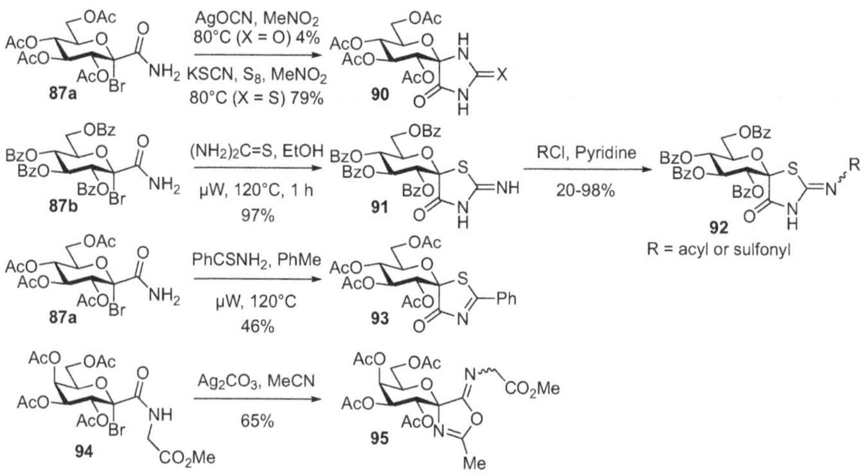

Scheme 29 Five-membered ring spiro-bicyclic glucosides as potential glycogen phosphorylase inhibitors

Spiro-cyclization of the *C*-(glucosyl)formamide **87a** was achieved with silver cyanate to obtain the spiro-hydantoin **90** (X = O) in only 4% yield with a large portion of its spiro-epimer (25%) and hydrolysis product hemiacetal (53%) (Scheme 29) [88]. The deacetylated spiro-(thio)hydantoins **90** displayed micromolar inhibition toward glycogen phosphorylase. The same reaction performed with potassium isothiocyanate afforded the spiro-thiohydantoin **91** (X = S) as a single stereoisomer in good yield (79%) although the presence of catalytic amount of elemental sulfur (S_8) was required to overcome radical-mediated pathways. When the benzoylated *C*-(glucosyl)formamide **87b** was condensed with thiourea, the corresponding spiro-iminothiazolidinone **91** was obtained in nearly quantitative yield with complete stereocontrol [89]. Further acylation of the imine moiety afforded a series of *N*-substituted derivatives **92** as a mixture of isomers on the imine bond position. The debenzoylated compounds **91** and **92** were identified as micromolar inhibitors of glycogen phosphorylase. Similarly, when the acetylated *C*-(glucosyl) formamide **87a** was reacted with thiobenzamide, the corresponding spiro-thiazolinone **93** was obtained in moderate yield (46%) although with good stereocontrol [90]. The same synthesis performed from the benzoylated precursor **87b** provided lower yields (14–18%). Furthermore, deacetylation could only be performed in moderate yield (32%) using lithium hydroxide. It was also observed that water addition to the spiro-thiazolinone **93** moiety modified the structure and prevented the evaluation as potential glycogen phosphorylase inhibitors. Finally, addition of nitrile to the *N*-substituted *C*-(glucosyl)formamide **94** in a Ritter-type reaction afforded spiro-isoxazoline derivatives **95** which could then be further elaborated into anomeric glycosyl α-amino acids [91]. Stereochemistry at the imine position could not be controlled, while the stereocontrol at the anomeric position was very good.

5 Conclusion and Perspectives

Spiro-bicyclic compounds can be generated through a multitude of synthetic strategies within the organic chemist's arsenal. The present review has compiled a selection of examples highlighting the typical strategic routes for the creation of a spiro-junction at the anomeric position of hexopyranoses. The main strategies can be divided into intramolecular and intermolecular approaches, and a specific attention to the control of the stereochemistry at the anomeric center will be crucial. Cyclizations are typically performed through radical reactions, ring-closing metathesis (RCM), C-H activation, and [3 + 2] or [2 + 2 + 2] cycloaddition reactions but also through acetalation under thermodynamic control.

Most of the synthetic strategies for which the configuration of the anomeric center is affected have moderate to very poor stereocontrol. If the anomeric configuration is set in the starting material, then no epimerization is usually observed. But if the anomeric position has to be activated (e.g., to oxocarbenium) under other synthetic pathways, the stereocontrol is typically moderate. Nevertheless, a good diastereoisomeric excess can be obtained when a thermodynamic equilibrium can be reached. Mixtures of epimers could be most often separated by chromatography or crystallization.

The synthesis of spiro-bicyclic glycosides has provided access to a large diversity of chemical scaffolds which proved useful in the modulation of biological targets' activity such as in the context of type 2 diabetes (SGLT2 and GP) or papulacandins as natural products with antifungal properties. Among these pharmaceutically relevant molecules, the gliflozins have reached the market, and tofogliflozin is now a drug marketed and used to treat hyperglycemia for type 2 diabetes patients.

The next developments for the synthesis of spiro-bicyclic glycopyranosides will most probably rely on the stereocontrol of the anomeric configuration. New and creative strategies are needed to result in even further complicated structures reaching different natural products in which the spiro-junction is a key feature. This will allow the discovery of unexpected molecular scaffolds to inhibit proteins toward drug candidates.

Acknowledgments The authors thank the Université Claude Bernard Lyon 1 and the CNRS for financial support. MP is grateful to the Ministère de l'Enseignement supérieur et de la Recherche for a PhD stipend.

References

1. Vidal S (ed) (2019) Protecting groups: strategies and applications in carbohydrate chemistry. Wiley-VCH, Weinheim
2. Demchenko AV (ed) (2008) Handbook of chemical glycosylation: advances in stereoselectivity and therapeutic relevance. Wiley-VCH, Weinheim

3. Zulueta MML, Hung S-C (eds) (2016) Glycochemical synthesis: strategies and applications. Wiley-VCH, Weinheim
4. Chen G-R, Fei Zhong B, Huang X-T, Xie Y-Y, Xu J-L, Gola J, Steng M, Praly J-P (2001). Eur J Org Chem:2939–2946
5. Lambu MR, Hussain A, Sharma DK, Yousuf SK, Singh B, Tripathi AK, Mukherjee D (2014). RSC Adv 4:11023–11028
6. John Pal AP, Gupta P, Suman Reddy Y, Vankar YD (2010). Eur J Org Chem:6957–6966
7. Haudrechy A, Sinaÿ P (1992). Carbohydr Res 216:375–379
8. Yamanoi T, Oda Y, Muraishi H, Matsuda S (2008). Molecules 13:1840
9. Dondoni A, Marra A (2009). Tetrahedron Lett 50:3593–3596
10. Lin H-C, Chen Y-B, Lin Z-P, Wong FF, Lin C-H, Lin S-K (2010). Tetrahedron 66:5229–5234
11. Chen Y-B, Liu S-H, Hsieh M-T, Chang C-S, Lin C-H, Chen C-Y, Chen P-Y, Lin H-C (2016). J Org Chem 81:3007–3016
12. John Pal AP, Vankar YD (2010). Tetrahedron Lett 51:2519–2524
13. John Pal AP, Kadigachalam P, Mallick A, Doddi VR, Vankar YD (2011). Org Biomol Chem 9:809–819
14. Martín A, Salazar J, Suárez E (1995). Tetrahedron Lett 36:4489–4492
15. Betancor C, Dorta RL, Freire R, Prangé T, Suárez E (2000). J Org Chem 65:8822–8825
16. Martín A, Quintanal LM, Suárez E (2007). Tetrahedron Lett 48:5507–5511
17. Martín A, Pérez-Martín I, Suárez E (2009). Tetrahedron 65:6147–6155
18. Martín A, Pérez-Martín I, Suárez E (2005). Org Lett 7:2027–2030
19. Probst N, Grelier G, Ghermani N, Gandon V, Alami M, Messaoudi S (2017). Org Lett 19:5038–5041
20. Pezzotta J, Urban D, Guillot R, Doisneau G, Beau J-M (2014). Synlett 25:375–380
21. Briner K, Vasella A (1989). Helv Chim Acta 72:1371–1382
22. Blüchel C, Linden A, Vasella A (2001). Helv Chim Acta 84:3495–3502
23. Mangholz SE, Vasella A (1991). Helv Chim Acta 74:2100–2111
24. Vasella A, Waldraff CAA (1991). Helv Chim Acta 74:585–593
25. Somsák L, Praly J-P, Descotes G (1992). Synlett:119–120
26. Praly JP, El Kharraf Z, Descotes G (1990). Tetrahedron Lett 31:4441–4442
27. Blüchel C, Ramana CV, Vasella A (2003). Helv Chim Acta 86:2998–3036
28. Brand C, Rauch G, Zanoni M, Dittrich B, Werz DB (2009). J Org Chem 74:8779–8786
29. Vasella A, Witzig C, Waldraff C, Uhlmann P, Briner K, Bernet B, Panza L, Husi R (1993). Helv Chim Acta 76:2847–2875
30. Vasella A, Dhar P, Witzig C (1993). Helv Chim Acta 76:1767–1778
31. Lay L, Nicotra F, Panza L, Russo G (1995). Synlett:167–168
32. Schweizer F, Inazu T (2001). Org Lett 3:4115–4118
33. Zhang K, Schweizer F (2005). Synlett:3111–3115
34. Zhang K, Wang J, Sun Z, Nguyen D-H, Schweizer F (2007). Synlett:0239–0242
35. Zhang K, Mondal D, Zhanel GG, Schweizer F (2008). Carbohydr Res 343:1644–1652
36. Zhang K, Schweizer F (2009). Carbohydr Res 344:576–585
37. Praly JP, Brard L, Descotes G (1988). Tetrahedron Lett 29:2651–2654
38. Praly J-P, Kharraf ZE, Corringer P-J, Brard L, Descotes G (1990). Tetrahedron 46:65–75
39. Buchanan JG, Clelland APW, Wightman RH, Johnson T, Rennie RAC (1992). Carbohydr Res 237:295–301
40. Baddeley KL, Cao Q, Muldoon MJ, Cook MJ (2015). Chem Eur J 21:7726–7730
41. Zhang D, Ye D, Feng E, Wang J, Shi J, Jiang H, Liu H (2010). J Org Chem 75:3552–3557
42. McDonald FE, Zhu HYH, Holmquist CR (1995). J Am Chem Soc 117:6605–6606
43. Yamamoto Y, Yamashita K, Hotta T, Hashimoto T, Kikuchi M, Nishiyama H (2007). Chem Asian J 2:1388–1399
44. Bartolozzi A, Capozzi G, Falciani C, Menichetti S, Nativi C, Bacialli AP (1999). J Org Chem 64:6490–6494
45. Wrodnigg TM, Kartusch C, Illaszewicz C (2008). Carbohydr Res 343:2057–2066

46. Gallas K, Pototschnig G, Adanitsch F, Stütz AE, Wrodnigg TM (2012). Beilstein J Org Chem 8:1619–1629
47. Denmark SE, Regens CS, Kobayashi T (2007). J Am Chem Soc 129:2774–2776
48. Denmark SE, Kobayashi T, Regens CS (2010). Tetrahedron 66:4745–4759
49. Song KS, Lee SH, Kim MJ, Seo HJ, Lee J, Lee SH, Jung ME, Son EJ, Lee M, Kim J, Lee J (2011). ACS Med Chem Lett 2:182–187
50. Harada N, Inagaki N (2012). J Diabetes Investig 3:352–353
51. Madaan T, Akhtar M, Najmi AK (2016). Eur J Pharm Sci 93:244–252
52. Washburn WN (2012) SGLT2 inhibitors in development. In: Jones RM (ed) New therapeutic strategies for type 2 diabetes: small molecule approaches. The Royal Society of Chemistry, Cambridge, pp 29–87
53. Bokor É, Kun S, Goyard D, Tóth M, Praly JP, Vidal S, Somsák L (2017). Chem Rev 117:1687–1764
54. Aguillón AR, Mascarello A, Segretti ND, de Azevedo HFZ, Guimaraes CRW, Miranda LSM, de Souza ROMA (2018). Org Proc Res Dev 22:467–488
55. Poole RM, Prossler JE (2014). Drugs 74:939–944
56. Murakata M, Ikeda T, Kimura N, Kawase A, Nagase M, Yamamoto K, Takata N, Yoshizaki S, Takano K (2009) Crystal of spiroketal derivative, and process for production thereof. WO2009154276
57. Kobayashi T, Sato T, Nishimoto M (2005) Spiroketal derivative and use thereof as diabetic medicine. US2009030006
58. Kobayashi T, Sato T, Nishimoto M (2006) Preparation of 1,1-anhydro-1-C-[2-(hydroxyalkyl) aryl]-β-D-glucopyranose compounds as SGLT2 inhibitors. WO2006080421A1
59. Ohtake Y, Emura T, Nishimoto M, Takano K, Yamamoto K, Tsuchiya S, Yeu SY, Kito Y, Kimura N, Takeda S, Tsukazaki M, Murakata M, Sato T (2016). J Org Chem 81:2148–2153
60. Yang X-D, Pan Z-X, Li D-J, Wang G, Liu M, Wu R-G, Wu Y-H, Gao Y-C (2016). Org Process Res Dev 20:1821–1827
61. Ohtake Y, Sato T, Kobayashi T, Nishimoto M, Taka N, Takano K, Yamamoto K, Ohmori M, Yamaguchi M, Takami K, Yeu SY, Ahn KH, Matsuoka H, Morikawa K, Suzuki M, Hagita H, Ozawa K, Yamaguchi K, Kato M, Ikeda S (2012). J Med Chem 55:7828–7840
62. Ross SA, Gulve EA, Wang M (2004). Chem Rev 104:1255–1282
63. Morral N (2003). Trends Endocrinol Metab 14:169–175
64. Baker DJ, Greenhaff PL, Timmons JA (2006). Expert Opin Ther Pat 16:459–466
65. Khan M (2007). Top Heterocycl Chem 9:33–52
66. Somsák L, Czifrák K, Tóth M, Bokor E, Chrysina ED, Alexacou KM, Hayes JM, Tiraidis C, Lazoura E, Leonidas DD, Zographos SE, Oikonomakos NG (2008). Curr Med Chem 15:2933–2983
67. Praly J-P, Vidal S (2010). Mini-Rev Med Chem 10:1102–1126
68. Henke BR (2012) Inhibition of glycogen phosphorylase as a strategy for the treatment of type 2 diabetes. In: Jones RM (ed) New therapeutic strategies for type 2 diabetes: small molecule approaches. The Royal Society of Chemistry, Cambridge, pp 324–365
69. Gaboriaud-Kolar N, Skaltsounis A-L (2013). Expert Opin Ther Pat 23:1017–1032
70. Donnier-Maréchal M, Vidal S (2016). Expert Opin Ther Pat 26:199–212
71. Somsák L, Nagy V, Hadady Z, Docsa T, Gergely P (2003). Curr Pharm Des 9:1177–1189
72. Somsák L (2011). C R Chim 14:211–223
73. Praly J-P, Boyé S, Joseph B, Rollin P (1993). Tetrahedron Lett 34:3419–3420
74. Elek R, Kiss L, Praly J-P, Somsák L (2005). Carbohydr Res 340:1397–1402
75. Somsák L, Nagy V, Vidal S, Czifrák K, Berzsényi E, Praly J-P (2008). Bioorg Med Chem Lett 18:5680–5683
76. Nagy V, Benltifa M, Vidal S, Berzsényi E, Teilhet C, Czifrák K, Batta G, Docsa T, Gergely P, Somsák L, Praly J-P (2009). Bioorg Med Chem 17:5696–5707
77. RajanBabu TV, Reddy GS (1986). J Org Chem 51:5458–5461

78. Enderlin G, Taillefumier C, Didierjean C, Chapleur Y (2005). Tetrahedron Asymmetry 16:2459–2474
79. Benltifa M, Vidal S, Gueyrard D, Goekjian PG, Msaddek M, Praly J-P (2006). Tetrahedron Lett 47:6143–6147
80. Zhang P-Z, Li X-L, Chen H, Li Y-N, Wang R (2007). Tetrahedron Lett 48:7813–7816
81. Benltifa M, Hayes JM, Vidal S, Gueyrard D, Goekjian PG, Praly JP, Kizilis G, Tiraidis C, Alexacou KM, Chrysina ED, Zographos SE, Leonidas DD, Archontis G, Oikonomakos NG (2009). Bioorg Med Chem 17:7368–7380
82. Goyard D, Kónya B, Chajistamatiou AS, Chrysina ED, Leroy J, Balzarin S, Tournier M, Tousch D, Petit P, Duret C, Maurel P, Somsák L, Docsa T, Gergely P, Praly J-P, Azay-Milhau J, Vidal S (2016). Eur J Med Chem 108:444–454
83. Tite T, Tomas L, Docsa T, Gergely P, Kovensky J, Gueyrard D, Wadouachi A (2012). Tetrahedron Lett 53:959–961
84. Benltifa M, Kiss MD, Garcia-Moreno MI, Mellet CO, Gueyrard D, Wadouachi A (2009). Tetrahedron Asymmetry 20:1817–1823
85. Toumieux S, Compain P, Martin OR (2005). Tetrahedron Lett 46:4731–4735
86. Somsák L, Kovács L, Gyóllai V, Ősz E (1999). Chem Commun 7:591–592
87. Páhi A, Czifrák K, Kövér KE, Somsák L (2015). Carbohydr Res 403:192–201
88. Somsák L, Kovács L, Tóth M, Ősz E, Szilágyi L, Györgydeák Z, Dinya Z, Docsa T, Tóth B, Gergely P (2001). J Med Chem 44:2843–2848
89. Czifrák K, Páhi A, Deák S, Kiss-Szikszai A, Kövér KE, Docsa T, Gergely P, Alexacou K-M, Papakonstantinou M, Leonidas DD, Zographos SE, Chrysina ED, Somsák L (2014). Bioorg Med Chem 22:4028–4041
90. Szabó KE, Kun S, Mándi A, Kurtán T, Somsák L (2017). Molecules 22:1760
91. Czifrák K, Gyóllai V, Kövér KE, Somsák L (2011). Carbohydr Res 346:2104–2112

Top Heterocycl Chem (2019) 57: 27–50
DOI: 10.1007/7081_2019_28

Published online: 8 February 2019

Cycloaddition Reactions of Sugar-Based Olefins, Nitrones and Nitrile Oxides: En Route to Saccharidic Spiroisoxazoli(di)nes

Nadia Pellegrini-Moïse and Mylène Richard

Contents

Abstract Isoxazoline- and isoxazolidine-containing compounds are privileged structures of interest, notably in synthetic and medicinal chemistry. These heterocycles can be obtained by 1,3-dipolar cycloaddition reactions between an olefin and a nitrile oxide or a nitrone. This reaction generates one C–C and one C–O bond and up to three chiral centres in one step. In the present chapter, we aim to summarize and discuss reports of these cycloadditions on sugar olefins, with a focus on *exo*-methylene sugars or activated *exo*-glycals, leading to saccharidic spiroisoxazoli(di) nes with high regio- and stereocontrol. Additional examples of cycloaddition reactions involving chiral nitrone, sugar nitrile oxide, sugar nitrone and two chiral sugar partners will also be discussed. Due to the importance of the spiro structure in several biologically active compounds, these spiroheterocycles can be regarded as

N. Pellegrini-Moïse (✉) and M. Richard (✉)
Université de Lorraine-CNRS, UMR 7053 L2CM, Nancy, France
e-mail: nadia.pellegrini@univ-lorraine.fr; mylene.richard@cea.fr

spironucleoside analogues, mimics of natural building blocks or multicyclic sugar scaffolds suitable for selective derivatization. Some of them thus showed promising biological properties as antibacterial agents or enzyme inhibitors. Moreover, the labile nature of the N–O bond in the isoxazolidine ring makes it an attractive target for synthetic chemists. The reactivity of this scaffold has therefore been widely studied, and the cycloadducts have been converted to other classes of compounds of interest. Examples of the biological relevance and synthetic use and reactivity of these spiro-sugars will be given in this chapter.

Keywords Cycloaddition · Nitrile oxides · Nitrones · Spiroisoxazoli(di)nes · Sugar-based olefins

1 Introduction

The five-membered heterocycle isoxazolidine is part of several natural products and can be a useful synthetic intermediate due to the labile nature of the N–O bond. As reported in an excellent review by Parrot and Martinez [1], these compounds can be considered as analogues of natural building blocks, i.e. nucleosides and carbohydrates, and can also be incorporated into peptidic sequences for the design of peptidomimetics. From a medicinal point of view, these derivatives provided many biological applications as cytotoxic, antiviral or antimicrobial compounds. These heterocycles are commonly obtained through the well-known 1,3-dipolar cycloaddition between an olefin and a nitrone, thus resulting in the formation of one C–C and one C–O bond [2–5].

Isoxazoline derivatives, the unsaturated parent of isoxazolidine, are also a class of heterocyclic compounds of significance, since they are frequently found in many compounds that showed significant biological activities [6]. Besides, they can be easily transformed into other functional groups, making them valuable compounds in synthetic chemistry. Nitrile oxide cycloaddition with olefin represents an easy route to this type of heterocycle [7, 8].

Reactivity of carbohydrate-derived olefins has been tested in this type of cycloadditions [9–12]. In the present chapter, we aim to summarize and discuss reports of these cycloadditions on sugar olefins, with a focus on *exo*-methylene sugars or activated *exo*-glycals, leading to saccharidic spiroisoxazoli(di)nes with high regio- and stereocontrol. *Exo*-glycals or *C*-glycosylidenes have an exocyclic double bond on the carbon derived from the anomeric carbon of a parent saccharide [13, 14]. These compounds can be efficiently prepared by various olefination processes starting from sugar lactones and taking advantage of Wittig-type olefination or other organometallic reagents.

The first part of the present chapter is dedicated to *exo*-methylene sugars (with the exocyclic CH_2 group attached to the former C-1 or elsewhere in the sugar ring) as dipolarophiles in cycloaddition reactions with a nitrile oxide or a nitrone. We then focus on *exo*-glycals substituted by an electron-withdrawing group, labelled

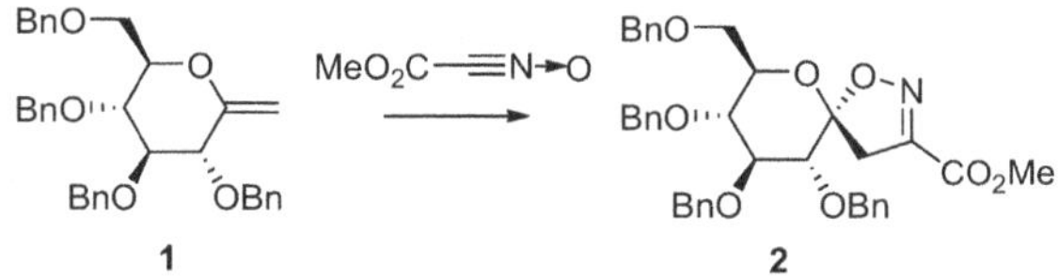

Scheme 1 Synthesis of the first saccharidic spiro-isoxazoline described by RajanBabu

"activated *exo*-glycals" and widely studied in our group with regard to their singular reactivity. Finally, the last part of this chapter describes some representative examples of 1,3-dipolar cycloaddition reactions of carbohydrate-derived nitrones and nitrile oxides.

2 *Exo*-Methylene Sugars as Dipolarophiles

In this first part, we report 1,3-dipolar cycloadditions with *exo*-methylene sugars or activated *exo*-glycals as dipolarophile partners.

2.1 C-1 Exo-Methylene Sugars as Dipolarophiles

RajanBabu et al. described the first example of 1,3-dipolar cycloaddition of methylene *exo*-glycals with a nitrile oxide [15]. The *O*-perbenzylated derivative **1** efficiently reacted with carbomethoxynitrile oxide to give spiro-isoxazoline **2** with full stereoselectivity, considering steric hindrance on the β-face (Scheme 1).

Few years later, Lieberknecht and Bravo gave full experimental accounts of this reaction and applied it to *Z*- or *E*-configurated *exo*cyclic enolethers **3** (Scheme 2) [16]. Starting from the *E*-configurated olefin, the cycloaddition reactions were performed with mesitonitrile oxide (stable monomer) in dichloromethane, and *trans*-configurated cycloadducts **4** were obtained in excellent yields. With ethoxycarbonylnitrile oxide generated in situ, microwave irradiations were the most efficient for the cycloaddition on these trisubstituted alkenes **3**, in particular for the less reactive 1-phenylenitol (R^1 = Ph). The diastereoisomeric ratio determined by NMR experiments confirmed the proposed structures for compounds **4** and **5**, in agreement with the attack of the nitrile oxide to the less hindered face of the *exo*-glycal. Starting from *E*-configurated *exo*-glycals, the *trans* cycloadducts were mainly obtained (95:5), and the *cis*-configurated cycloadducts were the main compounds when the reaction was performed on the *Z*-isomers. The regioselectivity was in line with trends found in nitrile oxide cycloadditions to trisubstituted alkenes [4]. The reaction was carried out in different conditions (irradiation time, irradiation power) explaining the 35–45% yield range obtained.

As part of their project concerning glucose derivatives as glycogen phosphorylase (GP) inhibitors, Praly et al. deeply developed the 1,3-dipolar cycloaddition of various

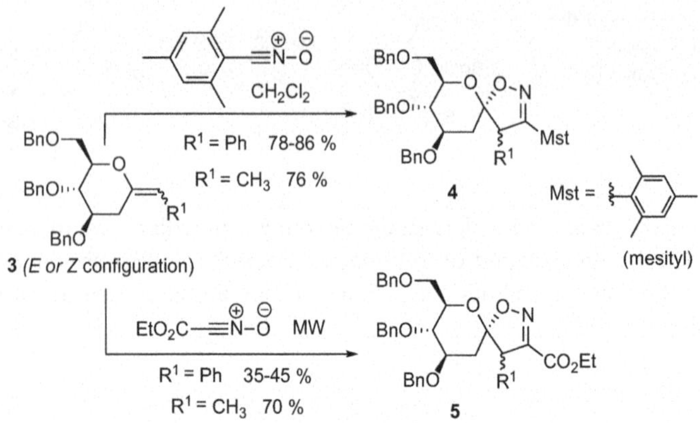

Scheme 2 Cycloadditions on *Z*- or *E*-configurated *exo*cyclic enolethers

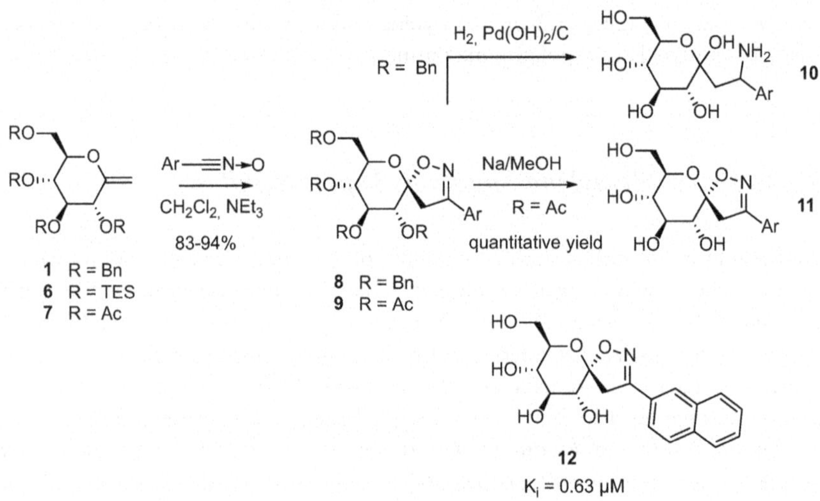

Scheme 3 Water-soluble spiro-isoxazolines as GP inhibitors

aryl nitrile oxides to differently protected *exo*-glucals (Scheme 3) [17]. They demonstrated that the benzylated *exo*-glucal **1** afforded the corresponding spiro-isoxazoline **8** in high yields (83–94%) with complete regio- and stereoselectivity established by 1D, 2D and NOESY NMR experiments. However, the reductive cleavage of the N–O bond during hydrogenolysis of benzyl-protecting group prompted them to envisage another hydroxyl-protecting group like the triethylsilyl group (TES). The attempted cycloaddition of the aryl nitrile oxide on derivative **6** did not give the expected cycloadduct. Finally, the acetylated methylene *exo*-glucal **7** was efficiently transformed in the corresponding adduct **9** subsequently deprotected by the Zemplén method. The obtained water-soluble spiro-isoxazolines were evaluated

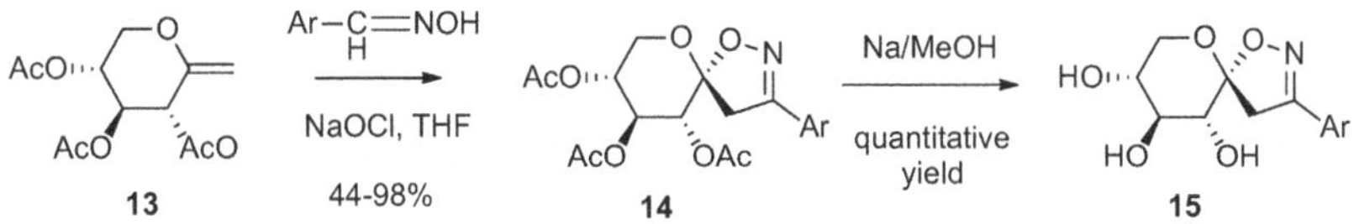

Scheme 4 Xylose-derived spiro-isoxazolines

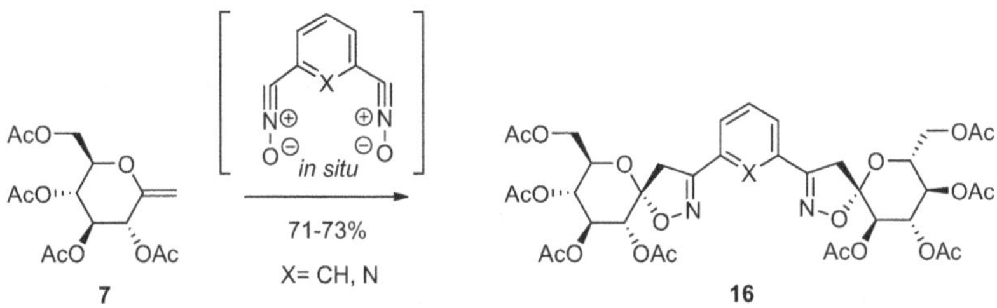

Scheme 5 Carbohydrate-based bis-spiro-isoxazolines as ligands for asymmetric catalysis

as glycogen phosphorylase inhibitors and showed IC_{50} values in the micromolar range. The X-ray structures of the enzyme-ligand complexes showed that this new family of glucose-based spiro-isoxazolines preferentially bound at the catalytic site of the enzyme retaining the less active T-state conformation [18]. Compound **12**, prepared by this methodology, presented the best biological activity with a K_i value of 0.63 μM for inhibition of GPb and was therefore deconstructed and used in a fragment-based binding evaluation [19]. Furthermore, 2-naphthyl-substituted glucopyranosylidene-spiro-isoxazoline **12** was selected for further in vitro and in vivo evaluation in a Zucker fa/fa rat model, and the promising results indicated that this type of compounds can be considered as anti-hyperglycemic agents in the context of type 2 diabetes treatment [20].

More recently, new spiro-isoxazolines **14** from D-xylose were synthesized by cycloaddition of *O*-peracetylated *exo*-xylal **13** with various nitrile oxides prepared in situ by oxidation of the corresponding aromatic aldoximes (Scheme 4) [21]. Here again, the configuration of the anomeric spiro-carbon was determined by NOE, and similar stereoselectivities were observed in the formation of glucose- and xylose-derived spiro-isoxazolines. The fully deprotected compounds **15** were evaluated towards GPb inhibition and showed moderate activity, highlighting the importance of the glucose moiety.

In the field of asymmetric catalysis, carbohydrates are excellent starting materials for the preparation of chiral auxiliaries. In this context, Vidal et al. efficiently used 1,3-dipolar cycloaddition of a bis(nitrile oxide) and methylene *exo*-glucal for the preparation of carbohydrate-based bis-spiro-isoxazoline ligands like **16** (Scheme 5) [22]. These ligands were evaluated in a Pd-catalysed Tsuji-Trost reaction but did not provide good results. Ring opening and formation of isoxazole, detected in the reaction mixture after addition of Pd(II), probably explain the low yields.

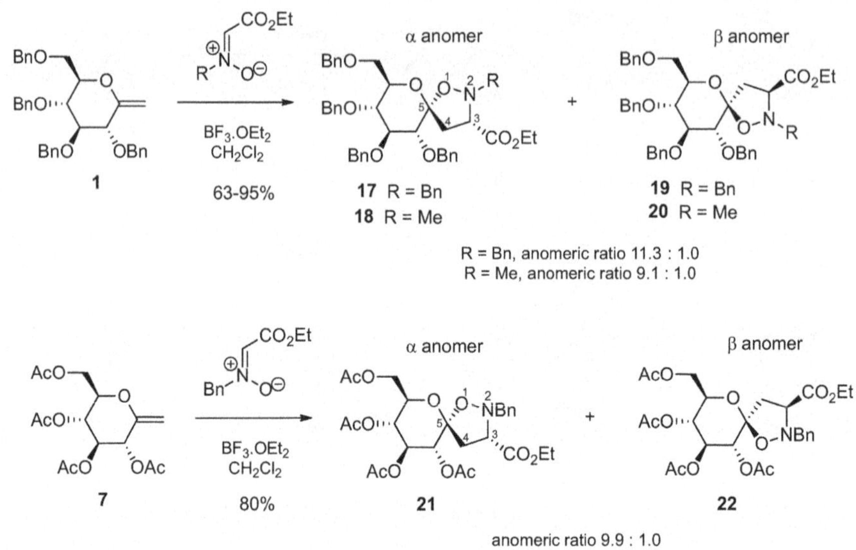

Scheme 6 Synthesis of spiro-oxazolidines by $BF_3{\cdot}OEt_2$ catalysis with differently substituted nitrones

The carbohydrate-based bis-spiro-isoxazoline ligands were also evaluated in Cu(I)-catalysed asymmetric imine alkynylation, and, in this case, the addition products were obtained in good yields.

1,3-Dipolar cycloaddition of nitrones with alkenes is widely used in organic synthesis, and Ikegami et al. described this reaction on 1-methylenesugar **1** under $BF_3{\cdot}OEt_2$ catalysis. At low temperature, the cyclization took place in a diastereo-selective manner and afforded the corresponding spiro-isoxazolidines in good to excellent yields (Scheme 6) [23, 24]. The yields and the reaction stereoselectivities were slightly affected by the nature of the solvent, the temperature, the number of equivalents of catalyst and the steric hindrance of the nitrone. Although four diastereomers can be potentially obtained (two anomers with R and S configuration on C-3), only two anomers **17** and **19** or **18** and **20** with R-configuration on C-3 were obtained, with α-isomers **17** and **18** as the predominant compounds.

The 1,3-dipolar cycloaddition reactions were also performed on benzyl-protected *exo*-glycals of D-*manno* and D-*galacto* configuration, and similar results were obtained [23]. Concerning the *manno* derivative bearing an axial benzyloxy group in position 2, an increase of β-stereoselectivity was observed (α/β ratio, 5.4:1 to 1.9:1 depending on experimental conditions and on the nitrone substitution). In the same manner, the 1,3-dipolar cycloaddition on the acetyl-protected 1-methylenesugar **7** was also performed under $BF_3{\cdot}OEt_2$ catalysis, and the two anomers **21** and **22** were obtained in a 9.9:1 ratio.

Interested by the development of glycosyl amino acids, Ikegami et al. described a new access to *C*-glycosyl amino acids from spiro-isoxazolidines by reductive cleav-age of the N–O bond (Scheme 7) [23]. The treatment of spiro-isoxazolidine **18** with

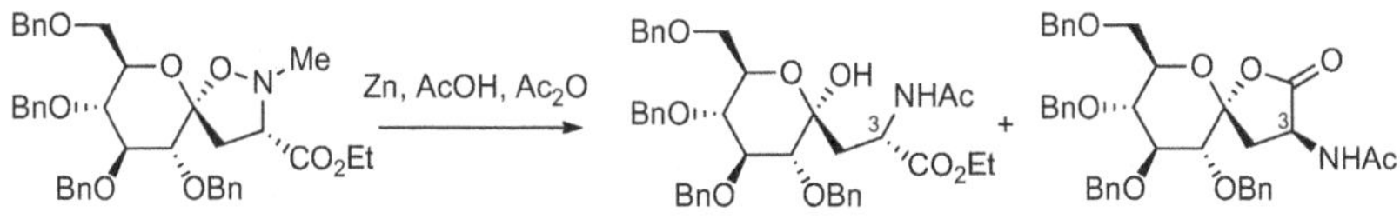

Scheme 7 Access to *C*-glycosyl amino acid by cleavage of N–O bond

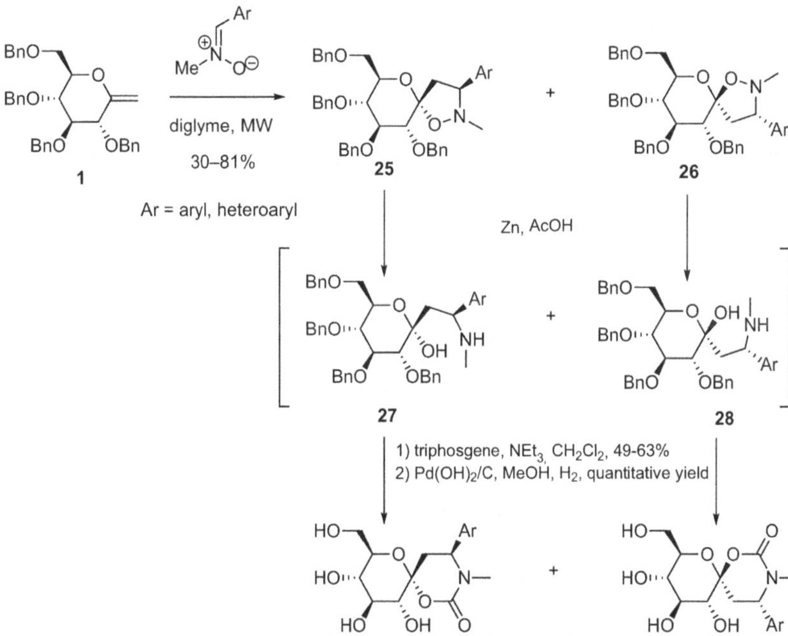

Scheme 8 Preparation of oxazinanone derivatives containing a sugar moiety

Zn in acetic acid and acetic anhydride solution resulted in the formation of *C*-glycosyl amino acid **23** with retention of configuration at C-3. Compound **24** resulting from intramolecular lactonization was also obtained.

This efficient approach was employed by Li et al. for the preparation of novel spiro-isoxazolidines by 1,3-dipolar cycloaddition with various nitrones [25]. Amino-*C*-glycosides **27** and **28** obtained as an anomeric mixture were sometimes difficult to separate because of their similar polarities (Scheme 8). Nevertheless, these compounds were useful intermediates for the preparation of oxazinanone derivatives containing sugar moiety and obtained by cyclization with triphosgene. As spironucleoside analogues, the inhibitory activities against glycosidases and cytotoxicities of deprotected compounds **29** and **30** were evaluated, and two derivatives showed selective inhibition against β-glucosidase and a slight antitumor activity.

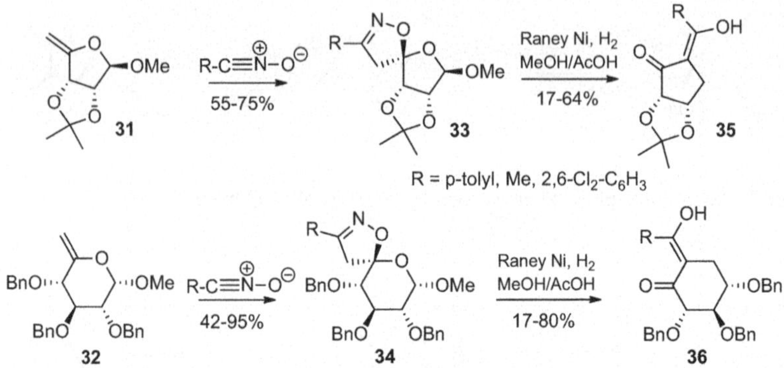

Scheme 9 Spiro-isoxazolines as intermediates for the synthesis of functionalized carbocycles

2.2 Other Exo-Methylene Sugars as Dipolarophiles

Carbohydrates are useful building blocks for the preparation of highly functionalized cyclopentane and cyclohexane rings found in biologically interesting compounds. In this context, Gallos et al. described the elegant synthesis of carbocycles **35** and **36** starting from hex-5-enopyranoside and pent-4-enofuranoside (Scheme 9) [26–28]. Cycloadditions of nitrile oxides with 4- or 5-*exo*-methylene derivatives **31** and **32**, respectively, gave spiro-isoxazolines **33** and **34** in good yields. These compounds were efficiently transformed into the corresponding carbocycles by action of Ni Raney in a mixture of methanol and acetic acid. Functionalized carbocycles **35** and **36** were obtained by reductive opening of the heterocyclic ring and spontaneous intramolecular aldol-like condensation. The hydroxylated six- and five-membered cyclic enaminones were the main products of the reaction, but the authors reported that by-products could be formed in some cases depending on the reaction conditions, the nature of substituents and the ring size.

2.3 Activated Exo-Glycals as Dipolarophiles

Sugar olefins labelled "activated" possess a C-1 or C-5 exocyclic double bond substituted by an electron-withdrawing group, giving them a capto-dative structure and a singular reactivity. Our group, headed by Yves Chapleur until 2014, developed a high-yielding, one-step method to prepare a series of activated *exo*-glycals. The saccharidic olefins were obtained in excellent yields as an easily separable mixture of *Z*- and *E*-isomers, via Wittig reaction on sugar lactones [13, 29]. We investigated their behaviour in cycloaddition reactions with nitrile oxides and nitrones several years ago, and we recently explored the potential biological activities of spiro-isoxazolidine-based carbohydrates.

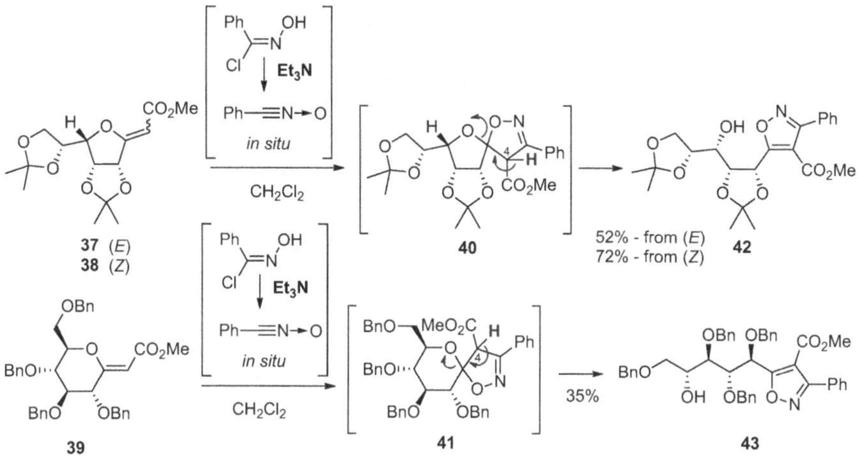

Scheme 10 Cycloadditions of activated pyranoid and furanoid *exo*-glycals with nitrile oxide

Cycloaddition of activated *exo*-glycals **37**, **38** and **39** was firstly tested with benzonitrile oxide (Scheme 10) [30, 31]. The reaction proceeded smoothly at room temperature, but the expected spiro-isoxazolines **40** and **41** were not obtained, and careful NMR examination indicated the formation of open-chain compounds **42** and **43** in 52% or 72% and 35% yields, respectively. Indeed, due to the basic conditions needed for in situ nitrile oxide formation, abstraction of the acidic proton H-4 promotes the aromatization into the corresponding isoxazoles via β-elimination of the sugar ring oxygen. Jimenez et al. had previously observed such rearrangement of activated olefins when reacted with nitrile oxides [32].

In a similar fashion, the reactivity of these activated sugar olefins in cycloadditions with nitrones was also explored in our group (Scheme 11). It is noteworthy that these reactions exclusively proceeded under microwave activation, no reaction occurring in standard thermal conditions. Cycloadducts were obtained in average to excellent yields depending on sugar and double-bond configuration or nitrone substitution. On furanoid *E*-*exo*-glycals **37** and **44**, cycloaddition with unsubstituted nitrone ($R^2 = H$) generated only one diastereoisomer, implying a complete facial control due to the stereochemistry of the sugar moiety and the steric hindrance on α face. When considering more substituted nitrones ($R^2 \neq H$), two diastereoisomers at C-3 were obtained, *trans* compounds **46** and **47** being the major ones as expected. However, the *trans*/*cis* ratio was also dependant on the nature of R^3 substituent. Indeed, bulky substituents like a benzyl group gave a 1:1 ratio, whereas a small substituent like a methyl group led to a 4:1 *trans*/*cis* ratio in the case of **37** (*gulo* derivative). Similar results were obtained on furano-*Z*-*exo*-glycals **38** and **45**, except that *trans* selectivity was greatly enhanced with D-*ribo* olefin **45**. However, when considering pyrano-*exo*-glycal **39**, a different behaviour was observed. Cycloaddition with unsubstituted nitrone ($R^2 = H$) led to both anomers, and four adducts

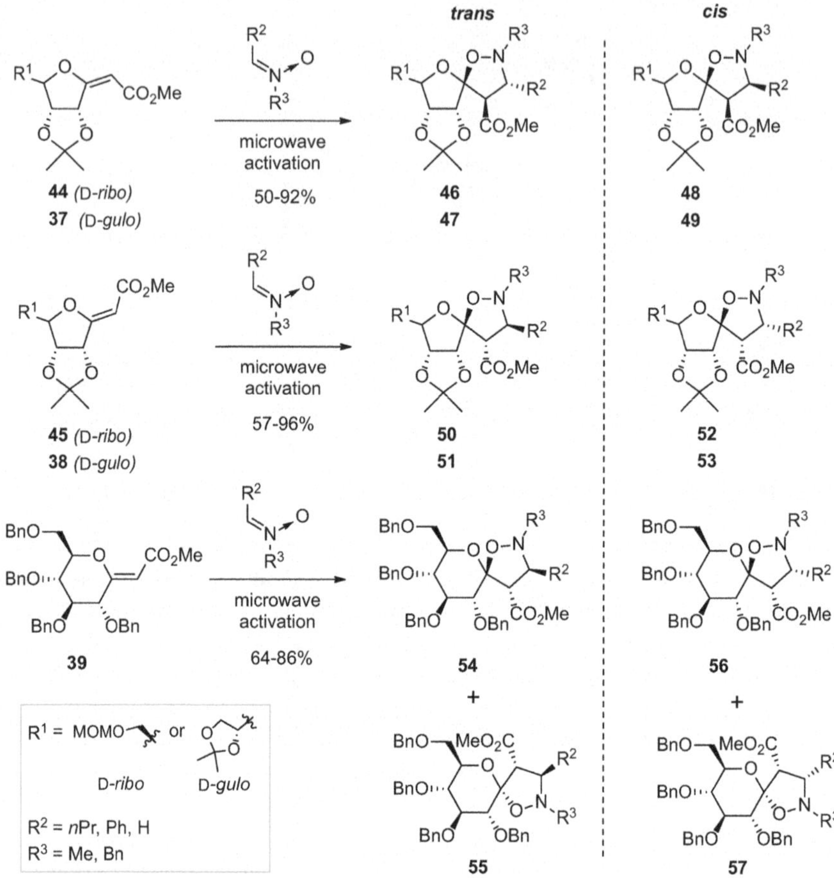

Scheme 11 Cycloadditions of activated pyranoid and furanoid *exo*-glycals with nitrones

54–57 were obtained in case of reaction with substituted nitrones ($R^2 \neq H$), indicating an overall poor facial selectivity of the sugar ring.

The stereochemistry of the cycloadducts and the *trans*/*cis* ratios were rationalized through postulated transition states (TS, Scheme 12). In *endo* TS case, steric hindrance between the methyl ester and the R^2 group is minimized for both *E*- and *Z*-*exo*-glycals, whereas *exo* TS is congested, and formation of the corresponding *cis* compounds is therefore disfavoured.

These spiro-isoxazolidines, easily deprotected in acidic media, can be considered as spironucleoside analogues and have thus potential biological properties. Indeed, the natural spironucleoside (+)-hydantocidin displays both a plant growth regulatory function and a potent herbicidal activity harmless to microorganisms and mammals [33]. As an example, the deprotection of representative spirosaccharide **49** was carried out with trifluoroacetic acid in water, giving **58** in quantitative yield (Scheme 13). It was established that these compounds are stable

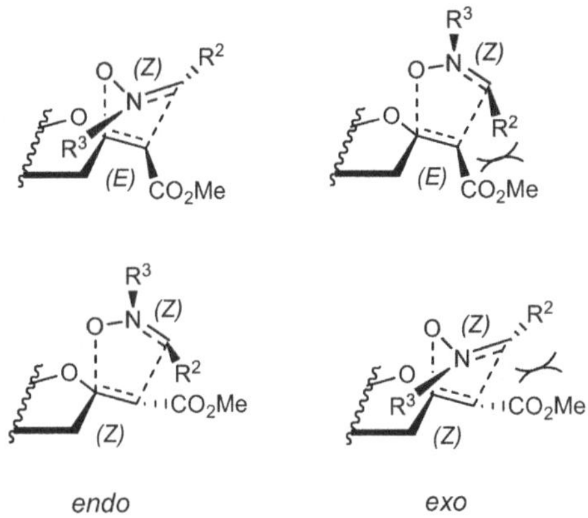

Scheme 12 Postulated transition states

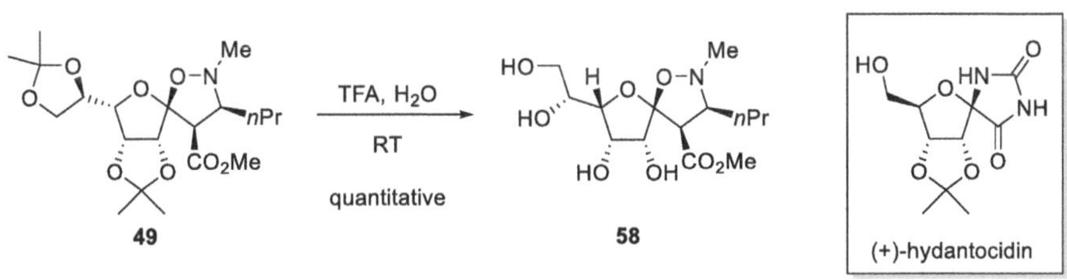

Scheme 13 Acidic deprotection of spiro-isoxazolines

in acidic conditions and do not undergo furano to pyrano isomerization and that anomeric configuration is maintained through deprotection.

To explore the potential biological activities of spiro-isoxazolidine-based carbohydrates, we prepared a library of peptidomimetics to target neuropilin-1 (NRP-1) (Scheme 14) [34]. NRP-1 is a transmembrane receptor of vascular endothelial growth factor-A (VEGF-A) implicated in various biological processes and notably angiogenesis. Our group has been interested in NRP-1 targeting for a few years with the aim of building a sugar-based peptidomimetic of the natural ligand of NRP-1 [35]. Using a sugar as a template provides a selected chirality and enables a precise spatial orientation of the substituents. In this context, we opted for a saccharidic spiro-isoxazolidine template presenting several anchoring points. Following previous studies on NRP-1/VEGF-A interactions, we functionalized the spiro scaffolds at two sites with arginine or arginine-like groups via copper-catalysed azide-alkyne cycloaddition (compounds **59** and **60**) or peptidic coupling (**61** and **62**) and tested their binding properties. However, given the moderate affinity of these peptidomimetics for NRP-1, the central bicyclic structure was probably too bulky for an appropriate binding to the receptor. Molecular modelling studies corroborated this hypothesis and helped us design more appropriate ligands [36, 37].

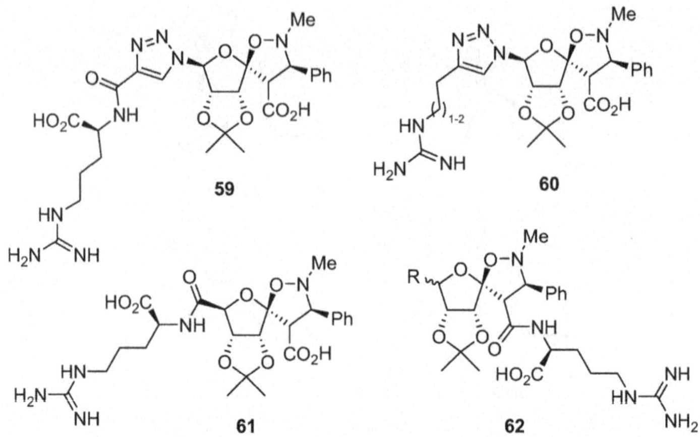

Scheme 14 Spiro-isoxazolidine based peptidomimetics for NRP-1 targeting

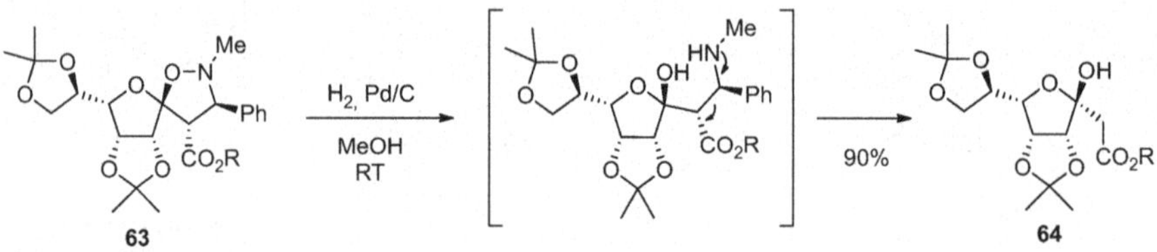

Scheme 15 Isoxazolidine cleavage by hydrogenolysis

The labile nature of the isoxazolidine N–O bond makes it a valuable target for synthetic chemists, and, as described by Li et al. [25, 38] and Gallos et al. [28], it can be easily cleaved with Zn in acetic acid and acetic anhydride or in Raney nickel hydrogenation conditions (cf. I.1. above). During our work related to synthesis of NRP-1 peptidomimetics, we noted a similar cleavage of the N–O bond during hydrogenolysis of **63**, leading to isoxazolidine ring opening and loss of N-benzylmethyl amine moiety to produce **64** (Scheme 15).

Very few examples are reported on cycloadditions between two sugar partners, although this reaction could lead to highly chiral analogues of natural products like hikosamine or tunicamycins (Scheme 16) [39]. Indeed, this synthetic approach would benefit of the easiness of isoxazolidine ring opening combined to the facial selectivity and chirality of the sugar moieties.

In this context, we more recently examined the cycloaddition of activated *exo*-glycals with carbohydrate-derived nitrones (Scheme 17). Nitrones were beforehand prepared by Swern oxidation of protected L-threitol followed by reaction with appropriate hydroxylamines. Then, starting from D-*manno*-configurated olefins **65** or **66**, cycloaddition with nitrone **67** led to a single product **68** and **69**, respectively. This indicates an excellent facial selectivity induced by steric hindrance on

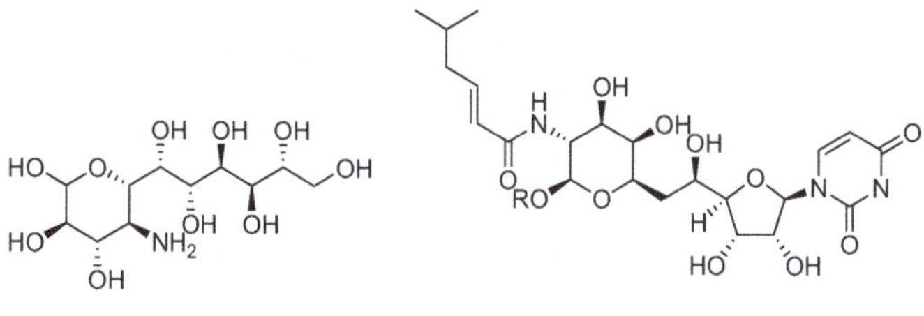

Scheme 16 Long-chain sugars natural products

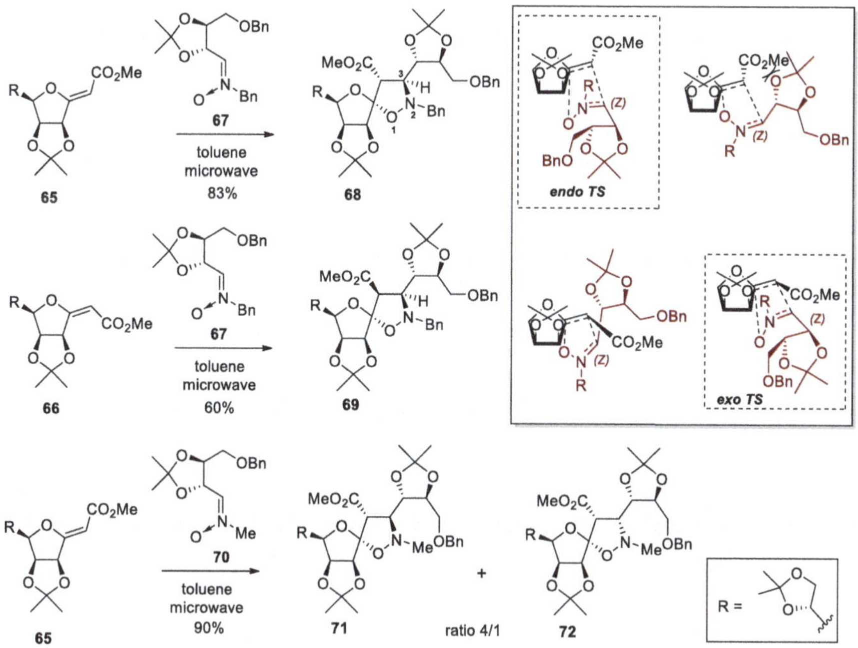

Scheme 17 Cycloadditions with open-chain sugar-derived nitrones

β-face of *exo*-glycals. Moreover, due to the bulkiness of benzyl group on nitrone **67**, only one C-3 diastereoisomer was obtained. Configurations at C-3/C-4 were established from the coupling constants between protons H-4 and H-3 and indicated a *trans* configuration for adduct **68** and a *cis* configuration for **69**. These observations may be rationalized by an analysis of postulated transition states. In the case of *Z*-*exo*-glycal **65**, the *endo* TS minimizes the steric hindrance, leading to compound **68** exclusively. For *E*-*exo*-glycal **66**, interactions of the nitrone benzyl group with the sugar ring destabilizes the *endo* transition state and formation of **69** probably arose from an *exo* TS. When performed with less hindered nitrone **70** on

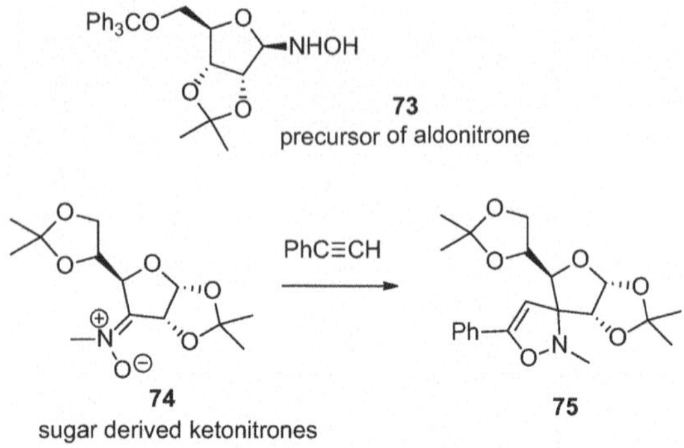

Scheme 18 1,3-Dipolar cycloaddition reactions of carbohydrate-derived aldonitrones and ketonitrones

exo-glycal **65**, the same reaction led to the formation of two C-3 diastereoisomers **71** and **72** in excellent yield, compound **71** (*trans* configuration) being the major product.

As seen in these examples, cycloaddition between carbohydrate-based nitrones and enoates is a valuable way to connect two sugars via generation of three new chiral centres, leading to complex carbohydrates and up to nine chiral centres. The remarkable selectivity of sugar-based nitrones has been explored on cycloaddition reactions with various olefins, and the last part of this chapter focuses on some representative examples.

3 Carbohydrate-Derived Nitrones and Oximes as 1,3-Dipoles

This part is dedicated to sugar-based nitrones and nitrile oxides involved in cycloaddition reactions with various olefins and suppress spiro-isoxazolines.

3.1 *Non-anomeric Sugar Nitrones or Nitrile Oxides*

The 1,3-dipolar cycloaddition of aldonitrones [8] and in particular sugar-derived aldonitrones prepared in situ from hydroxylamine derivatives (e.g. compound **73**) was fully described by Vasella et al. [40, 41] and allowed the synthesis of several interesting biologically active molecules. Similar reactions with more hindered sugar-derived ketonitrones like compound **74** were first described by Tronchet and Mihaly, but the yields were not reported (Scheme 18) [42].

Later, nitrone **76** was prepared from the corresponding 4-oxo-mannopyranose derivative and the cycloaddition reaction with ethyl vinyl ether afforded in a

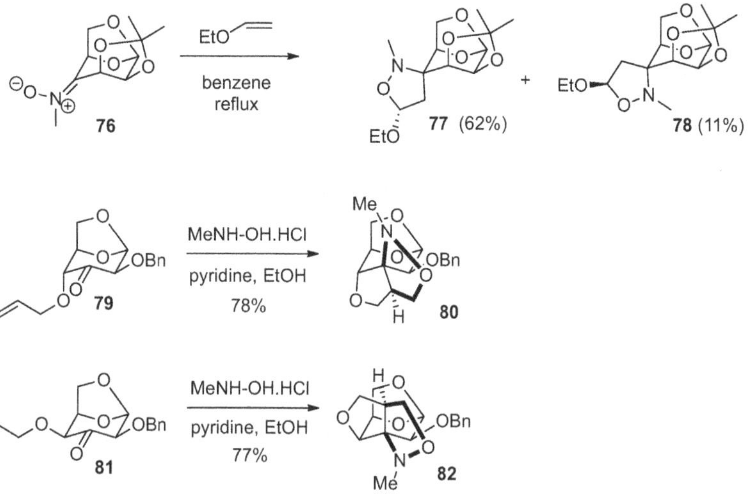

Scheme 19 Examples of heterocyclic systems obtained by 1,3-dipolar cycloaddition reactions

regioselective manner cycloadducts **77** and **78** in a good yield (73%) (Scheme 19) [43]. It is noteworthy that the quaternary centre was formed with a modest stereoselectivity (**77**:**78** ratio 6:1). An intramolecular version of this reaction was next investigated on various pyranoid or furanoid derivatives bearing a 4-*O*-allyl group as the internal dipolarophile. For example, with compound **79** (axial C-4 hydroxyl), the reaction proceeded with good yield and in this case a single stereoisomer **80** was isolated. Starting from **81** isomer (equatorial C-4 hydroxyl), the exclusive formation of the adduct **82** was observed. These results clearly showed that the stereochemistry of the reaction depends on the C-4 configuration. This methodology could thus be useful for the preparation of precursors of compounds of interest like (−)-tetrodotoxin and (+)-lactacystin requiring a nitrogenated quaternary centre [44, 45].

Nucleoside analogues with C-2′ or C-3′ spiro substitutions were prepared via intramolecular cycloaddition reactions between an electron-rich olefin and a vicinal nitrone (Scheme 20) [46, 47]. As an example, for synthesis of 3′-substituted spiro-isoxazolidine thymine derivative **84**, the reaction of 3′-ketone with *N*-methylhydroxylamine hydrochloride gave the 3′-methylnitrone **83** subsequently treated with ethyl vinyl ether. The cycloaddition reaction occurred in a fully diastereoselective manner, giving only compound **84** in 95% yield. The stereochemistry of the cycloadduct was explained by a postulated transition state, and the C-3′-(or C-2′-) configuration was proved by 1D NOE difference spectroscopy. Other 2′ or 3′- spiro-isoxazolidine derivatives were efficiently prepared by this methodology and constitute a new class of nucleoside analogues.

The 1,3-dipolar cycloaddition of sugar ketonitrones with differently substituted dipolarophiles was studied by Alonso et al. [48]. The aim of this work was to take advantage of this useful synthetic procedure to get original compounds with

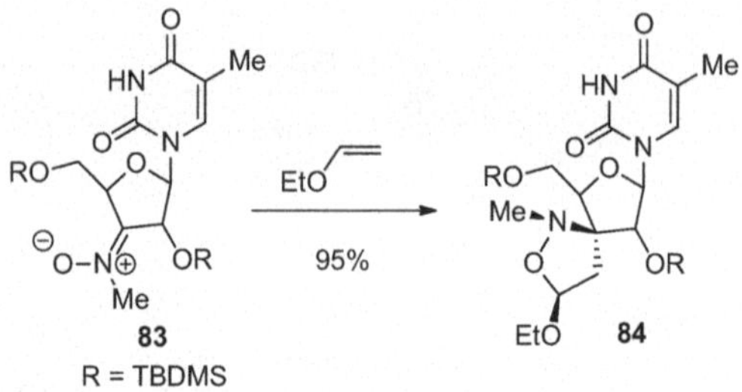

Scheme 20 Example of a spiro-isoxazolidine derivative of thymine obtained by 1,3-dipolar cycloaddition reactions

nitrogenated quaternary centres. The authors firstly developed highly efficient intramolecular procedures with mono-, di- and trisubstituted olefins. They then turned to an intermolecular version with ketonitrone **86** prepared by treatment of the corresponding lactone with methylhydroxylamine hydrochloride in pyridine (Scheme 21). When heated with ethyl vinyl ether in refluxing benzene, nitrone **86** led to cycloadducts **87** and **88** in 73% overall yield. The reaction occurred in a complete regioselective manner but with a moderate stereoselectivity as attested by the **87/88** ratio (6:1). To go further, the authors studied the reactivity of ketonitrone **90**, directly available from D-glucurono-3,6-lactone **89**. This tricyclic compound is an attractive system due to its high facial selectivity and can thus lead to only one stereoisomer. When **90** was reacted with phenylacetylene in refluxing benzene, the expected spiro-isoxazolidine **91** was obtained in a total regio- and stereoselective manner. A complete regio- and stereocontrol was also observed with allyl alcohol, leading to cycloadduct **92** in excellent yield. Only one stereoisomer was obtained, but the configuration of the newly created asymmetric carbon bearing the hydroxymethyl group was not established. It is worthy to note that the use of substituted dipolarophiles like 2-methyl-2-propen-1-ol gave low diastereoselectivity.

3.2 C-1 Sugar-Based Nitrones

Chiral isoxazolidines bearing a sugar moiety have been extensively developed by Yokoyama et al. via a tandem Michael addition-1,3-dipolar cycloaddition [11, 49, 50]. These authors described the preparation of sugar nitrone **94** by reaction of sugar oxime **93** with methyl acrylate. The intermediate nitrone **94**, prepared according to Griggs's reaction [51], was not isolated and reacted with another methyl acrylate molecule in a 1,3-dipolar cycloaddition reaction (Scheme 22). For this purpose, the mixture was heated in dry toluene in a sealed tube and gave the spiro-isoxazolidine **95** in a regio- and stereoselective manner. The cycloadducts **96** and **97** were obtained in 48% and 37% yield, respectively, by reaction with methyl vinyl ketone. The stereochemistry of compound **96** at C-3 was unambiguously determined by X-ray analysis. When reacted with acrylonitrile, an inseparable mixture of C-3 stereoisomers **98** was obtained.

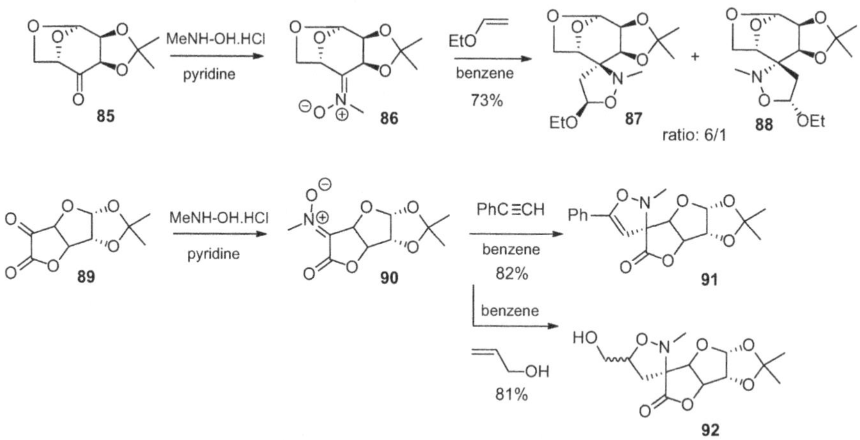

Scheme 21 1,3-Dipolar cycloaddition of sugar ketonitrones

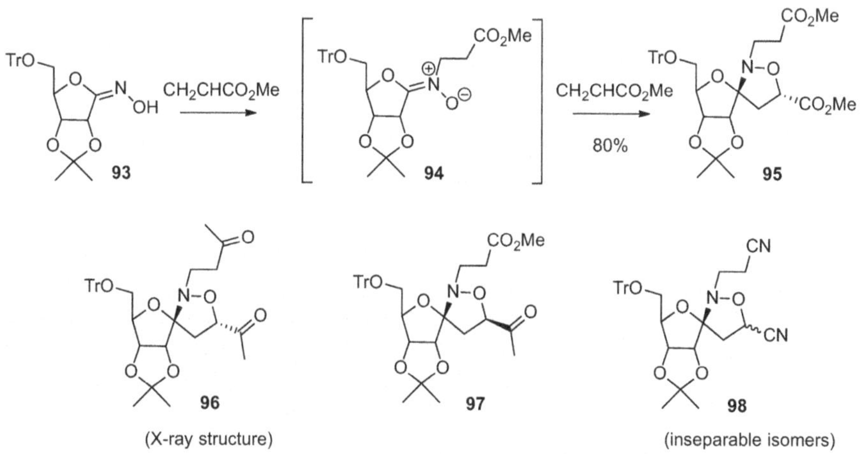

Scheme 22 Sugar hydroximolactones in a tandem Michael addition-1,3-dipolar cycloaddition

Yokoyama et al. also reported the preparation of two new bicyclic nucleosides **100** and **101** [52]. The formation of these original fused systems is presumed to proceed through a spiro-isoxazoline intermediate **99** prepared by cycloaddition of dimethyl acetylenedicarboxylate and sugar oxime **93**. Instability of spiro adduct **99** prompted the opening and consecutive closure of the sugar ring and **100** and **101** were obtained in a 2.7:1 ratio as a diastereoisomeric mixture in 85% overall yield (Scheme 23).

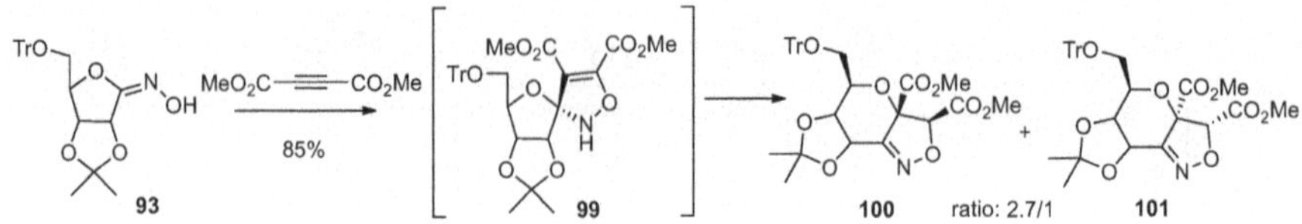

Scheme 23 New bicyclic nucleosides from hydroxyiminolactone

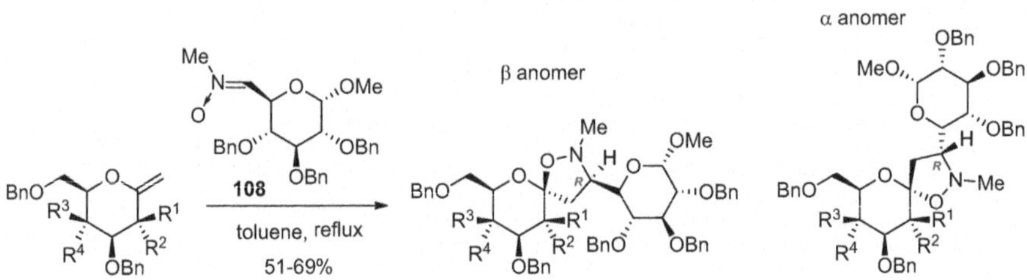

1 (D-*gluco*) $R^1 = R^3 = H$, $R^2 = R^4 = OBn$ — 104 — anomeric ratio 3:1 — 107
102 (D-*galacto*) $R^1 = R^4 = H$, $R^2 = R^3 = OBn$ — 105 — -
103 (D-*manno*) $R^2 = R^3 = H$, $R^1 = R^4 = OBn$ — 106 — -

Scheme 24 Reaction between *exo*-glycals and sugar nitrone

3.3 Reactions of Carbohydrate-Derived Nitrones and Nitrile Oxides with Exo-Methylene Sugars

In connection to their work on *exo*-methylene sugars (cf. I.1), Ikegami and coworkers carried out the diastereoselective cycloaddition of sugar nitrone **108** on *exo*-glycals **1**, **102** and **103** (Scheme 24) [38]. With glucose derivative **1**, the reaction afforded α and β spiro-isoxazolidines **104** and **107** with a 3:1 ratio in a 69% overall yield. NOE correlations analysis enabled assignment of absolute configuration of C-6 as *R* for both anomers. In the case of D-*galacto*- and D-*manno-exo*-glycals **102** and **103**, only β anomers **105** and **106** were obtained in average yields.

Similarly, cycloadditions between *exo*-glycals **1**, **102** and **103** and nitrile oxides were reported by Zhang et al. (Scheme 25) [53]. Preparation of sugar nitrile oxides **a–c** involved formation of oximes from aldehydes followed by chlorination and dehydrochlorination steps. Handling of nitrile oxides **a–c** proved complex, and they were used in the next step without workup or purification. When the cycloaddition reaction was carried out on glucose and mannose derivatives **1** and **103**, only α anomers **109a–c** and **111a–c** were obtained in all cases, indicating a diastereoselective addition. Concerning olefin **102**, the reaction was stereoselective with nitrile oxides **b** and **c**, but a 3:2 anomeric mixture was obtained with nitrile oxide **a**. It should be noted that a nitrile oxide dimer side product was also formed in all cases. After deprotection of the saccharidic spiro-isoxazolines by catalytic hydrogenation, their biological properties as glycosidase inhibitors were tested. Although the compounds showed a low inhibitory effect, they were selective

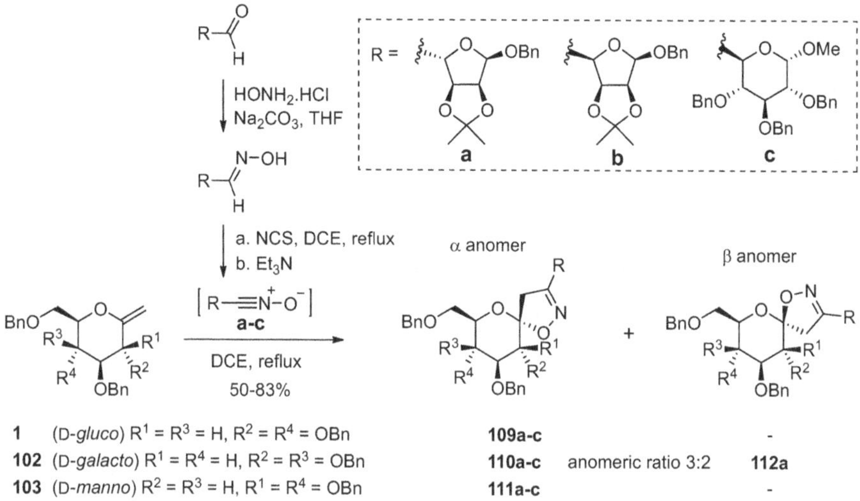

Scheme 25 Reactions between *exo*-glycals and sugar nitrile oxides

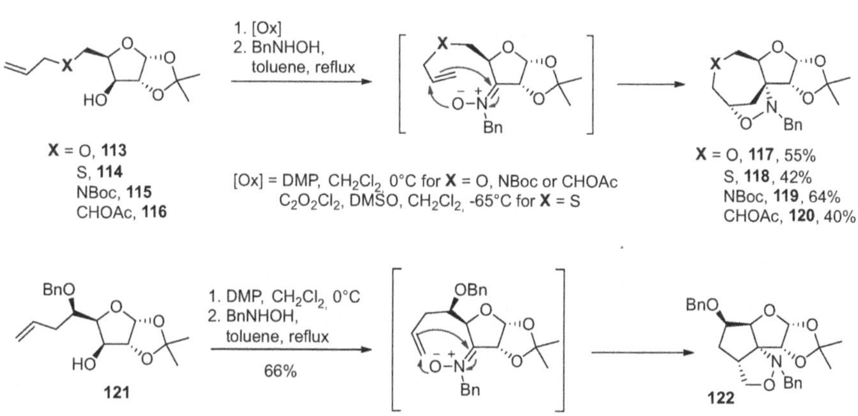

Scheme 26 Intramolecular nitrone – olefin cycloaddition

between glycosidases, with a higher inhibitory activity against α-amylase. As an example, at 2.6 mmol/mL, compound **111c** showed a 40% inhibitory activity against α-amylase, whereas its inhibitory activity against β-glucosidase was evaluated at 12% and barely reached 2% against α-glucosidase.

3.4 *Intramolecular Cycloadditions*

Recently, Das et al. published some examples of intramolecular cycloadditions on unsaturated sugar nitrones (Scheme 26) [12]. Starting from sugar olefins **113–116**,

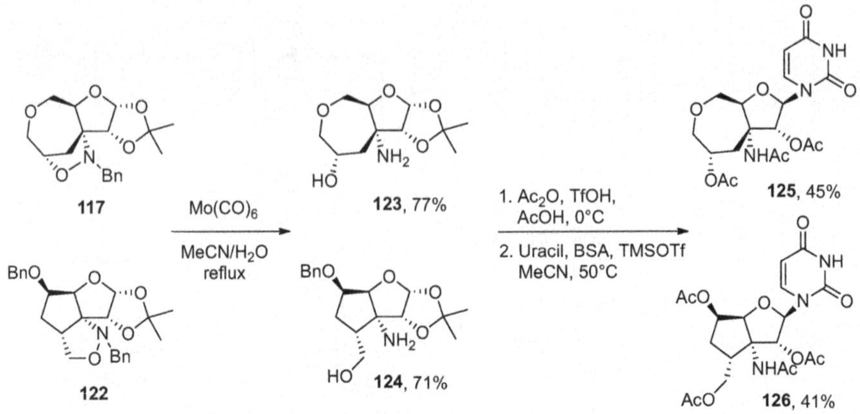

Scheme 27 Preparation of bicyclic nucleosides

the alcohol function was oxidized by Dess-Martin periodinane (DMP) or oxalyl chloride and dimethyl sulfoxide, followed by condensation of *N*-benzyl hydroxylamine on the ketone. The unstable nitrones underwent intramolecular cyclization to furnish exclusively spiro-isoxazolidine-fused seven-membered rings **117–120**. Stereochemistry of the cycloadducts was assigned by NMR analysis and X-ray crystallography. This method thus enables the formation of furanose-fused oxepane (**117**), thiepane (**118**), azepane (**119**) and cycloheptane (**120**). In the case of olefin **121**, attack of the nitrone occurred on the unsubstituted carbon of the double bond, leading to five-membered carbocycle **122** in good yield.

Furthermore, they took advantage of the labile nature of the N–O bond to prepare original bicyclic nucleoside derivatives (Scheme 27). Cleavage of the isoxazolidine ring of **117** and **122** was carried out with molybdenum hexacarbonyl in refluxing aqueous acetonitrile, leading to **123** and **124** in 77% and 71% yields, respectively. After deprotection of the sugar 1,2-diol of compounds **123** and **124** and subsequent acetylation of the free amino and hydroxyl groups, the fully protected compounds underwent glycosylation in Vorbrüggen conditions to give nucleosides **125** and **126** in 45% and 41% yields.

4 Conclusion

The 1,3-dipolar cycloaddition reaction between an olefin and a nitrile oxide or a nitrone is an efficient way to prepare isoxazolines and isoxazolidines, respectively. The adaptation of this reaction on saccharidic systems has been studied in the literature for several decades. In this chapter, we put the emphasis on the preparation of saccharidic spiro derivatives containing these heterocycles, an unexplored subject until the preliminary results of Rajanbabu et al. in 1986. Since then, the syntheses of saccharidic spiroisoxazoli(di)nes have been expanded on various types of olefins, notably *exo*-methylene sugars or activated *exo*-glycals. The cycloaddition proceeds in many cases with high regio- and stereocontrol, and the stereochemistry of the adducts can be established by NMR experiments and/or X-ray analyses. Elegant examples involving sugar-based nitrile oxides and nitrones and also intramolecular cycloadditions were reported and paved the way to the preparation of original bicyclic systems and various carbocycles. In several examples, the properties of the obtained saccharidic cycloadducts were investigated for various biological applications such as spironucleoside analogues, antibacterial agents or enzyme inhibitors. In addition to their biological properties and taking into account the originality of their structures, these systems were utilized as sugar scaffolds and proved suitable for selective derivatization.

References

1. Berthet M, Cheviet T, Dujardin G, Parrot I, Martinez J (2016) Isoxazolidine: a privileged scaffold for organic and medicinal chemistry. Chem Rev 116:15235–15283
2. Tufariello JJ (1984) Nitrones. 1,3-Dipolar cycloadditions in chemistry. Wiley, New York, pp 83–168
3. Confalone PN, Huie EM (2004) The [3 + 2] nitrone–olefin cycloaddition reaction. Organic reactions. American Cancer Society, Atlanta, pp 1–173
4. Padwa A, Pearson WH (eds) (2002) Synthetic applications of 1,3-dipolar cycloaddition chemistry toward heterocycles and natural products. Wiley, New York
5. Gothelf KV (2001) Asymmetric metal-catalyzed 1,3-dipolar cycloaddition reactions. Cycloaddition reactions in organic synthesis. Wiley, Weinheim, pp 211–247
6. Kaur K, Kumar V, Sharma AK, Gupta GK (2014) Isoxazoline containing natural products as anticancer agents: a review. Eur J Med Chem 77:121–133
7. Frederickson M (1997) Optically active isoxazolidines via asymmetric cycloaddition reactions of nitrones with alkenes: applications in organic synthesis. Tetrahedron 53:403–425
8. Gothelf KV, Jørgensen KA (1998) Asymmetric 1,3-dipolar cycloaddition reactions. Chem Rev 98:863–910
9. Giuliano RM (1992) Cycloaddition reactions in carbohydrate chemistry. An overview. ACS Symp Ser 494:1–23
10. Paton RM (1998) The nitrile oxide/isoxazoline route to C-disaccharides. In: Chapleur Y (ed) Carbohydrate mimics. Wiley, Hoboken, pp 49–66
11. Osborn HMI, Gemmell N, Harwood LM (2002) 1,3-Dipolar cycloaddition reactions of carbohydrate derived nitrones and oximes. J Chem Soc Perkin Trans 1:2419–2438
12. Das SN, Chowdhury A, Tripathi N, Jana PK, Mandal SB (2015) Exploitation of in situ generated sugar-based olefin keto-nitrones: synthesis of carbocycles, heterocycles, and nucleoside derivatives. J Org Chem 80:1136–1148
13. Taillefumier C, Chapleur Y (2004) Synthesis and uses of *exo*-glycals. Chem Rev 104:263–292
14. Pellegrini-Moïse N, Richard M, Chapleur Y (2014) Chapter 6. *Exo* -glycals as useful tools for anomeric functionalization of sugars. In: Pilar Rauter A, Lindhorst T, Queneau Y (eds) Carbohydrate chemistry, vol 40. Royal Society of Chemistry, Cambridge, pp 99–117
15. Rajan Babu TV, Reddy GS (1986) 1-Methylene sugars as C-glycoside precursors. J Org Chem 51:5458–5461
16. Colinas PA, Jäger V, Lieberknecht A, Bravo RD (2003) Nitrile oxide cycloadditions to olefinated sugars. Tetrahedron Lett 44:1071–1074

17. Benltifa M, Vidal S, Gueyrard D, Goekjian PG, Msaddek M, Praly J-P (2006) 1,3-Dipolar cycloaddition reactions on carbohydrate-based templates: synthesis of spiro-isoxazolines and 1,2,4-oxadiazoles as glycogen phosphorylase inhibitors. Tetrahedron Lett 47:6143–6147
18. Benltifa M, Hayes JM, Vidal S, Gueyrard D, Goekjian PG, Praly JP, Kizilis G, Tiraidis C, Alexacou KM, Chrysina ED, Zographos SE, Leonidas DD, Archontis G, Oikonomakos NG (2009) Glucose-based spiro-isoxazolines: a new family of potent glycogen phosphorylase inhibitors. Bioorg Med Chem 17:7368–7380
19. Krimm I, Lancelin J-M, Praly J-P (2012) Binding evaluation of fragment-based scaffolds for probing allosteric enzymes. J Med Chem 55:1287–1295
20. Goyard D, Kónya B, Chajistamatiou AS, Chrysina ED, Leroy J, Balzarin S, Tournier M, Tousch D, Petit P, Duret C, Maurel P, Somsák L, Docsa T, Gergely P, Praly JP, Azay-Milhau J, Vidal S (2016) Glucose-derived spiro-isoxazolines are anti-hyperglycemic agents against type 2 diabetes through glycogen phosphorylase inhibition. Eur J Med Chem 108:444–454
21. Somsák L, Bokor É, Czibere B, Czifrák K, Koppány C, Kulcsár L, Kun S, Szilágyi E, Tóth M, Docsa T, Gergely P (2014) Synthesis of C-xylopyranosyl- and xylopyranosylidene-spiro-heterocycles as potential inhibitors of glycogen phosphorylase. Carbohydr Res 399:38–48
22. Goyard D, Telligmann SM, Goux-Henry C, Boysen MMK, Framery E, Gueyrard D, Vidal S (2010) Carbohydrate-based spiro bis(isoxazolines): synthesis and evaluation in asymmetric catalysis. Tetrahedron Lett 51:374–377
23. Li X, Takahashi H, Ohtake H, Ikegami S (2003) Synthesis of ketosyl spiro-isoxazolidine by 1,3-dipolar cycloaddition of 1-methylenesugars with nitrones - a new access to C-glycosyl amino acids. Heterocycles 59:547–571
24. Ikegami S, Namme R (2009) Chemistry of 1-methylenesugars: synthetic utilizations to 1′-C-methyl-saccharides and related carbohydrates. Heterocycles 78:19–44
25. Li X, Wang R, Wang Y, Chen H, Li Z, Ba C, Zhang J (2008) Stereoselective synthesis and biological activity of novel spiro-oxazinanone-C-glycosides. Tetrahedron 64:9911–9920
26. Gallos JK, Koftis TV, Koumbis AE, Moutsos VI (1999) A novel carbocyclic ring closure of hex-5-enopyranosides and pent-4-enofuranosides. Synlett 1999:1289–1291
27. Gallos JK, Demeroudi SC, Stathopoulou CC, Dellios CC (2001) Expeditious synthesis of seven-membered iminocyclitols. Tetrahedron Lett 42:7497–7499
28. Gallos JK, Koftis TV (2001) Carbocyclic ring closure of hex-5-enopyranosides and pent-4-enofuranosides: a nitrile oxide approach. J Chem Soc Perkin Trans 1:415–423
29. Lakhrissi M, Chapleur Y (1996) Wittig olefination of lactones. Angew Chem Int Ed Engl 35:750–752
30. Enderlin G, Taillefumier C, Didierjean C, Chapleur Y (2005) Cycloaddition reactions on activated exo-glycals. Tetrahedron Asymmetry 16:2459–2474
31. Taillefumier C, Enderlin G, Chapleur Y (2005) Cycloaddition of nitrones and nitrile oxides to activated exo-glycals. Lett Org Chem 2:226–230
32. Salgado H, Jiménez R, Pérez L, Tamariz J, Salgado H (1993) 1,3-Dipolar cycloaddition reactions involving captodative olefins. Heterocycles 35:591–598
33. Nakajima M, Itoi K, Takamatsu Y, Kinoshita T, Okazaki T, Kawakubo K, Shindo M, Honma T, Tohjigamori M, Haneishi T (1991) Hydantocidin: a new compound with herbicidal activity from Streptomyces hygroscopicus. J Antibiot (Tokyo) 44:293–300
34. Richard M, Chapleur Y, Pellegrini-Moïse N (2016) Spiro sugar-isoxazolidine scaffold as useful polyfunctional building block for peptidomimetics design. Carbohydr Res 422:24–33
35. Novoa A, Pellegrini-Moïse N, Bechet D, Barberi-Heyob M, Chapleur Y (2010) Sugar-based peptidomimetics as potential inhibitors of the vascular endothelium growth factor binding to neuropilin-1. Bioorg Med Chem 18:3285–3298
36. Richard M, Chateau A, Jelsch C, Didierjean C, Manival X, Charron C, Maigret B, Barberi-Heyob M, Chapleur Y, Boura C, Pellegrini-Moïse N (2016) Carbohydrate-based peptidomimetics targeting neuropilin-1: synthesis, molecular docking study and in vitro biological activities. Bioorg Med Chem 24:5315–5325

37. Gong C, Valduga J, Chateau A, Richard M, Pellgrini-Moïse N, Barberi-Heyob M, Chastagner P, Boura C (2018) Stimulation of medulloblastoma stem cells differentiation by a peptidomimetic targeting neuropilin-1. Oncotarget 9:15312–15325
38. Li X, Takahashi H, Ohtake H, Ikegami S (2004) 1,3-Dipolar cycloaddition of exo -methylenesugars with nitrone: approach to new amino-C-ketosyl disaccharides. Tetrahedron Lett 45:4123–4126
39. Oukani H, Pellegrini-Moïse N, Jackowski O, Chrétien F, Chapleur Y (2013) The 1,3-dipolar cycloaddition reaction of chiral carbohydrate-derived nitrone and olefin: towards long-chain sugars. Carbohydr Res 381:205–214
40. Vasella A, Voeffray R (1982) Total Synthese von Nojirimycin. Helv Chim Acta 65:1134–1144
41. Vasella A, Voeffray R, Pless J, Huguenin R (1983) Synthesis of D- and L-5-oxaproline and of a new captopril analogue. Helv Chim Acta 66:1241–1252
42. Tronchet JMJ, Mihaly E (1972) C-glycosylic derivatives. VII. Synthesis and reactions of sugar nitrones. Helv Chim Acta 55:1813–1815
43. Torrente S, Noya B, Paredes MD, Alonso R (1997) Intramolecular 1,3-dipolar cycloadditions of sugar ketonitrones: a convenient method for stereoselective formation of nitrogenated quaternary centers. J Org Chem 62:6710–6711
44. Kishi Y, Aratani M, Fukuyama T, Nakatsubo F, Goto T, Inoue S, Tanino H, Sugiure S, Kakoi H (1972) Synthetic studies on tetrodotoxin and related compounds. IV. Stereospecific total syntheses of DL-tetrodotoxin. J Am Chem Soc 94:9219–9221
45. Fenteany G, Standaert RF, Lane WS, Choi S, Corey EJ, Schreiber SL (1995) Inhibition of proteasome activities and subunit-specific amino-terminal threonine modification by lactacystin. Science 268:726–731
46. Hossain N, Papchikhin A, Plavec J, Chattopadhyaya J (1993) Synthesis of 2′- and 3′-spiro-isoxazolidine derivatives of thymidine & their conversions to 2′,3′-dideoxy-2′,3′-didehydro-3′-C-substituted nucleosides by radical promoted fragmentation. Tetrahedron 49:10133–10156
47. Rong J, Roselt P, Plavec J, Chattopadhyaya J (1994) The synthesis and conformation of 2′- and 3′-hypermodified tricyclic nucleosides and their use in the synthesis of novel 2′- or 3′-isomeric 4 (7)-substituted isoxazolidine-nucleosides. Tetrahedron 50:4921–4936
48. Torrente S, Noya B, Branchadell V, Alonso R (2003) Intra- and intermolecular 1,3-dipolar cycloaddition of sugar ketonitrones with mono-, di-, and trisubstituted dipolarophiles. J Org Chem 68:4772–4783
49. Yokoyama M, Yamada N, Togo H (1990) Synthesis of spiro sugar isoxazolidines via tandem michael addition-1,3-dipolar cycloaddition. Chem Lett 19:753–756
50. Yokoyama M, Sujino K, Irie M, Yamazaki N, Hiyama T, Yamada N, Togo H (1991) Additional reactions of sugar oximes, nitrile oxides and hydroximolactones. J Chem Soc Perkin Trans 1:2801–2809
51. Grigg R, Markandu J, Perrior T, Surendrakumar S, Warnock WJ (1992) X = Y - ZH systems as potential 1,3-dipoles part 35. Generation of nitrones from oximes. Class 3 processes. Tandem intramolecular Michael addition (1,3-azaprotio cyclotransfer) – intermolecular 1,3-dipolar cycloaddition reactions. Tetrahedron 48:6929–6952
52. Yokoyama M, Yamada N (1989) Synthesis of spiro and bicyclic nucleosides from ribose nitrile oxide with dimethyl acetylenedicarboxylate. Tetrahedron Lett 30:3675–3676
53. Zhang PZ, Li XL, Chen H, Li YN, Wang R (2007) The synthesis and biological activity of novel spiro-isoxazoline C-disaccharides based on 1,3-dipolar cycloaddition of exo-glycals and sugar nitrile oxides. Tetrahedron Lett 48:7813–7816

Top Heterocycl Chem (2019) 57: 51–104
DOI: 10.1007/7081_2019_30

Published online: 16 February 2019

Carbohydrate Spiro-heterocycles via Radical Chemistry

Angeles Martín and Ernesto Suárez

Contents

A. Martín and E. Suárez (✉)
Síntesis de Productos Naturales, Instituto de Productos Naturales y Agrobiología del CSIC, San Cristóbal de La Laguna, Tenerife, Spain
e-mail: angelesmartin@ipna.csic.es; esuarez@ipna.csic.es

Abstract This review summarizes the current status of the preparation of spiro-heterocycles fused with a pyranose or furanose carbohydrate skeleton, using free radical chemistry. A variety of heterospiro[*m.n*]alkane bicyclic structures ($m = 3$–5, $n = 4$–6) possessing one, two, or three heteroatoms (N, O, Si, S) have been collected, in addition to three different 1,6,8-trioxadispiro-tetradecane and 1,6,8-trioxadispiro-pentadecane tricyclic systems. C(sp^3)–H bond functionalization by 1,5- or 1,6-hydrogen atom transfer (HAT) initiated by C(sp^3)-, C(sp^2)-, *O*-, or *N*-radicals and 5-*exo*-trig or 6-*exo*-trig cyclization reactions are the most useful strategies employed for the construction of the heterocyclic rings. The intramolecular HAT promoted by photoexcited monoketones, α-diketones, furanones, and succinimides via a Norrish type II–Yang cyclization process has also been successfully applied.

Keywords Alkoxyl radicals · Carbohydrates · C-radicals · Hydrogen atom transfer (HAT) · N-radicals · Radical cascade · Radical cyclization

1 Introduction

Carbohydrates and radical chemistry have maintained a fruitful relationship over the last decades. As a result, a variety of new and more efficient methodologies have emerged for the synthesis of different sugar derivatives such as *C*-glycosides, *C*-ketosides, and polyhydroxylated carbocycles. For instance, the intramolecular hydrogen atom transfer (HAT) processes promoted by *C*-, *N*-, and *O*-centered radicals have been shown to be extremely effective for the remote and selective functionalization of unactivated C(sp^3)–H bonds of the sugar skeleton or even between units of contiguous sugars in di- and oligosaccharide systems. Moreover, the β-fragmentation of alkoxyl radicals permits unique synthetic transformations of alcohols and the preparation of structures that would otherwise be more difficult to synthesize by other methods.

The application of these radical processes in carbohydrate chemistry has been comprehensively reviewed covering the literature up to 2011 [1, 2]. Since then, an excellent book in two volumes on this subject has also been published [3], and some other related topics have been collected in more specific reviews [4–9].

This book chapter provides an up-to-date review of the construction, using specifically radical chemistry of spiro-heterocycles fused with a carbohydrate skeleton. To the best of our knowledge, this topic has never hitherto been reviewed. However, several more general reviews on the preparation of spiro-heterocycles are available [10, 11]. Additional reports on the synthesis of heterocycles from carbohydrate precursors [12, 13], the use of free-radical chemistry in the synthesis and functionalization of heterocycles [14], and on spirocyclic scaffolds present in natural products [15–19] have also been recently published.

The discussion has been classified into sections according to total number of atoms of the spiro-system, size of the smaller ring, number of heteroatoms involved, and finally molecular weight of the system (Table 1). A brief overview of the radical processes most frequently involved in the cascade reactions has also been included below.

1.1 Radical Chain Processes

As will be seen throughout this review, complex transformations can be achieved that in many cases surprise by their extraordinary regio- and stereoselectivity using radical processes in chain [20]. An excellent example is the new methodology developed independently by Tanaka and Chatgilialoglu for the transformation of 6-(2,2-dibromovinyl)uridine **I** into anomeric spironucleoside **V**, described in more detail in Sect. 3.2.1 (Fig. 1) [21, 22].

Hexabutylditin serves as an initiator, which undergoes photochemical cleavage to give tributylstannyl radicals. This reactive species goes into the propagation cycle that comprises a cascade of radical reactions involving bromine homolysis with formation of the vinyl radical **II**, 1,5-HAT of the γ-hydrogen at C1′ of the sugar moiety **III**, 5-*endo*-trig cyclization of the anomeric radical into the double bond of the tether at C6 **IV**, and finally bromine elimination to give the desired 1-oxa-6-azaspiro[4.4]nonane system **V**. The homolytic cleavage of the hexabutylditin by the bromine radical completes the chain cycle.

1.2 Hydrogen Atom Transfer (HAT)

Efficient 1,*n*-HAT ($n = 4$–8) initiated by C(sp^3)-, C(sp^2)-, alkoxyl, and amidyl radicals has been described, but 1,5-HAT is by far the most widely used and is present in almost all the sections of this chapter [23–34]. However, 1,6-HAT can also be of synthetic utility as noted in several sections of this review (see, e.g., Sects. 4.1.2 and 5.2.2).

Alkoxyl radicals, as electrophilic species, are highly efficient to perform this type of reactions. Computational work by Houk and coworkers has demonstrated that the six-membered transition state (TS) involved in the 1,5-HAT promoted by alkoxyl radicals resembles a five-membered ring in an envelope conformation, with a long bond constituted by C–H–O (2.5 Å) (Fig. 2) [35–37]. The angle of the C–H–O is close to linearity (153°) and quite different from that expected for a chair-like

Table 1 Main spirocyclic structures described

Name	Structure	Section
5-Oxaspiro[3.5]nonane	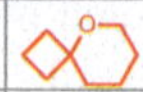	2.1
1,5-Dioxaspiro[3.5]nonane		2.2
1-Oxaspiro[4.4]nonane	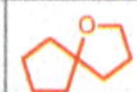	3.1
1-Oxa-6-azaspiro[4.4]nonane	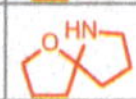	3.2
1-Oxa-7-azaspiro[4.4]nonane	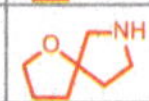	3.3
1,6-Dioxaspiro[4.4]nonane	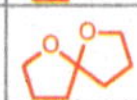	3.4
1,6-Dioxa-4-azaspiro[4.4]nonane	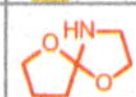	3.5
6-Oxa-1-aza-3-silaspiro[4.4]nonane		3.6
1,7-Dioxa-8-silaspiro[4.4]nonane	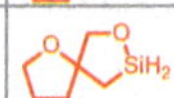	3.7
1-Oxa-6-azaspiro[4.5]decane	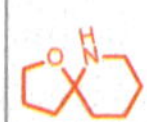	4.1
6-Oxa-1-azaspiro[4.5]decane	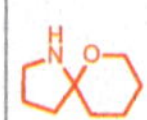	4.2
1,6-Dioxaspiro[4.5]decane		4.3
6-Oxa-1,3-diazaspiro[4.5]decane	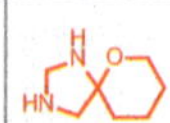	4.4
1,4,6-Trioxaspiro[4.5]decane	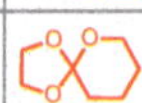	4.5
1,3,6-Trioxaspiro[4.5]decane		4.6
1,7-Dioxa-8-silaspiro[4.5]decane	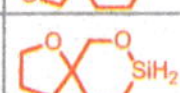	4.7
1,6-Dioxa-9-thiaspiro[4.5]decane	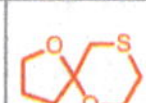	4.8
1-Oxa-7-azaspiro[5.5]undecane		5.1
1,7-Dioxaspiro[5.5]undecane		5.2
1,5,7-Trioxaspiro[5.5]undecane		5.3
7-Oxa-1,5-dithiaspiro[5.5]undecane	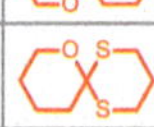	5.4
1-Oxa-8-azaspiro[5.6]dodecane	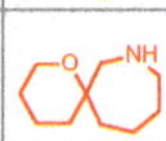	6.1
1,6,8-Trioxadispiro[4.1.4^{7}.3^{5}]tetradecane	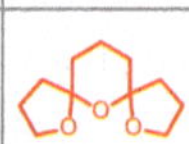	7.1
1,6,8-Trioxadispiro[4.1.5^{7}.2^{5}]tetradecane		8.1
1,6,8-Trioxadispiro[4.1.5^{7}.3^{5}]pentadecane	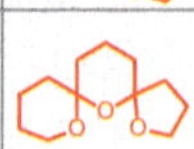	9.1

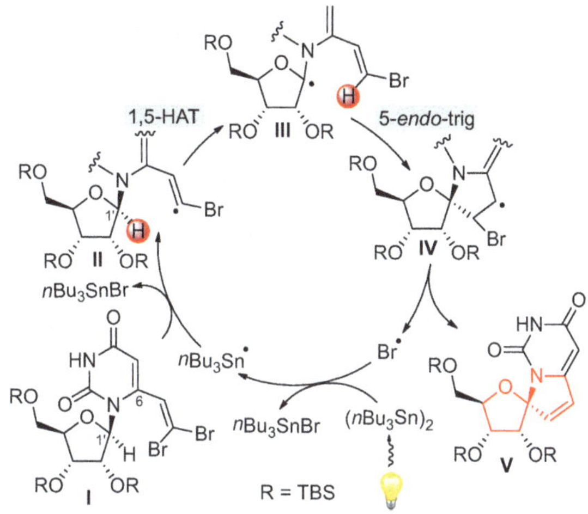

Fig. 1 Tanaka and Chatgilialoglu radical chain synthesis of spironucleosides [21, 22]

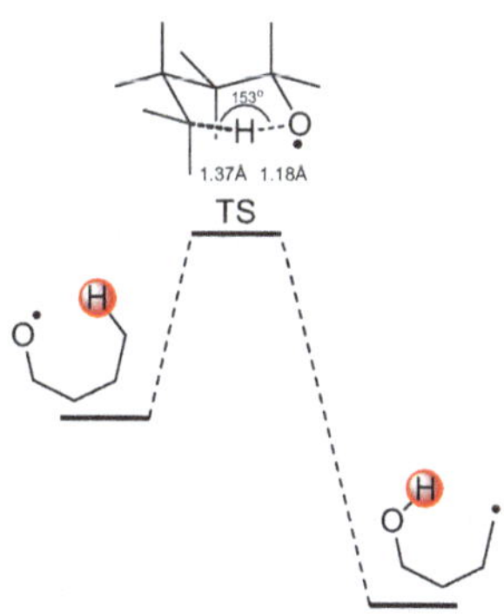

Fig. 2 Houk's computed transition state for 1,5-HAT by alkoxyl radicals

conformation of a cyclohexane ring (109°). Typically, the distance between the radical center and the hydrogen atom to be abstracted should be approximately 3 Å. The new methods that are now available for the generation of radicals under milder conditions, i.e., the use of hypervalent iodine reagents [38–40] and photoredox conditions involving a catalytic SET reaction [41–45], open a wider range of synthetic possibilities for the HAT processes [46–48].

Carbonyl compounds undergo photoinduced intramolecular hydrogen atom transfer (PHAT) to form intermediate biradicals which can be stabilized principally by two competitive processes: Norrish type II photoelimination and Yang photocyclization [49–52]. These 1,5- and 1,6-PHAT by excited carbonyls closely resemble the HAT by alkoxyl radicals commented above. The UV irradiation of 2-oxopropyl glycosides, schematically depicted in **VI** ($n = 1$) (Fig. 3), produces after intersystem crossing (ISC) a triplet ketone **VII** ($n = 1$) that undergoes 1,5-PHAT to form a 1,4-biradical **VIII** ($n = 1$) by a strongly favored abstraction of the γ-positioned anomeric hydrogen. The triplet 1,4-biradical mainly cleaves to photoenol and lactone **IX**, while the Yang cyclization is, in this particular case, a very minor process (see Sect. 2.2.1). In contrast, the *C*-glycosides with a

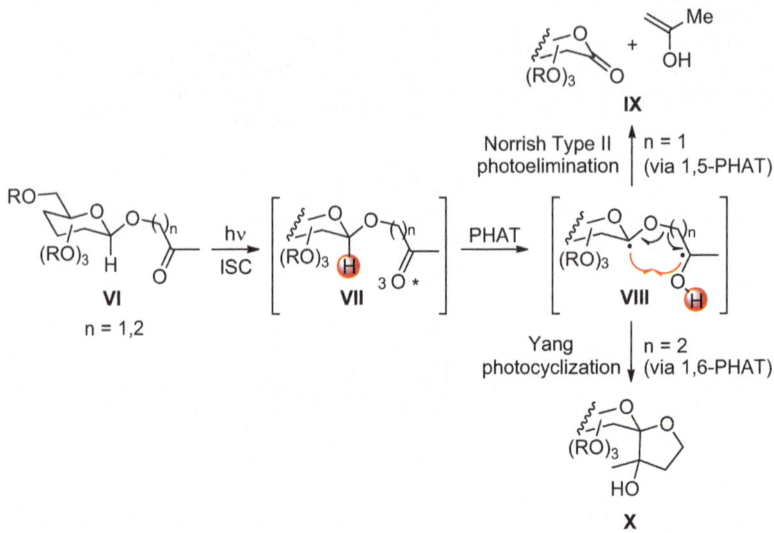

Fig. 3 Norrish type II–Yang cyclization mechanism

1,2-diketone tether described in Sect. 2.1.1 present a different behavior; after irradiation with visible light abstract the γ-hydrogen exclusively by the external carbonyl to give hydroxycyclobutanones. The 1,4-biradicals only cyclize; thus no cleavage products are detected.

The 2-oxobutyl glycosides **VI** ($n = 2$), in the absence of γ-hydrogens, abstracts the δ-anomeric hydrogen through a 1,6-PHAT reaction to form an intermediate 1,5-biradical **VIII** ($n = 2$). These biradicals cannot cleave and exclusively proceed via Yang photocyclization to give tetrahydrofuran rings **X** (see Sects. 4.3.1 and 7.1.1).

1.3 Radical Cyclization

The ability of free-radical chemistry to construct polyfunctionalized carbocyclic and heterocyclic frameworks by intramolecular cyclization of acyclic carbohydrates [53–57] has received considerable attention among synthetic organic chemists. By far, the most frequently used radical cyclizations involve the formation of five- and six-membered rings. Examples of this reaction initiated by C(sp^3)-, C(sp^2)-, or alkoxyl radicals with alkenes, alkynes, and aldehydes as acceptors can be found throughout this review. According to the Baldwin [58, 59] and Beckwith [60–62] guidelines, 4-pentenyl, 5-hexenyl, and 6-heptenyl radicals cyclize predominantly through favored *exo*-trig processes to give the smaller rings **XI**, **XII**, and **XIII**, respectively (Fig. 4). The alternative *endo*-trig cyclizations lead generally to minor products (i.e., **XV** and **XVI**), or in the case of 5-*endo*-trig **XIV**, the process is classified as disfavored. The 5-hexenyl and 6-heptenyl cyclization reactions proceed

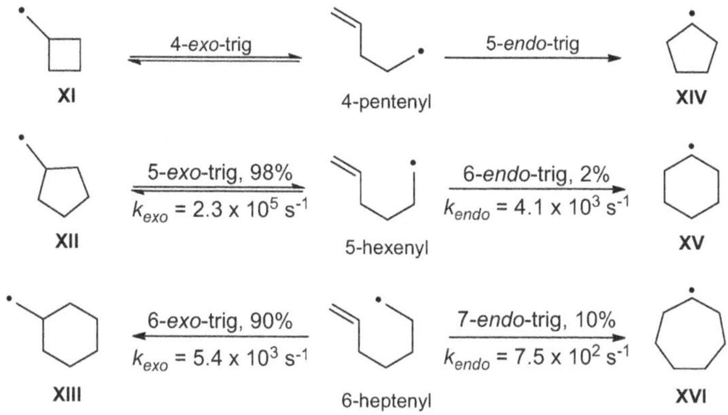

Fig. 4 Ring closure of ω-alkenyl radicals (rate constants taken from Ref. [62])

under kinetic control to make the less stable primary radicals **XII** and **XIII** much faster than the secondary radicals. The regioselectivity may change if the 5-hexenyl radical cyclization is reversible and the reaction is run under thermodynamic conditions; six-membered rings are formed predominantly by a 6-*endo*-trig process (i.e., **XV**).

Although the 5-*endo*-trig mode has been predicted by Baldwin and Beckwith as a disfavored process, a number of efficient and synthetically useful examples have been recently discovered (see Sects. 3.2.1 and 4.6.1) [63–69]. Kinetic and theoretical studies of the competition between 4-*exo*-trig and 5-*endo*-trig cyclization of carbamoylmethyl radicals have shown that, contrary to previous beliefs, the 5-*endo*-trig ring closure is not only thermodynamically but also kinetically favored [70]. The 4-*exo*-trig cyclization is not an easy process either; it is reversible, and on thermodynamic grounds, the isomerization equilibrium is shifted to the acyclic radical. The reaction occurs preferentially when the radical intermediates, after cyclization, are highly stabilized by adjacent groups. The stereoselective construction of substituted tetrahydrofurans from 4-penten-1-oxyl radical ring closure reactions has been reviewed [71]. An example where the alkoxyl radical adds intramolecularly to the alkene in a 5-*exo*-trig form can be found in Sect. 4.3.4.

2 Heterospiro[3.5]nonanes

2.1 5-Oxaspiro[3.5]nonanes

2.1.1 Photoinduced Norrish Type II–Yang Cyclization

The Norrish–Yang photocyclization of methyl nonopyranoside-7,8-diulose **1** with a 2,3-butanedione tether extending from C5 afforded 5-oxaspiro[3.5]nonane **2** with total stereoselectivity in high yield (Scheme 1) [72, 73]. The reaction proceeded

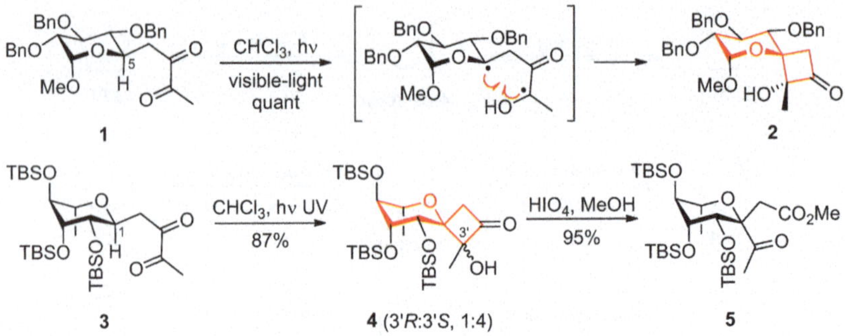

Scheme 1 5-Oxaspiro[3.5]nonanes via Norrish type II–Yang cyclization

under irradiation with visible or UV light through a 1,5-photoinduced hydrogen atom transfer (1,5-PHAT) of the axial hydrogen at C5, promoted by the external excited carbonyl, followed by 1,4-biradical cyclization and diastereoselective formation of the hydroxycyclobutanone. The reaction strongly depends on conformational and stereoelectronic factors, for example, the photocyclization of the TBS-protected *C*-glycoside **3** derived from L-rhamnose, which is in a $^{4}C_{1}$ conformation gave **4** also with high selectivity. Notwithstanding, the photoreaction of the same *C*-glycoside protected by acetyl groups, which is now in a $^{1}C_{4}$ conformation, is initiated by a 1,5-PHAT of the equatorial hydrogen at C1, to proceed with low selectivity, giving a mixture of the four possible isomers. The hydroxycyclobutanone moiety can be opened under oxidative conditions to give *C*-ketosides (e.g., **5**) which are valuable synthetic intermediates [74].

2.2 *1,5-Dioxaspiro[3.5]nonanes*

2.2.1 Photoinduced Norrish Type II–Yang Cyclization

The irradiation of 2-oxopropyl α- and β-glycosides (e.g., **6**) leading exclusively to aldonolactones **8** has been reported by Descotes and coworkers (Scheme 2) [75, 76]. The reaction goes through a Norrish type II photoelimination by β-cleavage of the 1,4-biradical intermediate and is clearly related to the photochemistry of *S*-phenacyl thioglycosides described by Vasella [77] or the Binkley [78] photolysis of 1-*O*-pyruvate ester of glucopyranose derivatives, which give aldonothiolactones and aldonolactones, respectively. The possible oxetane compounds produced by a Yang cyclization of the 1,4-biradical intermediate were not detected in any of these reactions, except in the particular case of phenacyl glycoside **7** where Brunckova and Crich [79] have been able to isolate and characterize **10** in low yield (20%), apart from the lactone **9** (56%). To the best of our knowledge, this is the only example of a 1,5-dioxaspiro[3.5]nonane derivative described in a carbohydrate framework.

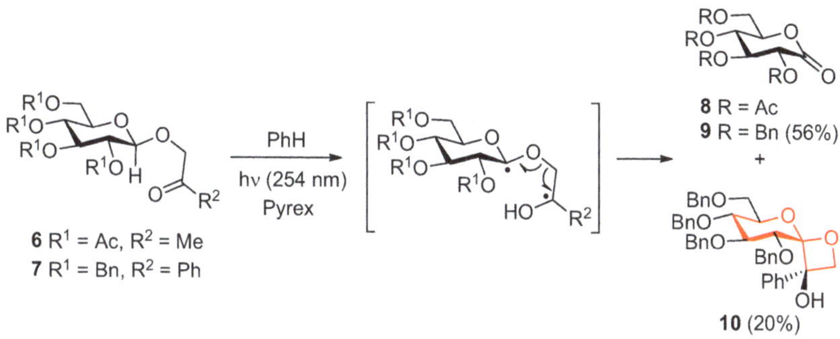

Scheme 2 1,5-Dioxaspiro[3.5]nonane via Norrish type II–Yang cyclization

3 Heterospiro[4.4]nonanes

3.1 1-Oxaspiro[4.4]nonanes

3.1.1 C(sp^3)-Radical-Initiated Intermolecular Addition–5-*exo*-trig Cyclization

Motherwell and coworkers have developed an efficient radical route for the preparation of the 1-oxaspiro[4.4]nonane system. Based on their synthesis of difluoromethylene-linked *C*-glycosides and *C*-disaccharides that implies an intermolecular addition of nucleophilic and electrophilic radicals to *gem*-difluoro *exo*-glycals [80], this group described an interesting extension to two spirofused difluorocarbocycles in a one-pot, two-step sequence (Scheme 3) [81]. The typical radical chain reaction mechanism is shown in the scheme; the intermolecular addition of diethyl allylmalonate radical to the exocyclic carbon atom of *exo*-glycals **11** and **13** generated the intermediate anomeric radical. Subsequent 5-*exo*-trig cyclization, from the least sterically hindered face, yielded, after $n\text{Bu}_3\text{SnH}$ reduction, the corresponding spirocycles **12** and **14** with total regio- and stereoselectivity.

3.2 1-Oxa-6-azaspiro[4.4]nonanes

3.2.1 C(sp^2)-Radical-Initiated 1,5-HAT–5-*endo*-trig Cyclization

In 1996, the Tanaka [21, 82, 83] and Chatgilialoglu [22, 84] groups independently reported the synthesis of a series of spironucleosides with a 1-oxa-6-azaspiro[4.4] nonane bicyclic framework, as shown in Scheme 4. The radical reaction of 6-(2,2-dibromovinyl)deoxyuridine **15** initiated by the $n\text{Bu}_3\text{SnH}$/AIBN or $(n\text{Bu}_3\text{Sn})_2$/h$\nu$ systems afforded the isomeric spirocyclic **16** and **17** in good yield. The first step of the reaction is the abstraction of a bromine atom by the stannyl

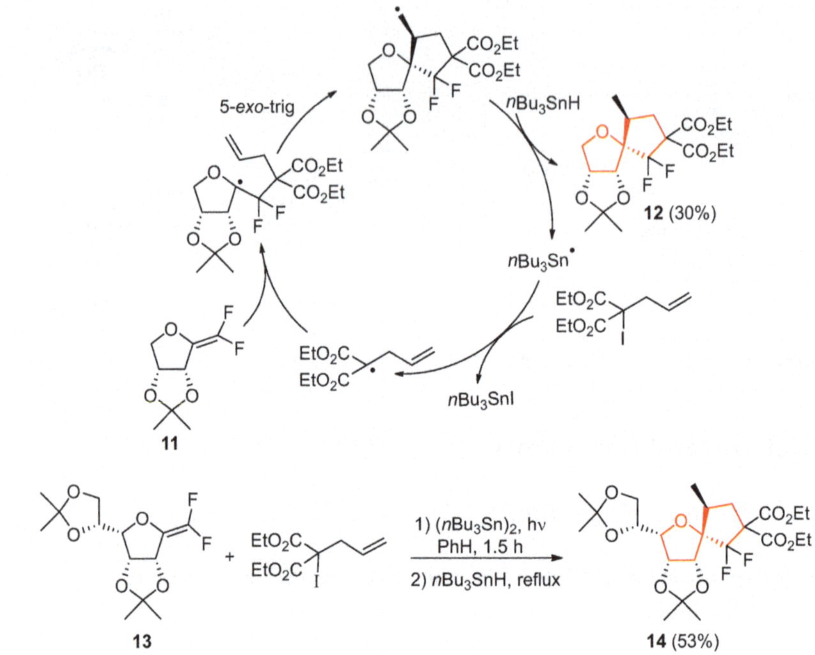

Scheme 3 1-Oxaspiro[4.4]nonanes via intermolecular addition–5-*exo*-trig cyclization

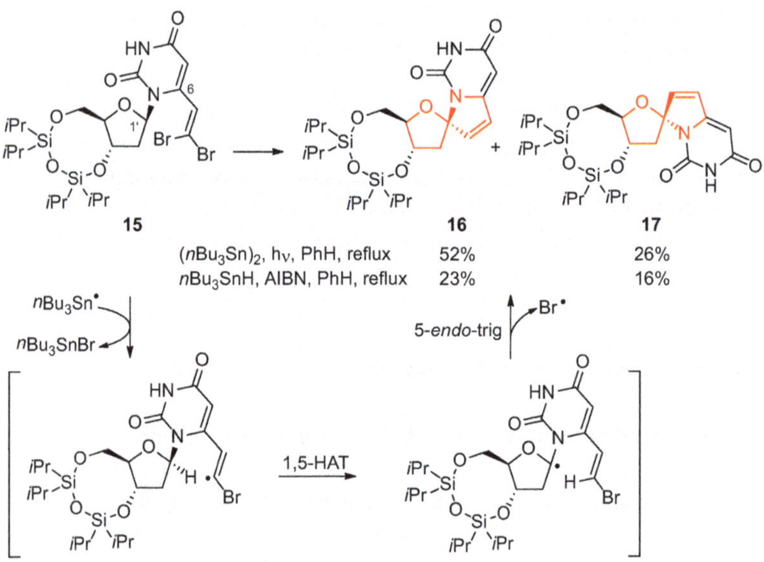

Scheme 4 1-Oxa-6-azaspiro[4.4]nonene via 1,5-HAT–5-*endo*-trig cyclization

radical to generate a monobromovinyl radical which through the equilibrated *E*-isomer abstracts regioselectively the anomeric hydrogen at C1′ via a 1,5-HAT process. A rare 5-*endo*-trig cyclization of the anomeric radical gives an adduct radical which is subsequently stabilized by bromine atom elimination. Several examples of this type of radical cyclization, classified as unfavorable by the Baldwin-Beckwith guidelines [58–62], have been appearing in recent years, and a theoretical study has supported that at least in certain cases, the process can be very efficient [70].

The stereoselectivity of the cyclization is strongly influenced by the stereochemistry and steric hindrance of the substituent at C2′, as can be deduced by the examples described in Table 2. The D-ribofuranosyl derivatives **18a–c** gave almost exclusively cyclized products **19a–c** by radical attack to the α-side of the molecule. The D-arabinofuranose derivatives **18d** and **18e**, notwithstanding, afforded principally β-cyclized products **20d** and **20e**.

To gain insight into the mechanism, (*E*)- and (*Z*)-6-(2-monobromovinyl)uridine (**21***E* and **21***Z*) were investigated (Scheme 5) [21]. Although the (*E*)-isomer **21***E* gave the spiro compound **23** in 31% yield, the (*Z*)-isomer **21***Z*, unexpectedly, afforded no cyclized product. The authors assume that the rate of attack by the

Table 2 1-Oxa-6-azaspiro[4.4]nonenes on D-*ribo* and D-*arabino* models

Substrate	Method[a]	Products		Reference
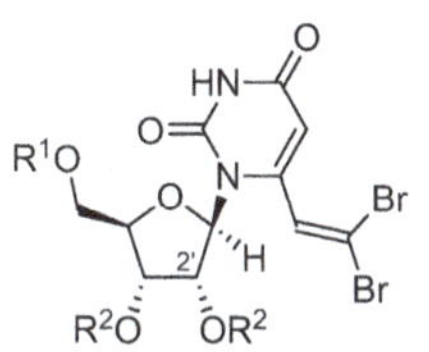 **18**		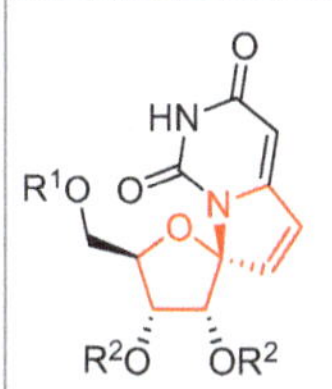 **19**	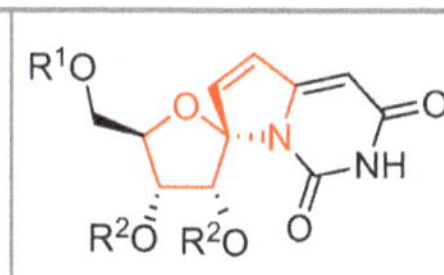 **20**	
a $R^1 = R^2 =$ TBS	A	37%	–	[22]
	B	50%	–	[21]
b $R^1 = R^2 =$ Ac	B	40%	7%	[21]
c $R^1 =$ TBS, R^2–$R^2 = CMe_2$	B	40%	3%	[21]
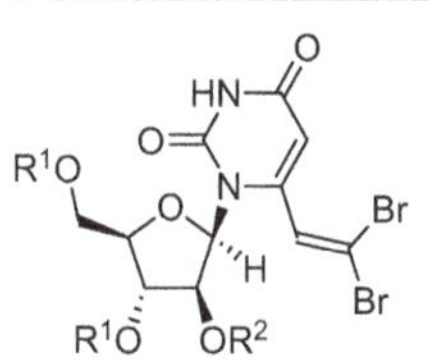		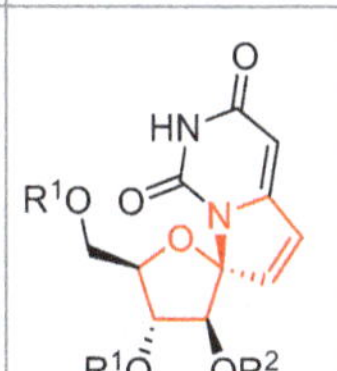	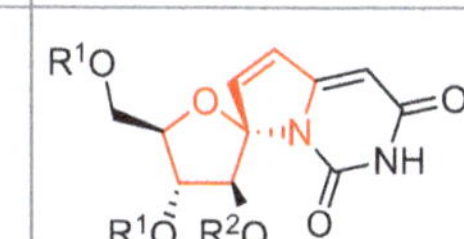	
d $R^1 =$ TBS, $R^2 =$ H	B	11%	26%	[21]
e $R^1 = R^2 =$ Ac	B	12%	26%	[21]

[a]Method A: $(nBu_3Sn)_2$ (2 equiv) in refluxing benzene and irradiation with 300 W tungsten-filament lamp. Method B: nBu_3SnH (2.5 equiv) and AIBN (0.5 equiv) in refluxing benzene

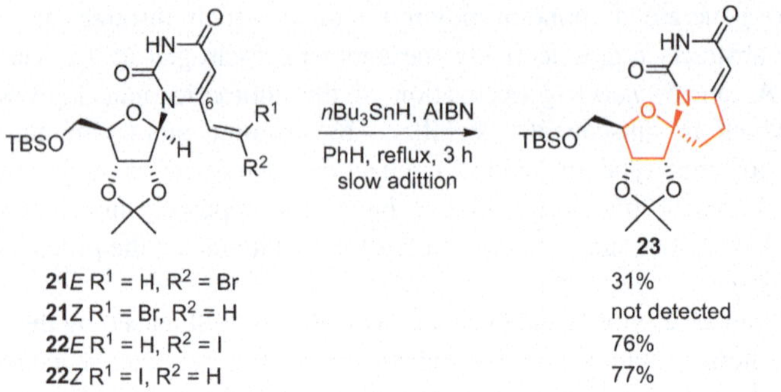

Scheme 5 1-Oxa-6-azaspiro[4.4]nonanes on (*E*)- and (*Z*)-6-(2-monohalovinyl)uridine models

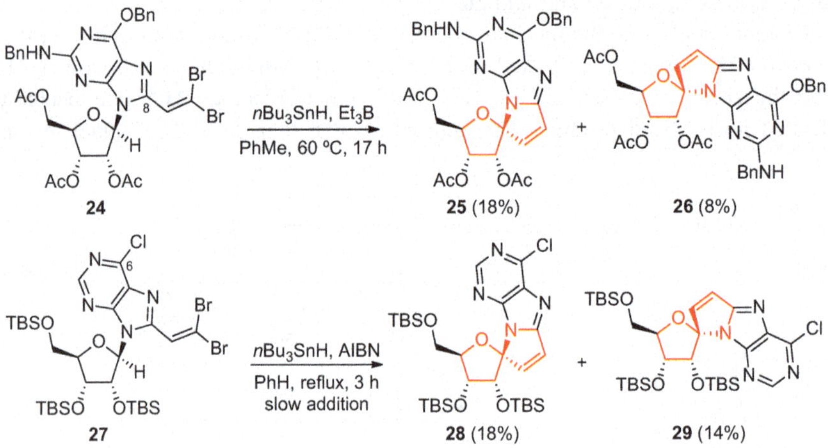

Scheme 6 1-Oxa-6-azaspiro[4.4]nonenes on D-ribofuranosylpurine nucleoside models

stannyl radical at the hindered bromine atom of (*Z*)-isomer **21***Z* should be extremely slow.

However, the reaction of both 6-(2-monoiodovinyl)uridine **22***E* and **22***Z* was highly effective, giving stereoselectively the spirocycle **23** in high yield. It is clear that once a vinyl radical is formed, the radical cascade proceeds with high efficiency. As far as we know, no attempt has been made to test the reaction with a 6-(2,2-diiodovinyl)uridine derivative.

To further extend the scope of the described methodology, two D-ribofuranosylpurine nucleosides were also investigated (Scheme 6): the guanosine derivative **24** containing a 2,2-dibromovinyl group tethered at the C8 position described by Diederichsen and coworkers [85] and the analogous 6-chloropurine nucleoside **27** prepared by Tanaka and coworkers [21]. Both compounds, under similar conditions, afforded the expected mixture of purine spironucleosides **25**, **26** and **28**, **29**, respectively, in low yields.

3.2.2 Amidyl-Radical-Initiated 1,5-HAT–SET Oxidation–5-*exo*-trig Cyclization

Resembling the intramolecular hydrogen atom transfer reaction promoted by nitrogen-centered radicals developed by Hofmann-Löffler-Freytag (HLF) [30, 86, 87], our group has described a protocol for the synthesis of 1-oxa-6-azaspiro[4.4] nonane skeletons which proceeds under mild conditions compatible with the more typical protecting groups used in carbohydrate chemistry [88, 89], thus avoiding the strong acids employed in the original HLF process. The reaction consists of the treatment of the suitable *N*-phosphoramidate derivative **30** with $PhI(OAc)_2$ and molecular iodine under irradiation with visible light. The *N*-iodoamide generated in situ by reaction with transient acetyl hypoiodite was homolyzed by heat or light to give the *N*-phosphoramidyl radical that abstracts the anomeric γ-hydrogen through a 1,5-HAT (Scheme 7) [90]. Subsequent radical-polar crossover step via SET oxidation and 5-*exo*-trig cyclization of the amide group onto the less sterically demanding face of the oxocarbenium ion produce the corresponding oxaazaspiroketal **31**. It is interesting to mention the umpolung reactivity of the nitrogen group during the reaction, acting first as an electrophilic radical in the hydrogen abstraction and later as a nucleophile in the cyclization step.

Moreover, the synthesis of the 1-oxa-6-azaspiro[4.4]nonane system can also be initiated by primary carboxamides (Scheme 8) [91]. We have developed an analogous tandem 1,5-HAT–SET oxidation–5-*exo*-trig cyclization process promoted by amidyl radicals generated from **32** or **33** that displays an *O*- and *N*-ambident nucleophilic character of the amide group. Thus, modulating the electrophilicity of

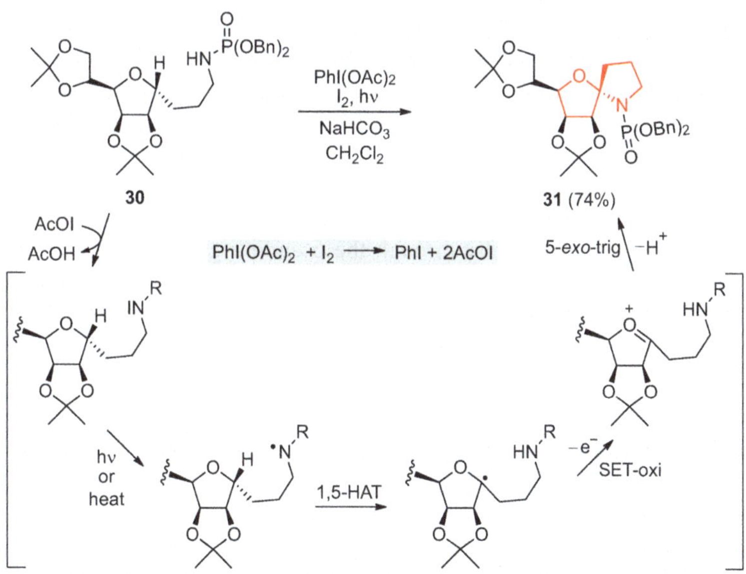

Scheme 7 1-Oxa-6-azaspiro[4.4]nonane via 1,5-HAT–SET oxidation–5-*exo*-trig cyclization

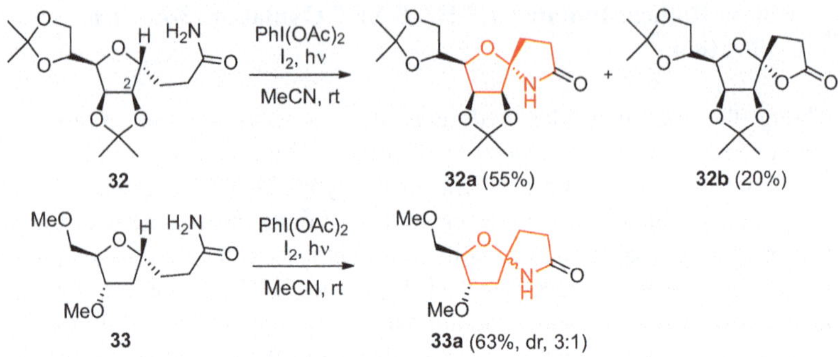

Scheme 8 1-Oxa-6-azaspiro[4.4]nonanes on carboxamide models

the oxocarbenium ion intermediate by tuning the electron-withdrawing nature of the C2 substituent can lead to the preferential or exclusive formation of lactones **32b** or lactams **32a** and **33a**. This ambident nucleophilic reactivity of the carboxamide group will be commented on in more detail during the synthesis of 6-oxa-1-azaspiro[4.5]decane models in Sect. 4.2.1.

3.2.3 C(sp^2)-Radical-Initiated 5-*exo*-trig Cyclization

The preparation of 1-oxa-6-azaspiro[4.4]nonane bicyclic nucleosides (e.g., **35** and **36**, Scheme 9) can also be achieved by the stannyl radical-promoted reaction of 1-(2-deoxy-D-*erythro*-pent-1-enofuranosyl)uracil derivative **34** possessing a 2,2-dibromovinyl group tethered at the C6 position of the base [92–94]. The vinyl radical is added to the sugar double bond by a 5-*exo*-trig mechanism to give an adduct radical, which is then reduced by the hydride. A small amount of a side product **37** produced by a competitive 6-*endo*-trig cyclization is also obtained as a mixture of isomers. When the reaction of **34** was carried out in an oxygen atmosphere, a tandem radical cyclization–peroxidation occurred, and the D-*arabino* alcohol **38** was obtained in low yield [93]. In a 6-chloro purine model **39**, the reaction proceeded preferentially through the 6-*endo*-trig ring closure to give **41**; only a minor amount of the alternative 5-*exo*-trig product **40** was formed [94]. The structures and stereochemistries have been confirmed by X-ray crystallographic analysis of **40** and the fully deprotected alcohols corresponding to **35**, **36**, and **38**.

3.2.4 C(sp^3)-Radical-Initiated Two Consecutive 5-*exo*-trig Cyclizations

During the preparation of 1′α-branched chain ribonucleosides from uridine, Shuto and Matsuda and coworkers discovered the reaction outlined in Scheme 10 [95]. The 1′α-phenylselenouridine derivative **42** reacts with $(TMS)_3SiH$/AIBN to give **43** in moderate yield.

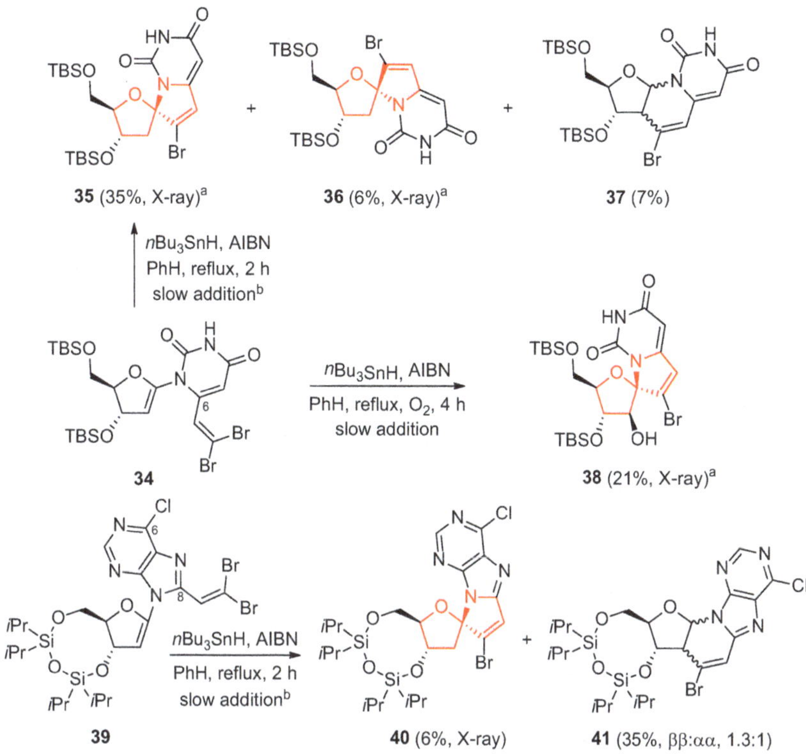

Scheme 9 1-Oxa-6-azaspiro[4.4]nonenes via 5-*exo*-trig cyclization; [a]X-ray of the completely unprotected compound; [b]reaction carried out under unspecified atmosphere

The 1-oxa-6-azaspiro[4.4]nonane skeleton must have been constructed by a mechanism involving a radical cascade initiated by the C1′-radical followed by two consecutive 5-*exo*-trig cyclizations. The stereochemistry depends on the configuration at C2′, since each radical cyclization must proceed to form a *cis* ring-fused cyclopentane.

3.3 *1-Oxa-7-azaspiro[4.4]nonanes*

3.3.1 C(sp^2)-Radical-Initiated 1,5-HAT–5-*exo*-trig or 1,5-HAT–5-*exo*-dig Cyclization

A novel synthetic strategy for the preparation of 4′-spironucleosides possessing a 1-oxa-7-azaspiro[4.4]nonane ring system has been developed by Wu and coworkers [96]. The starting 5′-uronamides **44** and **45** were prepared by reaction of *N*-allyl-2-bromoaniline with 5′-carboxylic nucleosides of the uridine and thymidine type, respectively (Scheme 11).

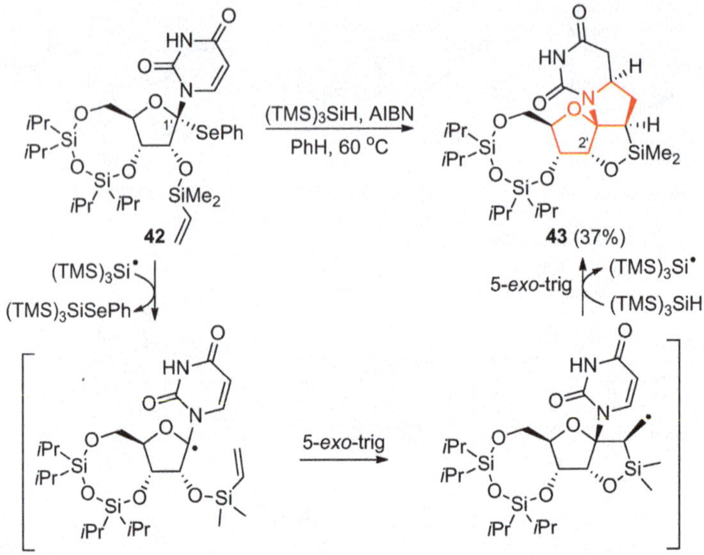

Scheme 10 1-Oxa-6-azaspiro[4.4]nonane via 5-*exo*-trig–5-*exo*-trig cyclizations

The synthetic approach to compounds **46** and **47** implies a radical chain reaction initiated by $n\text{Bu}_3\text{SnH}$/AIBN. The homolytic bromine abstraction generates an aryl radical intermediate that triggers a cascade reaction of two steps: 1,5-HAT of the H4′ followed by a 5-*exo*-trig addition to the double bond of the *N*-allyl group. The sequence proceeds with absolute regio- and stereoselectivity by retention of configuration at C4′ and formation of a single diastereomer at C8′. This radical cascade can also be applied to the *N*-propynyl analogue **48** where a 5-*exo*-dig cyclization is now the final step of the process. The spiro compound **49** was obtained analogously in good yield and diastereoselectivity.

3.4 1,6-Dioxaspiro[4.4]nonanes

3.4.1 Alkoxyl-Radical-Initiated 1,5-HAT–SET Oxidation–5-*exo*-trig Cyclization

In a related synthetic process to the one shown in Sect. 3.2.2, we have described the synthesis of 1,6-dioxaspiro[4.4]nonane substructure using an alkoxyl radical as key reactive intermediate (Scheme 12) [97, 98]. This oxygen-centered radical generated in situ from *C*-mannofuranosyl glycoside **50** with PhI(OAc)_2 and iodine performs the 1,5-HAT of the anomeric hydrogen delivering the C1 radical which undergoes SET oxidation and cyclization in a 5-*exo*-trig manner producing the spiroacetals **51**

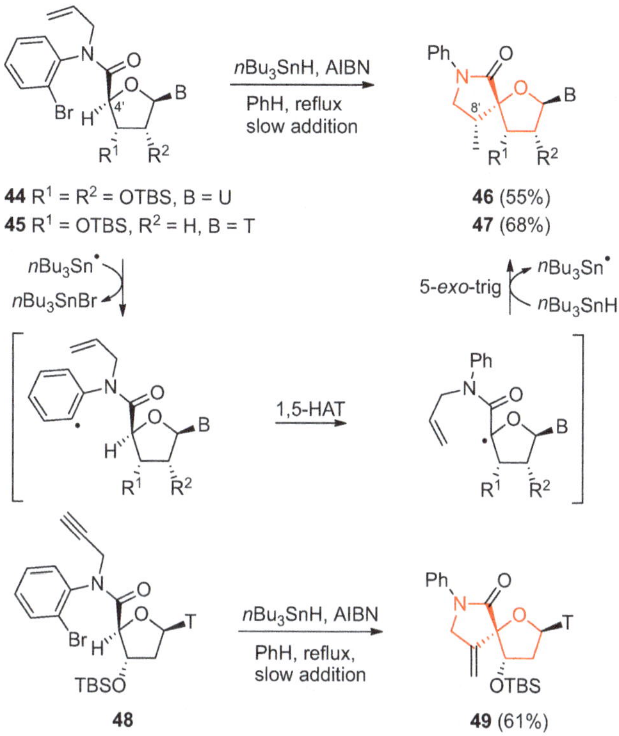

Scheme 11 1-Oxa-7-azaspiro[4.4]nonanes via 1,5-HAT–5-*exo*-trig or –5-*exo*-dig cyclization

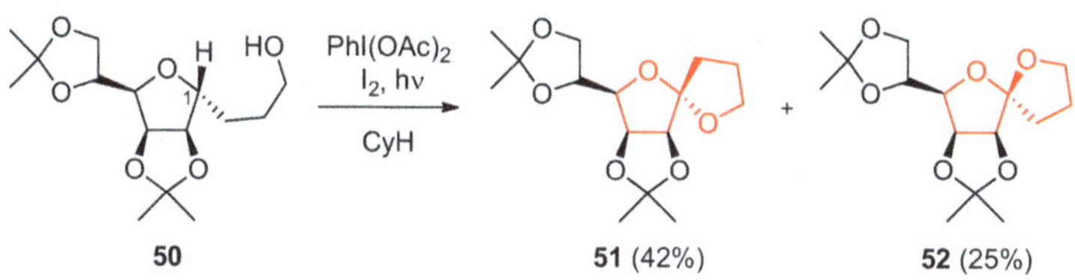

Scheme 12 1,6-Dioxaspiro[4.4]nonanes via 1,5-HAT–SET oxidation–5-*exo*-trig cyclization

and **52** in good overall yield. In this furanosyl system, stereocontrol in the cyclization step seems to be governed principally by steric factors.

3.4.2 Alkoxyl-Radical-Initiated 1,5-HAT–Radical-Polar Crossover–5-*exo*-trig Cyclization

Sartillo-Piscil and coworkers have developed another synthetic strategy to construct this 1,6-dioxaspiro[4.4]nonane framework inspired by the methodology reported by Crich and Newcomb on the rearrangements of β-phosphatoxyl radicals [99, 100]. An

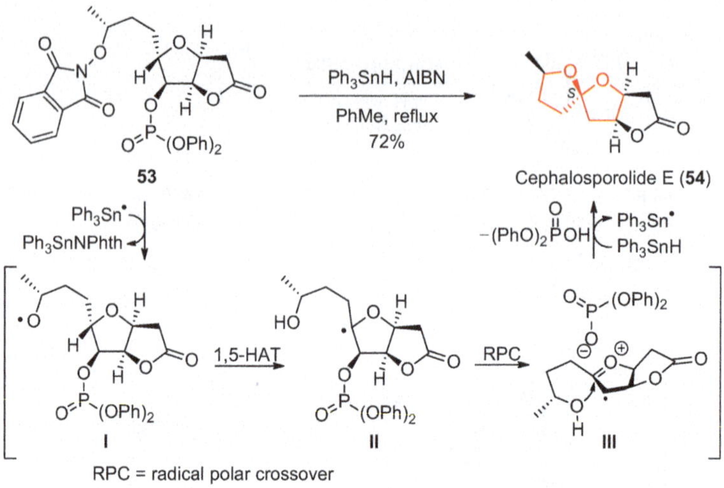

Scheme 13 1,6-Dioxaspiro[4.4]nonane via 1,5-HAT–RPC–5-*exo*-trig cyclization

alkoxyl radical **I** (Scheme 13), generated from the reaction of the appropriate *N*-alkoxyphthalimide **53** with a triphenylstannyl radical, promoted the 1,5-hydrogen transfer to give *C*-radical **II**, which undergoes a radical-polar crossover mechanism expelling the phosphate group and forming the radical cation **III** [101, 102]. Further 5-*exo*-trig cyclization by the hydroxyl group led to the exclusive formation of cephalosporolide E (**54**) in 72% yield. This is formally a 1,5-HAT–*cine*-substitution sequence where the oxygen atom acts first as an electrophilic radical and then as a nucleophile [103]. Furthermore, with the addition of base to the radical process, the authors were able to observe that the kinetic product, cephalosporolide F, the *R*-epimer at the spirocenter, is formed in the initial stage and is then transformed into thermodynamic cephalosporolide E via acid-catalyzed isomerization by the diphenyl hydrogen phosphate, generated in the reaction media.

3.4.3 C(sp^3)-Radical-Initiated 5-*exo*-trig or 5-*exo*-dig Cyclization

In an effort to develop new stereocontrolled strategies to naturally occurring spiroacetals, Sharma and coworkers reported an intramolecular regio- and stereoselective radical cyclization of α-bromoacetals derived from D-mannofuranosides to generate this 1,6-dioxaspiro[4.4]nonane framework, as depicted in Scheme 14 [104, 105]. Accordingly, the catalytic generation of nBu$_3$SnH with the nBu$_3$SnCl-NaCNBH$_3$ system [106] was successfully applied to 5-hexynyl **55a–b**, 5-hexenyl **57a–b** and 5-oxo-pentanyl **59** systems. The carbon-centered radical initially formed proceeds via 5-*exo*-dig or 5-*exo*-trig cyclizations, to yield exclusively the α-*C*-spiroacetals **56a–b**, **58a–b**, and **60**.

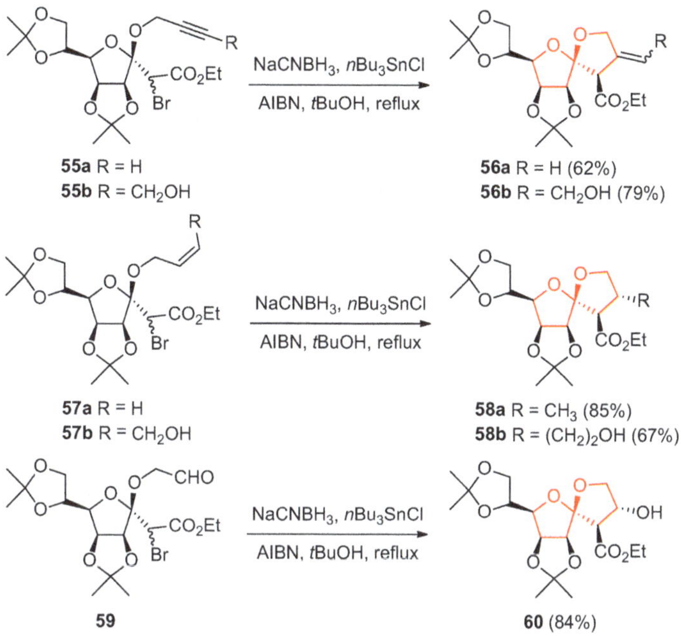

Scheme 14 1,6-Dioxaspiro[4.4]nonanes via 5-*exo*-trig or 5-*exo*-dig cyclization

3.5 *1,6-Dioxa-4-azaspiro[4.4]nonanes*

3.5.1 Alkoxyl-Radical-Initiated 1,5-HAT–SET Oxidation–5-*exo*-trig Cyclization

The preparation of anomeric spironucleosides with a 1,6-dioxa-4-azaspiro[4.4] nonane substructure has been independently studied by Tanaka [107, 108] and Chatgilialoglu (Table 3) [22, 109]. These compounds have a fixed *syn-anti* conformation around the *N*-glycosidic linkage and can be important models to determine the structure-activity relationship between glycosidic torsion angle and biological activity [110]. The introduction of the hydroxyalkyl group into the 6-position of conveniently protected 2′-deoxyuridines **61a–d** and uridines **61e–l** was accomplished by selective lithiation with LDA and subsequent treatment with the appropriate electrophile: DMF or ethyl formate, for the primary alcohols and acetaldehyde, and acetone for the secondary and tertiary alcohols, respectively.

The alkoxyl radicals were generated by two protocols: $Pb(OAc)_4/I_2/CaCO_3/h\nu$ and $PhI(OAc)_2/I_2/h\nu$. In general, it was found that for these cyclization reactions, PhI $(OAc)_2$ was a better oxidant than $Pb(OAc)_4$ in terms of chemical yield and facial selectivity [108]. The cyclization proceeds via 1,5-HAT by the radical-polar crossover mechanism described in Sect. 3.2.2, and in most cases a mixture of the two possible diastereomers **62a–l** and **63a–l** is obtained in high combined yield. The stereochemistries have been confirmed by X-ray crystallographic analysis of **61c**,

Table 3 1,6-Dioxa-4-azaspiro[4.4]nonanes via 1,5-HAT–SET oxidation–5-*exo*-trig cyclization

Substrate	Method[a]	Products		Reference
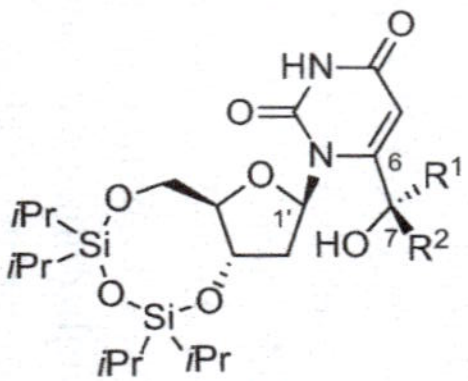 **61**		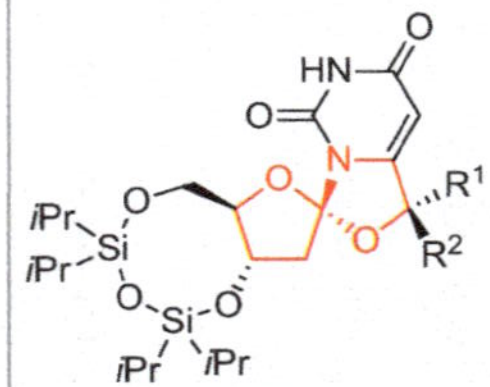 **62**	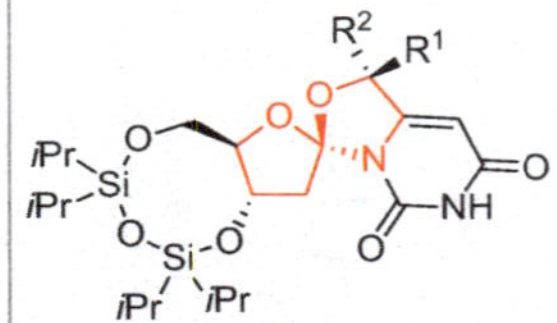 **63**	
a $R^1 = R^2 = H$	A	25% (X-ray)[b]	43% (X-ray)	[107]
	B	34%	37%	[22]
b $R^1 = H, R^2 = Me$	A	60%	9%	[107]
c $R^1 = Me, R^2 = H$ (X-ray)	A	2%	78%	[107]
d $R^1 = R^2 = Me$	A	44%	22%	[107]
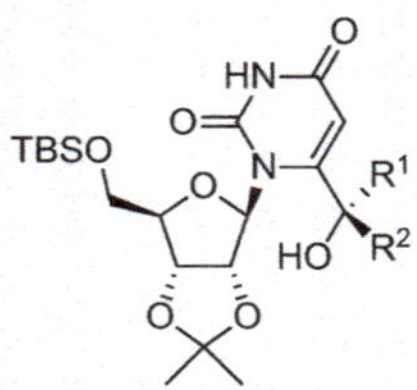		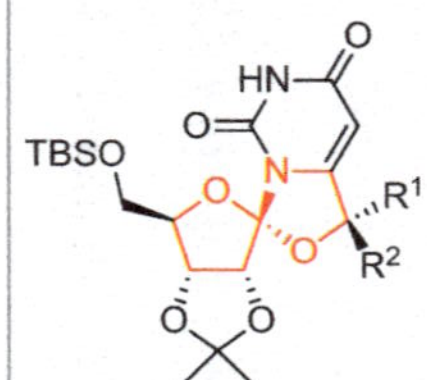	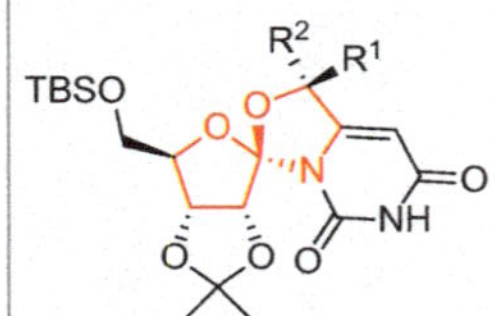	
e $R^1 = R^2 = H$	A	42%	1%	[107]
	B	49%	–	[22]
f $R^1 = H, R^2 = Me$	A	62%	2%	[107]
g $R^1 = Me, R^2 = H$	A	39%	14%	[107]
h $R^1 = R^2 = Me$	A	40%	19%	[107]

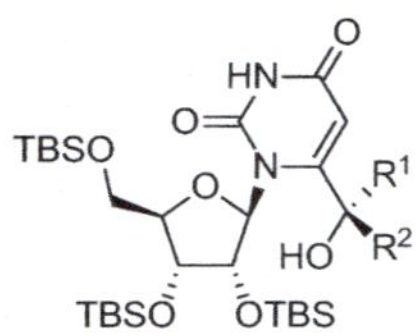		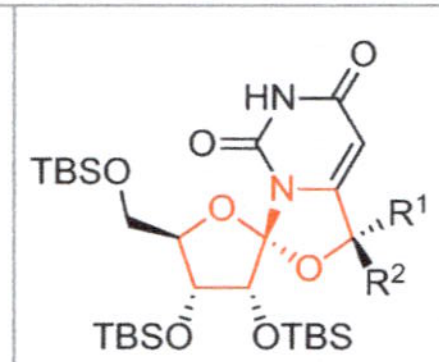	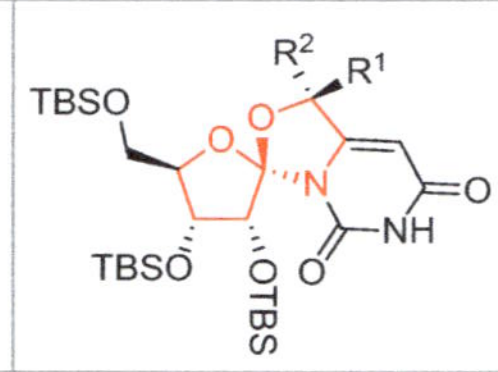	
i $R^1 = R^2 = H$	A	55% (X-ray)[b]	–	[107]
	B	36% (X-ray)[b]	–	[22]
j $R^1 = H, R^2 = Me$	A	61%	–	[107]
k $R^1 = Me, R^2 = H$	A	38%	19%	[107]
l $R^1 = R^2 = Me$	A	55% (X-ray)	10%	[107]

[a]Method A: $PhI(OAc)_2$ (2.4–3 equiv), I_2 (0.6 equiv), cyclohexane-CH_2Cl_2 (6:1 or 3:1), irradiation with a 250 W tungsten-filament lamp. Method B: $PhI(OAc)_2$ (3 equiv), I_2 (1.1 equiv), cyclohexane, irradiation with a 450 W tungsten-filament lamp

[b]X-ray of the compound fully deprotected

62l, and **63a** and the fully deprotected alcohols corresponding to **62a** and **62i** [107–109]. The diastereoselectivity strongly depends on the steric hindrance of the 2′α-*O*-substituent, especially in the primary alcohols. Therefore, a single diastereomer is produced in the uridine series (compare the reaction of 2′-deoxyuridine derivative **61a** with **61e** and **61i**). The comparison of the results of **61b** (7*R*) and **61c** (7*S*) reflects the sensitivity of the cyclization toward the steric demand of the methyl group in the secondary alcohols. Tertiary alcohols **61d**, **61h**, and **61l** gave spironucleosides in high yield but with lower selectivity.

3.6 6-Oxa-1-aza-3-silaspiro[4.4]nonanes

3.6.1 C(sp^3)-Radical-Initiated 5-*exo*-trig Cyclization

Reactions of 2′-substituted 6-(bromomethyl)dimethylsilyl-(2-deoxy-D-*erythro*-pent-1-enofuranosyl)uracils (**64a–d**) with $n\mathrm{Bu_3SnH}$ proceeded in preferential or exclusive 5-*exo*-trig cyclization mode to give the 6-oxa-1-aza-3-silaspiro[4.4]nonane systems (**65a–d**) in good yields (Scheme 15) [111]. The adducts were formed with high selectivity giving only addition products for the α-side of the furanose ring. In contrast, the 2′-unsubstituted **64e** undergoes exclusively 6-*endo*-trig cyclization. The introduction of a substituent seems to stabilize the adduct radical at C2′ and alter the cyclization bias in favor of a 5-*exo*-trig process. The oxidative cleavage of the cyclized products provides a new synthetic entry to biologically interesting 1′-*C*-hydroxymethyl branched ribonucleosides [95].

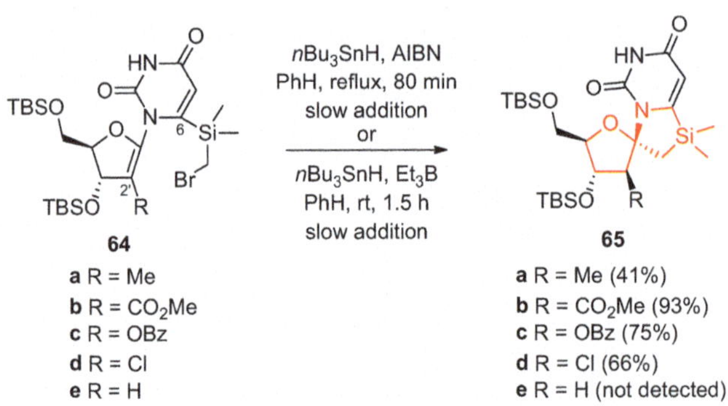

Scheme 15 6-Oxa-1-aza-3-silaspiro[4.4]nonanes via 5-*exo*-trig cyclization

3.7 *1,7-Dioxa-8-silaspiro[4.4]nonanes*

3.7.1 C(sp²)-Radical-Initiated 1,6-HAT–5-*exo*-trig Cyclization

The 2-bromobenzylidene group has been applied as protecting/radical-translocating (PRT) group to achieve the synthesis of the unusual 1,7-dioxa-8-silaspiro[4.4] nonane skeleton. Shuto and coworkers demonstrated that positions 2 or 3 of the D-*ribo*-derivative **66** are sterically inaccessible for 1,5-HAT from the aromatic radical. Instead the less common 1,6-HAT selectively occurs to produce a C4-radical **I** (Scheme 16) [112]. Then, the more favored β-face addition of this *C*-radical to the dimethylvinylsilyl tether through a 5-*exo*-trig cyclization produces the intermediate radical **II**, which under kinetic control, i.e., higher concentrations of nBu$_3$SnH and lower reaction temperatures, is quenched to generate **67** exclusively.

However, when the process is performed under thermodynamic conditions, i.e., lower concentrations of nBu$_3$SnH and higher reaction temperatures, the intermediate **II** rearranges into radical **III** to give the 1,7-dioxa-8-silaspiro[4.5]decane **68**, via a pentavalent silicon-bridging radical transition state [113]. A similar method has also been applied to $O^{3'}$,N^6,N^6-tribenzoyladenine derivative **69**. The C4′-radical generated by homolytic abstraction of the selenide precursor added analogously to the vinylsilane tether, but in this particular case, the primary radical intermediate **IV**

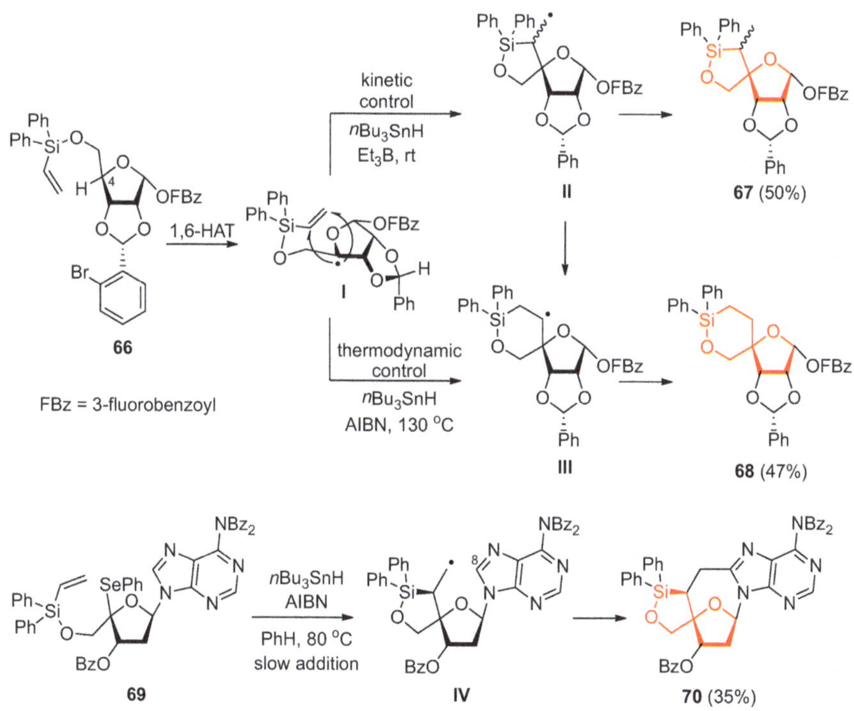

Scheme 16 1,7-Dioxa-8-silaspiro[4.4]nonanes via 1,6-HAT–5-*exo*-trig cyclization

does not react with nBu_3SnH but rapidly adds to the 8-position of the adenine to give **70** [114]. The spiro compounds **67**, **68**, and **70** which are supposed to be easily hydrolysable have not been fully characterized; the yields shown are of the diols obtained after Tamao oxidation [115] of the crude reaction mixtures.

4 Heterospiro[4.5]decanes

4.1 1-Oxa-6-azaspiro[4.5]decanes

4.1.1 C(sp^3)-Radical-Initiated 6-*exo*-trig Cyclization

The 1′-*C*-iodopropyl branched nucleoside **71** has been used by Matsuda and coworkers for the preparation of 6,1′-propanouridine **72** as described in Scheme 17 [116, 117]. The initially formed alkyl radical adds to the double bond by a favored 6-*exo*-trig radical cyclization to give a radical adduct which is subsequently reduced by tin hydride. In a second step, base-promoted dehydrochlorination afforded the required spironucleoside **72**, which possesses a fixed *syn* conformation around the *N*-glycosidic linkage. The *syn-anti* conformation of this bond appears to be important for the biological activity of nucleosides as enzyme inhibitors, and a circular dichroism (CD)-based methodology relates the magnitude of the Cotton effect to the glycosidic torsion angle.

4.1.2 Amidyl-Radical-Initiated 1,6-HAT–SET Oxidation–6-*exo*-trig Cyclization

Following the work described in Sect. 3.2.2, we have also extended this methodology to the synthesis of 1-oxa-6-azaspiro[4.5]decane systems by 1,6-HAT reactions promoted by amidyl radicals (Scheme 18) [88, 89]. The results showed that the reaction is highly dependent on the steric demand of the amine protector. Thus, treatment of the cyanamide derivative **73a** with $PhI(OAc)_2$ and iodine generated the spiroaminal **74a** with total regio- and stereoselectivity. In contrast, a similar model with a bulky phosphoramide group **73b** failed to undergo the 1,6-HAT process, and

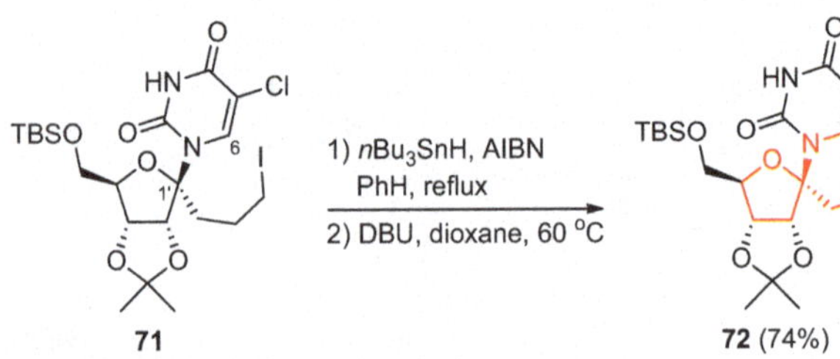

Scheme 17 1-Oxa-6-azaspiro[4.5]decane via 6-*exo*-trig cyclization

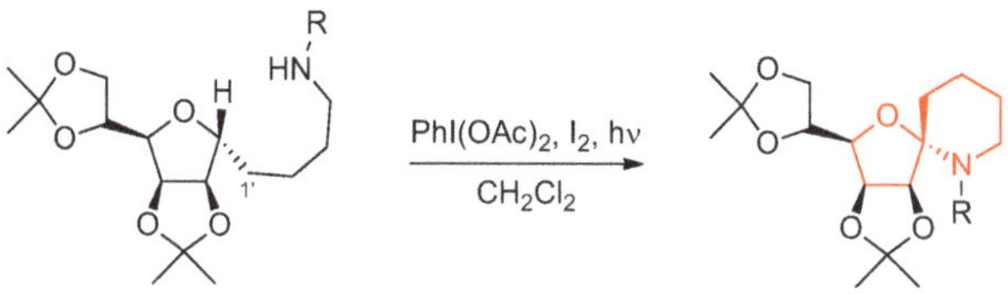

Scheme 18 1-Oxa-6-azaspiro[4.5]decane via 1,6-HAT–SET oxidation–6-*exo*-trig cyclization

the expected spirocycle **74b** was not detected; the only products that could be characterized were *C*-glycosides coming from 1,5-HAT reactions of hydrogens at C1′.

4.2 *6-Oxa-1-azaspiro[4.5]decanes*

4.2.1 Amidyl-Radical-Initiated 1,5-HAT–SET Oxidation–6-*exo*-trig Cyclization

The construction of the 6-oxa-1-azaspiro[4.5]decane skeleton can be achieved applying the same sequential process promoted by nitrogen-centered radical described in Sect. 3.2.2, using glycopyranoside derivatives as starting materials (Scheme 19) [88, 89]. Thus, the 1α-*C*-propylphosphoramidyl radical, generated from **75**, abstracts the anomeric hydrogen, through a six-membered transition state, from the β-side. Subsequent SET followed by nucleophilic cyclization from the less-hindered α-face delivered the pyrrolidine **76** in moderate yield. Moreover, when the propyl phosphoramide group is tethered to the C5 of the sugar, as occurs in **77**, a mixture of epimers **78** and **79** was obtained in 84% overall yield.

Additionally, we have also reported that this sequence can be initiated by primary carboxamidyl radicals [91]. As depicted in Scheme 20, with the 4,8-anhydro-2,3-dideoxynononamides **80** or **81**, the cyclization step can occur through the nitrogen or oxygen atom giving a mixture of lactams **80a** and **81a** and lactones **80b** and **81b**, suggesting that the configuration of C2 does not influence the course of the reaction. However, the electron density of the electrophilic oxocarbenium ion intermediate seems to be crucial for the regioselectivity. The absence of the group at C2 as in **82**, or replacement by a more electron-withdrawing group (OAc) as in **83**, leads to the exclusive formation of lactam **82a** or lactone **83a**, respectively. Moreover, the anomeric effect controls the stereoselectivity, and only the α-isomer is obtained in all cases.

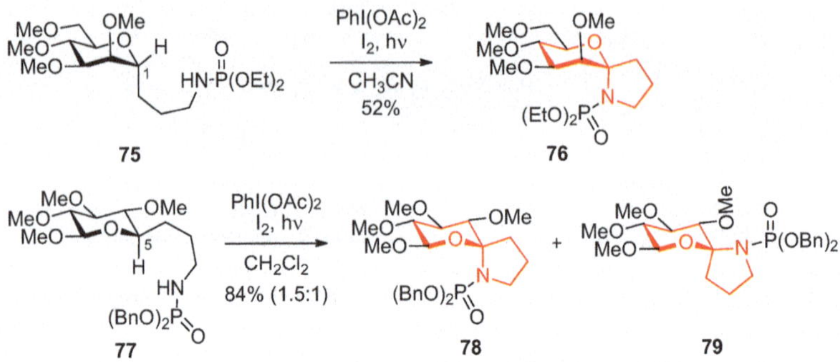

Scheme 19 6-Oxa-1-azaspiro[4.5]decanes via 1,5-HAT–SET oxidation–5-*exo*-trig cyclization

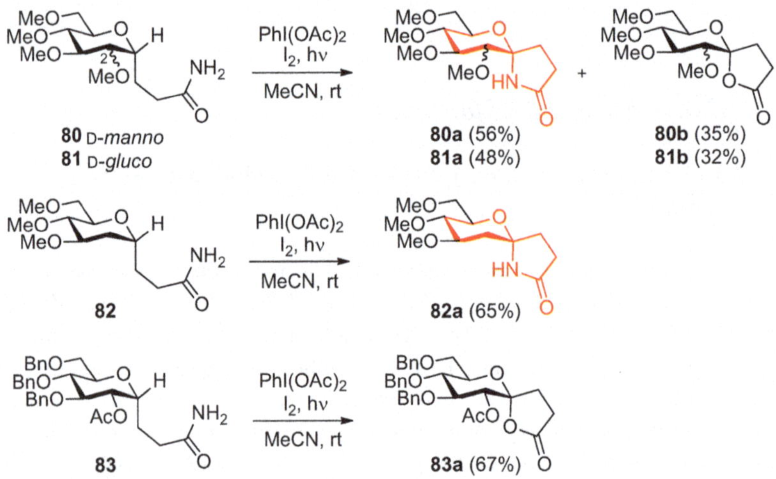

Scheme 20 6-Oxa-1-azaspiro[4.5]decanes and 1,6-dioxaspiro[4.5]decanes from 4,8-anhydro-2,3-dideoxynononamide models

4.3 1,6-Dioxaspiro[4.5]decanes

4.3.1 Photoinduced Norrish Type II–Yang Cyclization

In Scheme 21 a method developed by Descotes and coworkers for the construction of 1,6-dioxaspiro[4.5]decane derivatives using Norrish type II photocyclization of 3-oxobutyl β-D-glucopyranoside **84** is shown [118, 119]. The reaction proceeded via a 1,6-PHAT of the axial anomeric hydrogen and cyclization of the resulting 1,5-biradical to a mixture of spiroacetals **85***R* and **85***S* in moderate yield. As confirmed by X-ray crystallographic analysis, the cyclization occurs with retention of configuration, and the pyranose ring adopts a $^{1,4}B$ conformation [120].

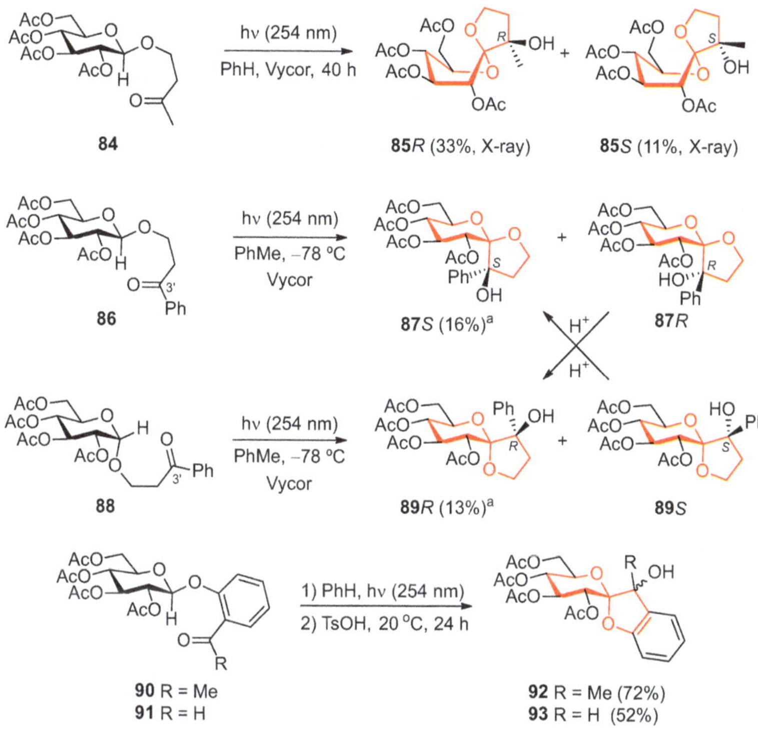

Scheme 21 1,6-Dioxaspiro[4.5]decanes via Norrish type II–Yang cyclization; [a]yields after acid-catalyzed isomerization

The cyclization is strongly dependent on the configuration of the anomeric center; the 3-oxobutyl α-D-glucopyranoside isomer does not react even after prolonged irradiation time (100 h). 1-*O*-(3-Oxobutyl)-2-deoxy- and 1-*O*-(3-oxobutyl)-2,3-dideoxy-β-D-glucopyranose and 1-*O*-(3-oxobutyl)-2,3-dideoxy-β-D-*erythro*-hex-2-enopyranose react with much less stereoselectivity, and a mixture of the four possible isomers is obtained in all cases [121]. This methodology has been extended to 3-oxobutyl β-D-mannopyranosides [122] and α-L-arabinopyranosides [123]. In the D-*manno* series, the reaction proceeds rapidly with retention of configuration at the anomeric center (two isomers, global yield, 33%). With the more flexible α-L-arabinopyranose, the stereoselectivity is lost (four isomers, global yield, 33%). Analogously to the D-*gluco* series, the corresponding axial isomeric glycosides 3-oxobutyl α-D-mannopyranosides and β-L-arabinopyranosides, where the hydrogen to be abstracted is in equatorial position, react much more slowly.

The influence of the nature of the carbonyl group has been studied with the models 1-*O*-(3-oxo-3-phenylpropyl)-D-glucopyranoses **86** and **88**. Both of them reacted at room temperature with poor diastereoselectivity to give a mixture of the four possible isomers **87** and **89**. Retention of the configuration is only achieved at low temperature (−78°C). Acid-catalyzed rearrangement of the mixture allows the transformation of **87***R* and **89***S* into the thermodynamically more stable **89***R* and **87***S*, respectively [124]. Irradiation of 1-*O*-(2-acetylphenyl)- and 1-*O*-(2-formylphenyl)-β-D-glucopyranoses **90** and **91** followed by acid-catalyzed isomerization afforded the spiroacetals **92** and **93**, respectively, in good overall yields and stereoselectivity [125, 126]. These photochemical reactions take place analogously to these described in Sect. 2.2.1 for 2-oxopropyl glycosides, with the difference that the 1,5-biradical intermediates cannot be cleaved and, consequently, cyclize.

4.3.2 Alkoxyl-Radical-Initiated 1,6-HAT–SET Oxidation–6-*exo*-trig or 1,5-HAT–SET Oxidation–5-*exo*-trig Cyclization

Two sequential processes were devised for the preparation of optically active 1,6-dioxaspiro[4.5]decane derivatives using the methodology employed in Sect. 3.4.1. In one of them, the D-mannofuranose derivative **94** is used as starting material, and the tetrahydropyran ring is synthesized by a 1,6-HAT initiated by the primary alkoxyl radical and 6-*exo*-trig cyclization generating the spiro compounds **94a**,**b** (Scheme 22) [98]. Products resulting from incorporation of iodine **94c**,**d** were also obtained, which could be explained by a competitive 1,5-HAT from C1′ via a more favored six-membered transition state.

The other route to these spiroacetals involves the use of D-glucopyranose derivatives **95** or **98**, and the tetrahydrofuran ring is obtained at C1 or C5 via an analogous sequence which implies 1,5-HAT reactions and the corresponding 5-*exo*-trig cyclizations [97, 98]. Both cases generate the corresponding kinetic isomers **96** and **99** as

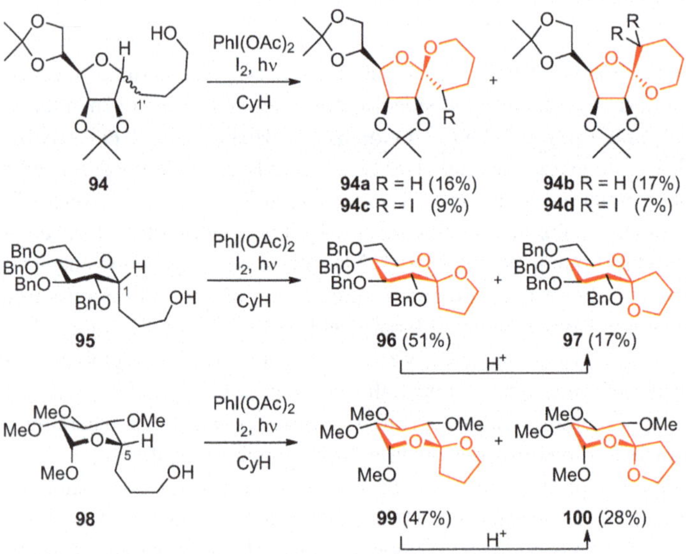

Scheme 22 1,6-Dioxaspiro[4.5]decanes via 1,6-HAT–SET oxidation–6-*exo*-trig or 1,5-HAT–SET oxidation–5-*exo*-trig cyclization

major products and can be transformed by acid-catalyzed isomerization to the thermodynamically more stable **97** and **100**, respectively. An interesting application of this methodology is the construction of the 1,6-dioxaspiro[4.5]decane system present in the ciguatoxin 3C [127].

4.3.3 C(sp^3)-Radical-Initiated 6-*exo*-dig, 5-*exo*-dig, or 5-*exo*-trig Cyclizations

Sharma and coworkers extended the methodology described in Sect. 3.4.3 to afford 1,6-dioxaspiro[4.5]decane systems [104, 105]. As depicted in Scheme 23, α-bromoacetals, derived from D-mannofuranosides and D-glucopyranosides, bearing homopropargyl **101a**, propargyl **101b**, or allyl chains **101c**, can undergo, upon treatment with nBu$_3$SnCl-NaCNBH$_3$, a radical homolytic dehalogenation and subsequent 6-*exo*-dig, 5-*exo*-dig, or 5-*exo*-trig cyclizations, respectively, to furnish highly functionalized spiroacetals **102a–c** with good regio- and stereoselectivity. Attempts to apply this radical cyclization on the corresponding homopropargyl derivatives from glucopyranosides to prepare 1,7-dioxaspiro[5.5]undecane skeletons failed, producing exclusively oct-3-ulopyranosonic acid derivatives from radical debromination.

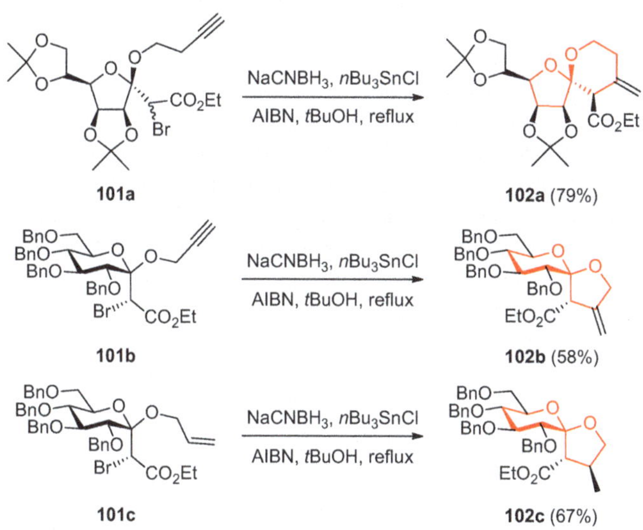

Scheme 23 1,6-Dioxaspiro[4.5]decanes via 6-*exo*-dig, 5-*exo*-dig, or 5-*exo*-trig cyclization

4.3.4 Alkoxyl-Radical-Initiated 5-*exo*-trig Cyclization and C(sp^3)-Radical-Initiated 5-*exo*-trig Cyclization

In 1987, Kraus and Thurston developed an approach for the synthesis of five-membered ring ethers based on the photolysis of unsaturated alcohols in the presence of HgO and iodine [128]. When a hemiketal such as tetra-*O*-benzyl-D-glucopyranoside derivative **103** is used, the alkoxyl radical generated can be added to the proximate olefin, through a 5-*exo*-trig cyclization to afford the corresponding spiroketal **104** as a single unassigned diastereomer (Scheme 24). Although the intermediacy of an alkoxyl radical is plausible under these conditions, an ionic mechanism involving a cyclic iodonium ion cannot be ruled out.

Later, Haudrechy and Sinaÿ described a new route to the 1,6-dioxaspiro[4.5] decane ring system [129]. The treatment of **105** with nBu$_3$SnH and catalytic amounts of AIBN triggered the formation of a carbon-centered radical and a subsequent stereoselective 5-*exo*-trig cyclization to yield spiro compound **106** (Scheme 24). As specified by the authors, a chair-like transition state for the cyclization step and the anomeric effect are probably responsible for the formation of a single 4′*S*-isomer.

4.3.5 C(sp^3)-Radical-Initiated β-Fragmentation–1,5-HAT–5-*exo*-trig Cyclization

Ferrier and Hall reported a direct procedure for the synthesis of carbohydrates with a 1,6-dioxaspiro[4.5]decane unit based on the radical rearrangement of 2′,3′-epoxy-4′-iodobutyl glycosides (Scheme 25) [130]. Thus, the treatment of iodoepoxide **107** with nBu$_3$SnH in benzene under UV irradiation produced a primary C4′-radical which triggered a chain reaction initiated by β-fragmentation of the adjacent C–O bond to form an alkoxyl radical. Then, a 1,5-HAT and further 5-*exo*-trig cyclization afforded the spiro compound **108** in good yield as a sole isomer. Interestingly, the reaction of the isomeric iodoepoxide **109** under the same conditions generated an unresolved mixture of four spiroacetals **110** in moderate yield. These differences in selectivity in both isomers are explained through the intramolecular hydrogen bonds

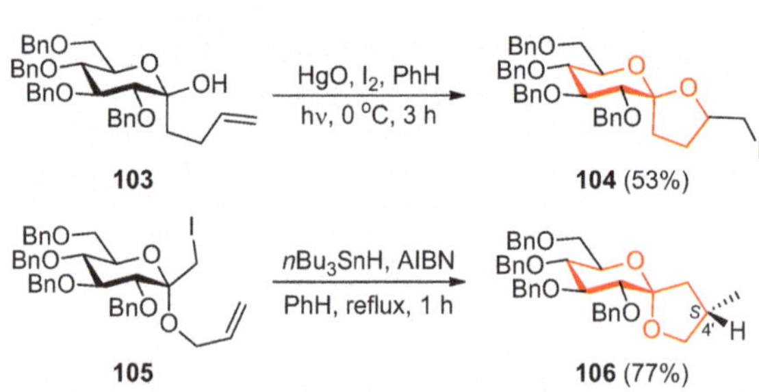

Scheme 24 1,6-Dioxaspiro[4.5]decanes via 5-*exo*-trig cyclization

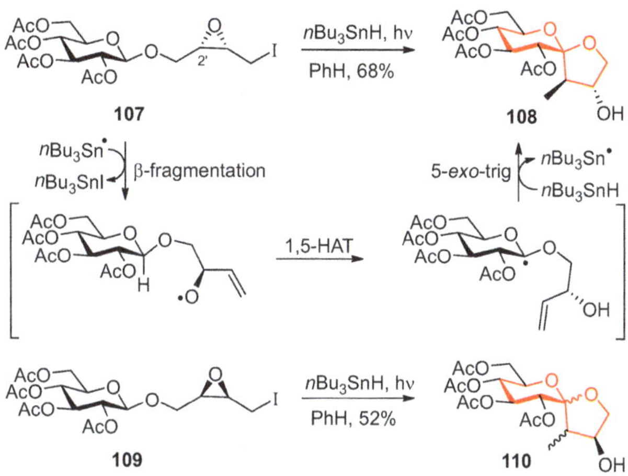

Scheme 25 1,6-Dioxaspiro[4.5]decanes via β-FRA–1,5-HAT–5-*exo*-trig cyclization

formed between the oxygen of the pyranose ring and the hydroxyl group in the intermediate anomeric radical.

4.4 6-Oxa-1,3-diazaspiro[4.5]decanes

4.4.1 Radical Anion-Initiated SET–Radical Coupling–Ring Closure

A number of 6-oxa-1,3-diazaspiro[4.5]decane systems have been prepared by Somsák and coworkers as biologically active surrogates of hydantocidin and 2-thiohydantocidin with a pyranose ring structure (Scheme 26) [131, 132]. Reaction of *C*-(1-bromo-1-deoxy-β-D-galactopyranosyl)formamide **111** with AgOCN in $MeNO_2$ in the presence of air gave the *epi*-hydantocidin analog **112** with retention of configuration at C1 and the alcohol **114** as a side product (89%, **112**:**114**, 1.5:1). Following the same procedure, AgSCN or KSCN afforded the inverted isomeric thiohydantocidin analog **113** together with **114** (90%, **113**:**114**, 2.9:1). The formation of the side product **114** can be substantially diminished or completely avoided under exclusion of oxygen. To rationalize these observations, an alternative mechanism to the classical S_N2 substitution has been proposed. The reaction can be started through SET via intermediate **I** which eventually leads to the corresponding free radical **II** [132]. A competitive radical trap by triplet oxygen may be the origin of the alcohol **114**. This methodology has also been applied to other glycopyranoses with D-*gluco*, D-*arabino*, and D-*xylo* configuration.

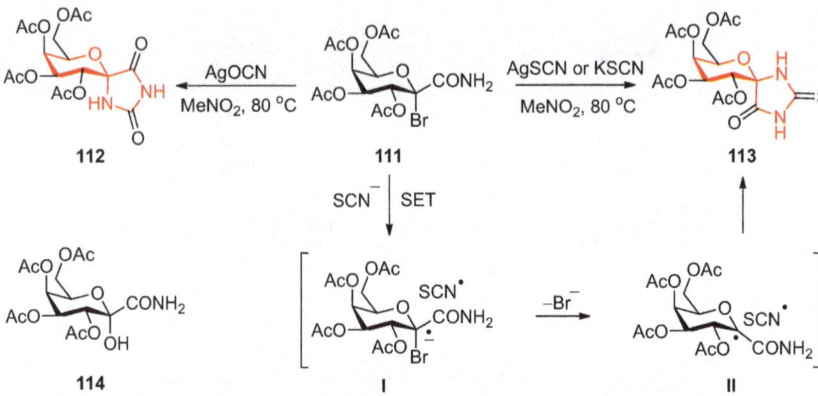

Scheme 26 6-Oxa-1,3-diazaspiro[4.5]decane via SET–radical coupling–ring closure

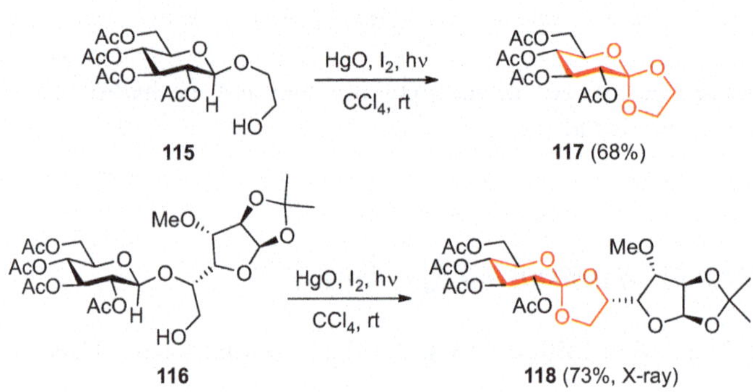

Scheme 27 1,4,6-Trioxaspiro[4.5]decanes via 1,5-HAT–SET oxidation–5-*exo*-trig cyclization

4.5 1,4,6-Trioxaspiro[4.5]decanes

4.5.1 Alkoxyl-Radical-Initiated 1,5-HAT–SET Oxidation–5-*exo*-trig Cyclization

The antibiotic activity exhibited by the orthosomycin family which contains 1,4,6-trioxaspiro[4.5]decane substructures has promoted methodologies to their synthesis [133–135]. Among them, Praly and Descotes described a radical approach for the generation of C1-spiro ortho esters in a one-pot procedure (Scheme 27) [136, 137].

β-D-Glucosides **115** and **116**, prepared from a Koenigs–Knorr reaction of glucosyl halides and the corresponding alcohols, can be irradiated in the presence of HgO and molecular iodine to produce, via a hypoiodite intermediate, an alkoxyl radical which promotes the radical sequence. 1,5-HAT of the anomeric hydrogen, followed by SET oxidation of the *C*-radical and further regio- and stereospecific

5-*exo*-trig cyclization of the alcohol, led to the spiro ortho esters **117** and **118** [138], respectively, in good yields.

4.6 1,3,6-Trioxaspiro[4.5]decanes

4.6.1 Alkoxyl-Radical-Initiated 1,8-HAT–5-*endo*-trig–SET Oxidation–10-*endo*-trig Cyclization

As a result of the expansion of the hydrogen atom transfer processes initiated by alkoxyl radical to disaccharide systems, our group reported an approach for the preparation of 1,3,6-trioxaspiro[4.5]decane framework based on our mild oxidative strategy [139, 140]. Under these conditions, the primary 6^{I}-*O*-pyranosyl radical **I**, generated from the D-maltose derivative **119**, abstracts regioselectively the hydrogen at $C5^{II}$, through an unusual 1,8-HAT to give a *C*-radical **II** (Scheme 28). Subsequent intramolecular addition to the carbonyl oxygen by a rare 5-*endo*-trig cyclization produces the intermediate **III**, which suffers a single-electron oxidation to finally collapse into the orthoacetate **120** via a 10-*endo*-trig process. An alternative mechanism involving SET-oxidation of radical **II** to an oxocarbenium ion and neighboring-group participation from the acetoxy group can also be considered.

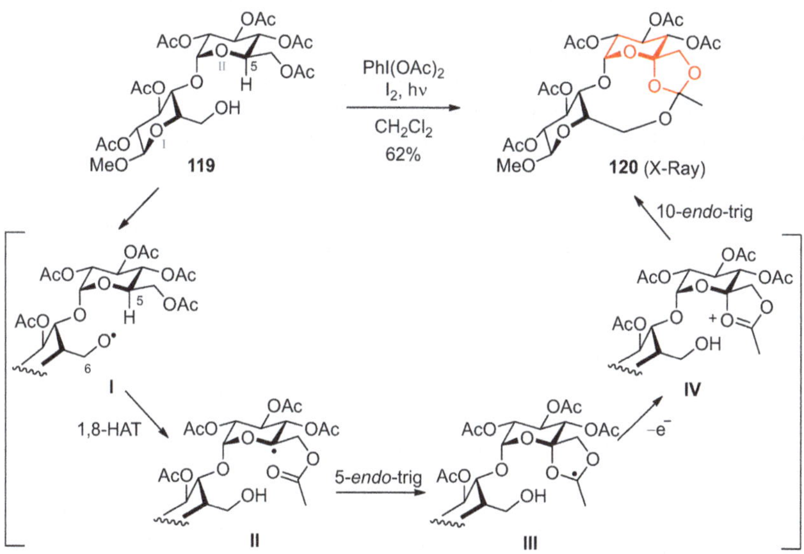

Scheme 28 1,3,6-Trioxaspiro[4.5]decane via 1,8-HAT–5-*endo*-trig–SET oxidation–10-*endo*-trig cyclization

4.7 *1,7-Dioxa-8-silaspiro[4.5]decanes*

4.7.1 C(sp^2)-Radical-Initiated 1,6-HAT–5-*exo*-trig–Cyclization–Ring Expansion

The synthesis of 1,7-dioxa-8-silaspiro[4.5]decane performed by Shuto and coworkers has been previously commented in Sect. 3.7.1 with the related 1,7-dioxa-8-silaspiro[4.4]nonane skeleton (Scheme 16) [112].

4.8 *1,6-Dioxa-9-thiaspiro[4.5]decanes*

4.8.1 Thiyl-Radical-Initiated Intermolecular Addition–*cine*-Substitution

During a study on the β-fragmentation mechanism of 4′-nucleotide radicals of considerable importance in the cleavage of DNA and RNA by hydrogen abstracting species and in the mode of action of several antitumor antibiotics, Crich and Huang discovered a new preparation of the 1,6-dioxa-9-thiaspiro[4.5]decane bicycle [141]. As illustrated in Scheme 29, the thiyl radical initiated from 2-mercaptoethanol added to the exomethylene derivative **121** generating a C4′-radical which triggered a *cine*-substitution with the alcohol acting as nucleophile [103]. The resulting C3′-radical was finally quenched with an excess of 2-mercaptoethanol. Compound **122**, obtained as a mixture of isomers, was unstable, and extensive decomposition was observed during the chromatographic purification, eventually enabling isolation in low yield (ca. 5%).

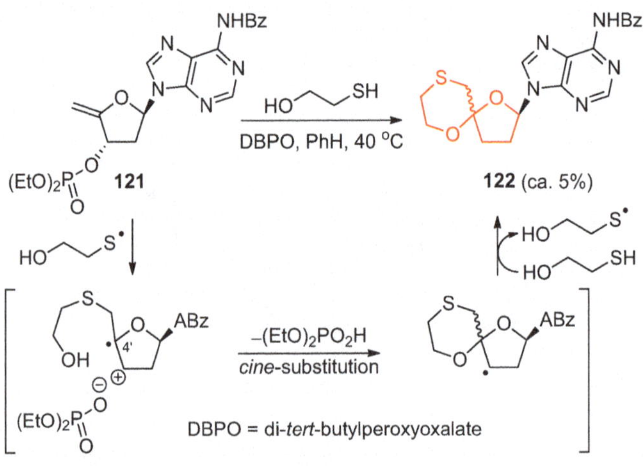

Scheme 29 1,6-Dioxa-9-thiaspiro[4.5]decane via intermolecular addition–*cine*-substitution

5 Heterospiro[5.5]undecanes

5.1 1-Oxa-7-azaspiro[5.5]undecanes

5.1.1 Amidyl-Radical-Initiated 1,6-HAT–SET Oxidation–6-*exo*-trig Cyclization

An analogous radical sequential process to that described in Sects. 3.2.2 and 4.2.1 can be applied to the synthesis of the 1-oxa-7-azaspiro[5.5]undecane framework. As depicted in Scheme 30, the *N*-phosphoramidylglucopyranose model **123** can react with the $PhI(OAc)_2$ and iodine system to generate the amidyl radical which initiates the radical sequence [88, 89]. Abstraction of the hydrogen at C5, via a seven-membered transition state, from the α-side, oxidation by SET to form the oxocarbenium ion, and subsequent 6-*exo*-trig cyclization of the amide group afforded the expected spiropiperidines **124** and **125** in moderate yield.

5.2 1,7-Dioxaspiro[5.5]undecanes

5.2.1 Photoinduced 1,5-PHAT–1,6-Biradical Cyclization

The stereo- and regioselectivity of triplet-sensitized radical reactions of α- and β-D-glucopyranosides possessing a 5-(hydroxymethyl)-2(5*H*)-furanone aglycon have been investigated by Abe and Hoffmann and coworkers (Scheme 31) [142]. β-Furanone **126** was excited to the 3ππ* state by triplet energy transfer from acetone. The reaction was then initiated by 1,5-PHAT of the anomeric hydrogen to the C3′ position of the furanone moiety. 1,5-Intramolecular cyclization of the C1-yl and oxoallyl radicals led finally to a mixture of diastereomers **127** and **128**. The overall process results in the glucopyranosyl group addition to the C4′ position of the furanone and formation of a new 1,7-dioxaspiro[5.5]undecane system. No transformation was observed by direct irradiation in the absence of acetone. Under the same conditions, the α-anomeric glucosyl derivative does not give the expected same mixture of stereoisomers. Instead, a 1,7-PHAT of the hydrogen at C5 of the sugar skeleton occurs regioselectively, and the subsequent cyclization of the

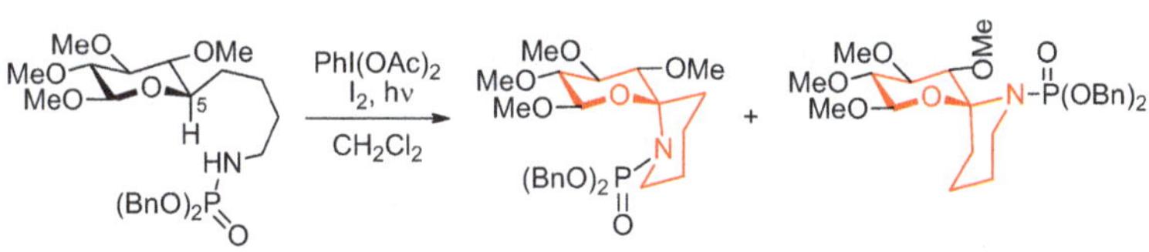

123 **124** (31%) **125** (10%)

Scheme 30 1-Oxa-7-azaspiro[5.5]undecanes via 1,6-HAT–SET oxidation–6-*exo*-trig cyclization

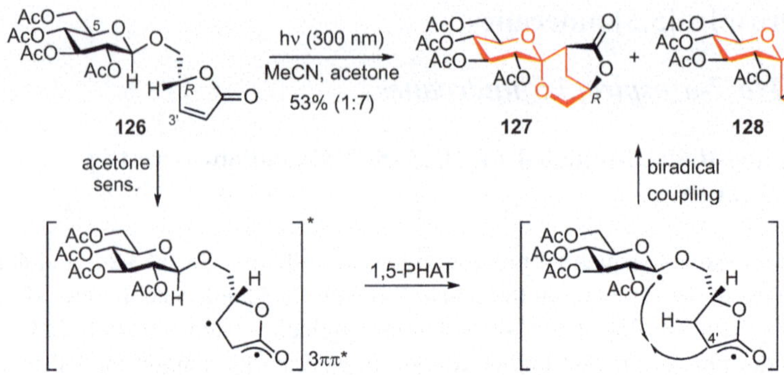

Scheme 31 1,7-Dioxaspiro[5.5]undecanes via 1,5-PHAT–1,6-biradical cyclization

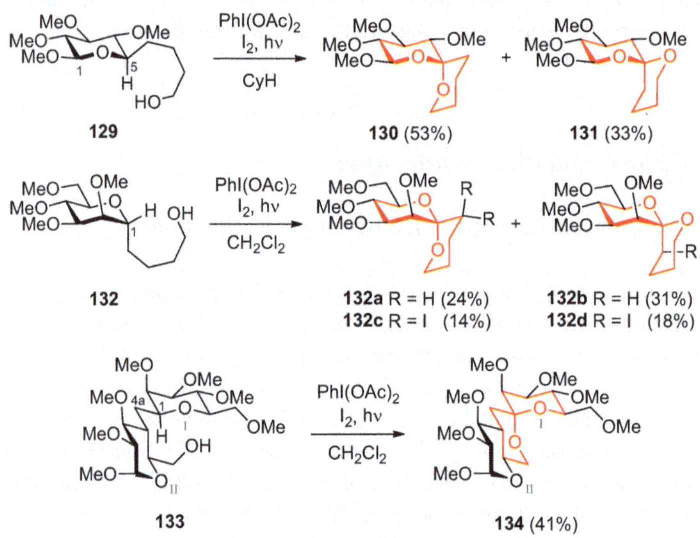

Scheme 32 1,7-Dioxaspiro[5.5]undecanes via 1,6-HAT–SET oxidation–6-*exo*-trig cyclization

resulting 1,8-diradical gives finally an interesting, although not spiroheterocyclic, 1,3-dioxocane ring.

5.2.2 Alkoxyl-Radical-Initiated 1,6-HAT–SET Oxidation–6-*exo*-trig Cyclization

We have also addressed the synthesis of spiroacetals of the [5.5]undecane series following a similar strategy to that described previously in Sect. 3.4.1 [97, 98]. Thus, the treatment of β-methyl glucopyranoside derivative **129** with $PhI(OAc)_2$ and iodine generated the spiro compounds **130** and **131** in good overall yield (Scheme 32). As can be observed, the major compound corresponds to the thermodynamically more stable one, in which each oxygen ring is axial with respect to the adjacent ring, thus obtaining maximum stability from the anomeric effect. Moreover, the use of other oxidant systems such as diphenylhydroxyselenium acetate/I_2 or HgO/I_2 led to poorer yields [143–145]. Interestingly, when the reaction was tested on methyl 6,7,8-trideoxy-2,3,4-tri-*O*-methyl-α-L-*gluco*-nonopyranoside (i.e., the corresponding α-anomer of **129**), a complex mixture was obtained [97]. The 1,3-diaxial steric interactions with the methoxyl group at C1 seem to disfavor the radical abstraction of the hydrogen at C5.

Additionally, this radical sequential process has also been applied to 1α-*C*-mannopyranosyl derivative **132** and *C*-disaccharide α-D-Man*p*-(1→4a)-4a-carba-α-D-Glc*p* **133** [146, 147]. In the former case, abstraction of the equatorial anomeric hydrogen is slower than the corresponding axial system, competing with 1,5-HAT to produce iodinated compounds **132c,d**, apart from the expected spiro compounds **132a,b**. In the case of the *C*-disaccharide **133**, the different conformational flexibility of the glycosidic bond allows products to be isolated from 1,5-, 1,6-, and 1,8-HAT processes, the spiroacetal **134**, arising from the 1,6-abstraction of the $H1^{I}$, being the major compound.

5.3 *1,5,7-Trioxaspiro[5.5]undecanes*

5.3.1 Alkoxyl-Radical-Initiated 1,6-HAT–SET Oxidation–6-*exo*-trig Cyclization

The formation of 1,5,7-trioxaspiro[5.5]undecane systems has also been explored during our investigations of HAT processes between the two pyranose units in (1→4)-*O*-disaccharide models. From these studies, we demonstrated that it is possible to modulate the regioselectivity of the abstraction step, 1,6- versus 1,8-HAT, by simply controlling the relative configuration of the four chiral centers involved in the cyclization step [140, 148]. Thus, the irradiation of the alcohol **135**, derived from α-L-Rham*p*-(1→4)-α-D-Glc*p* disaccharide, in the presence of PhI(OAc)$_2$ and iodine, generates a 6^{II}-*O*-yl radical which abstracts exclusively the anomeric hydrogen of the rhamnose unit via a 1,6-HAT (Scheme 33). The subsequent nucleophilic 6-*exo*-trig cyclization of the alcohol to the oxocarbenium ion intermediate led to the spiro ortho

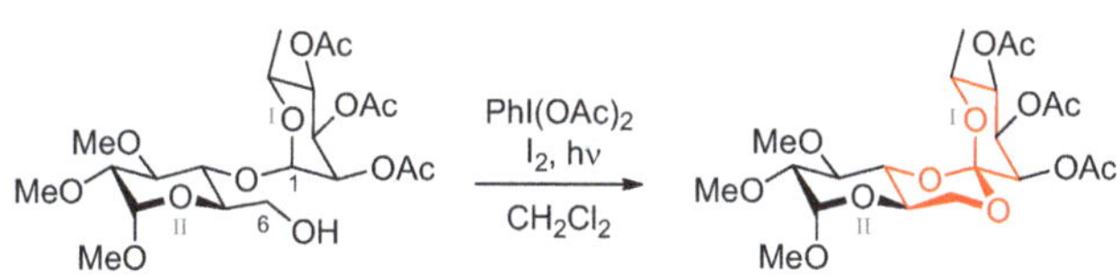

135 **136** (79%)

Scheme 33 1,5,7-Trioxaspiro[5.5]undecane via 1,6-HAT–SET oxidation–6-*exo*-trig cyclization

ester **136** in good yield. A few more examples of this radical sequential process have also been observed in a α-D-Lyx*p*-(1→4)-α-D-Glc*p* disaccharide containing a pentopyranose unit [149] as well as in one specific β-cyclodextrin model [150].

5.4 7-Oxa-1,5-dithiaspiro[5.5]undecanes

5.4.1 C(sp^3)-Radical-Initiated Intermolecular Addition–Radical Halogenation–Nucleophilic Substitution

Carbohydrates possessing a 7-oxa-1,5-dithiaspiro[5.5]undecane skeleton have been synthesized by Portella and coworkers from the radical trifluoromethylation of D-mannose-derived ketene dithioacetal **137** (Scheme 34) [151]. The trifluoromethyl radical was generated by SET from the sulfoxylate radical anion, which was obtained by reduction of sulfur dioxide with sodium formate.

The intermolecular addition proceeded stereoselectively, only one isomer of **138** being isolated with the trifluoromethyl group in equatorial position. The authors have proposed a mechanism involving a halogen atom transfer to a radical intermediate followed by intramolecular nucleophilic substitution for the cyclization step. One could also consider an alternative radical-polar crossover mechanism through a dithiinium ion, promoted by SET oxidation and subsequent 6-*exo*-trig cyclization (see related examples of radical-polar crossover mechanisms in Sects. 3.2.2 and 3.4.1).

Scheme 34 7-Oxa-1,5-dithiaspiro[5.5]undecane via intermolecular addition–radical halogenation–nucleophilic substitution

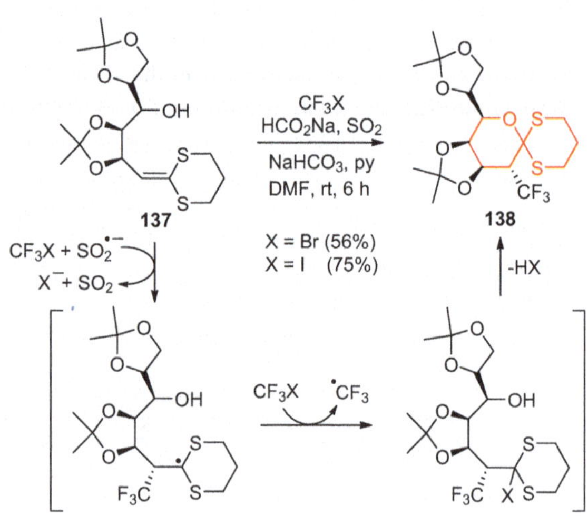

6 Heterospiro[5.6]dodecanes

6.1 1-Oxa-8-azaspiro[5.6]dodecanes

6.1.1 Photoinduced Norrish Type II–Yang Cyclization–Retro-Aldol Ring Expansion

Thiem and coworkers have described the Norrish–Yang photocyclization of a succinimide tether attached at C6 of a 6-deoxy-α-D-glucopyranose (e.g., **139** and **140**) (Scheme 35) [152, 153]. The intramolecular cyclization proceeded as expected via a 1,5-PHAT with regioselective abstraction of the hydrogen at C5 and subsequent collapse of the 1,4-biradical formed with retention of configuration at C5. The

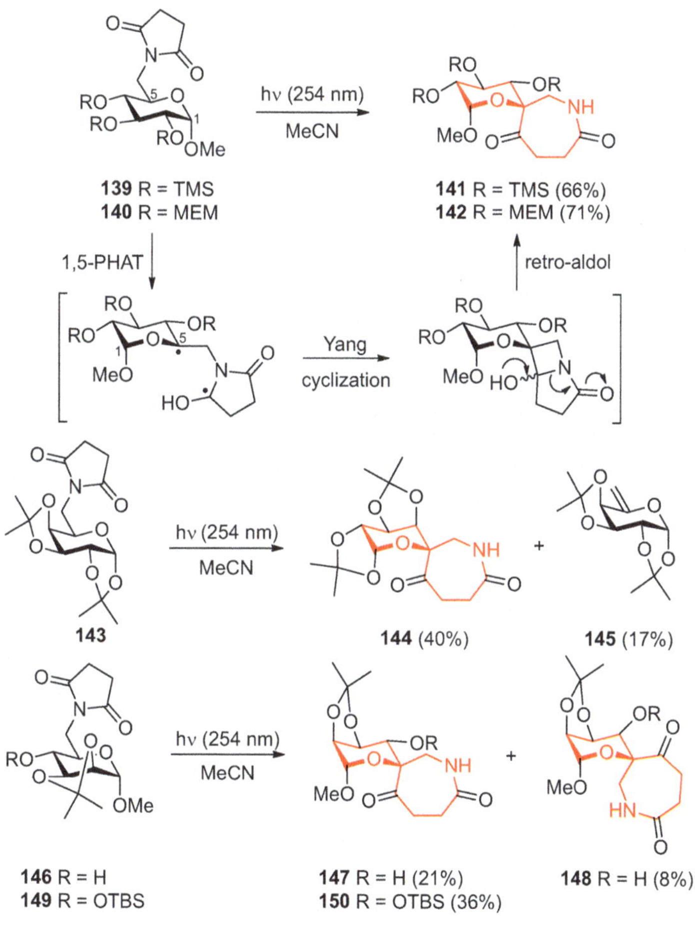

Scheme 35 1-Oxa-8-azaspiro[5.6]dodecanes via Norrish type II–Yang cyclization–retro-aldol ring expansion

azetidinol intermediate could not be isolated, and, due to an irreversible retro-aldol ring expansion, the 1-oxa-8-azaspiro[5.6]dodecane systems **141** and **142** were obtained instead. In the reaction of the D-*galacto* derivative **143**, apart from the spiro compound **144**, a small amount of the *exo*-glycal **145** was formed by a competitive Norrish Type II photoelimination process.

In the D-*manno* compound **146**, the cyclization proceeded with lower diastereoselectivity, and a mixture of isomers **147** and **148** was isolated. However, the fully protected mannopyranosyl compound **149** gave diastereomer **150** exclusively.

These examples have been selected from a more comprehensive Thiem study on the photochemistry of succinimide- and glutarimide-substituted carbohydrates [154–156].

7 Heterodispiro[4.1.4.3]tetradecanes

7.1 1,6,8-Trioxadispiro[$4.1.4^7.3^5$]tetradecanes

7.1.1 Photoinduced Two Consecutive Norrish Type II–Yang Cyclizations

The Descotes and coworkers synthesis of 1,6,8-trioxadispiro[$4.1.4^7.3^5$]tetradecane tricycles is outlined in Scheme 36 [157]. Although the authors start with a racemic product, the analogy with the methodology developed for carbohydrates in Sect. 4.3.1 is evident, and it seemed pertinent to include this work here.

The Norrish–Yang photocyclization of 6-methoxytetrahydro-2*H*-pyran **151** with a 4-hydroxy-2-butanone tether afforded 1,6-dioxaspiro[4.5]decane system **152** as a mixture of four isomers. Incorporation of another 4-hydroxy-2-butanone tether at C5 by acid-catalyzed transacetalization gave **153*R****** and **153*S****** as two separable

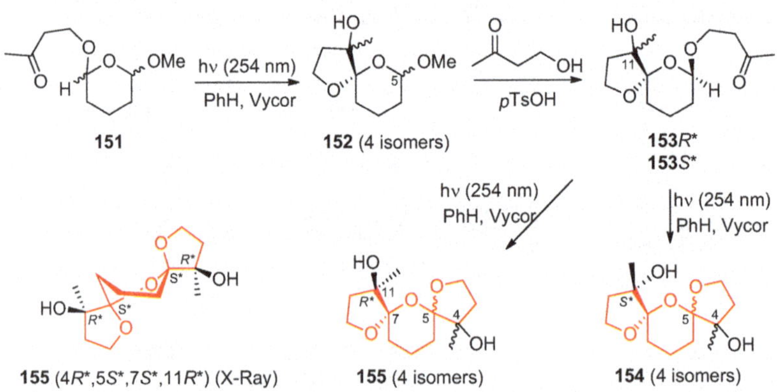

Scheme 36 1,6,8-Trioxadispiro[$4.1.4^7.3^5$]tetradecanes via two consecutive Norrish type II–Yang cyclization

compounds. The irradiation of each one separately allowed the preparation of dispirocycles **154** and **155**, obtained as mixtures of four isomers that were separated by chromatography and characterized. The structure and stereochemistry of the major component of **155** (4*R**,5*S**,7*S**,11*R**) was confirmed by X-ray diffraction analysis; the tetrahydropyrane ring adopts a skew boat ($^{5}S_{7}$) conformation that leaves the oxygens of the tetrahydrofuran rings in quasi-axial position [158]. The kinetic diastereomers were formed preferentially when the photolyses were conducted at low temperature, whereas acid-catalyzed isomerization produced the thermodynamic mixtures of products as determined by anomeric effects and 1,3-diaxial interactions [159].

8 Heterodispiro[4.1.5.2]tetradecanes

8.1 1,6,8-Trioxadispiro[4.1.5^{7}.2^{5}]tetradecanes

8.1.1 Alkoxyl-Radical-Initiated Two Consecutive 1,5-HAT–SET Oxidation–5-*exo*-trig Cyclization

The methodology developed in Sect. 3.4.1 has been applied successfully by Meilert and Brimble to the synthesis of 1,6,8-trioxadispiro[4.1.5^{7}.2^{5}]tetradecane skeleton which is present in spirolides A–F, a family of marine toxins of dinoflagellate origin [160]. Two iterative oxidative radical cyclization reactions of alcohols **156** and **157**, with the $PhI(OAc)_2/I_2$ system, allowed the construction of the bis-spiroacetal **158** as an equimolar mixture of four diastereomers in high yield (Scheme 37) [161–163].

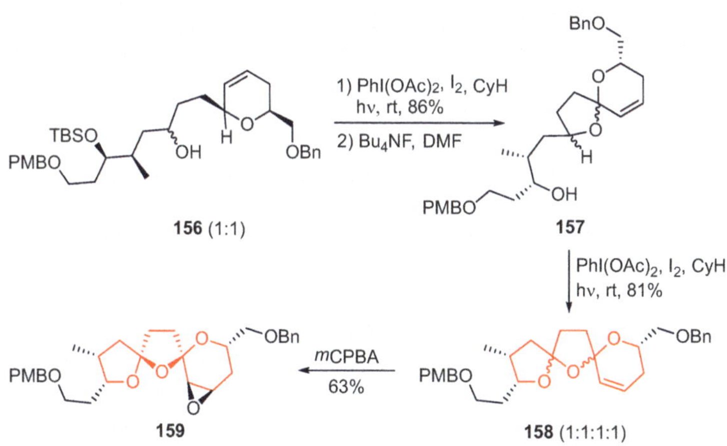

Scheme 37 1,6,8-Trioxadispiro[4.1.5^{7}.2^{5}]tetradecane via two consecutive 1,5-HAT–SET oxidation–5-*exo*-trig cyclization

Notably, the *m*CPBA treatment of **158** required in the next step of the synthesis afforded the β-epoxide **159** as a single diastereomer. The presence of *m*-chlorobenzoic acid and water may have affected equilibration of the mixture to the most thermodynamically favored isomer.

9 Heterodispiro[4.1.5.3]pentadecanes

9.1 1,6,8-Trioxadispiro[4.1.5^7.3^5]pentadecanes

9.1.1 Alkoxyl-Radical-Initiated 1,6-HAT–SET Oxidation–6-*exo*-trig Cyclization and 1,5-HAT–SET Oxidation–5-*exo*-trig Cyclization

Of special interest is the 1,6,8-trioxadispiro[4.1.5^7.3^5]pentadecane skeleton that is present in a diverse class of biologically active natural products, including polyether ionophores such as narasin [164], salinomycin [165], noboritomycin [166], CP44,661 [167], and X-14766A [168] isolated from fermentation cultures and the spirastrellolides A-G [169] obtained from the extract of a marine sponge.

Two different approaches to this dispiroacetal have been developed in our laboratory using the methodology displayed in Sects. 3.4.1 and 4.3.2. The first approach started with *C*-glycoside **160** prepared in several steps from D-galactose (Scheme 38) [170, 171]. Radical-initiated cyclization via 1,6-HAT with the PhI$(OAc)_2$/iodine system afforded 1,7-dioxaspiro[5.5]undecane derivatives **161** and

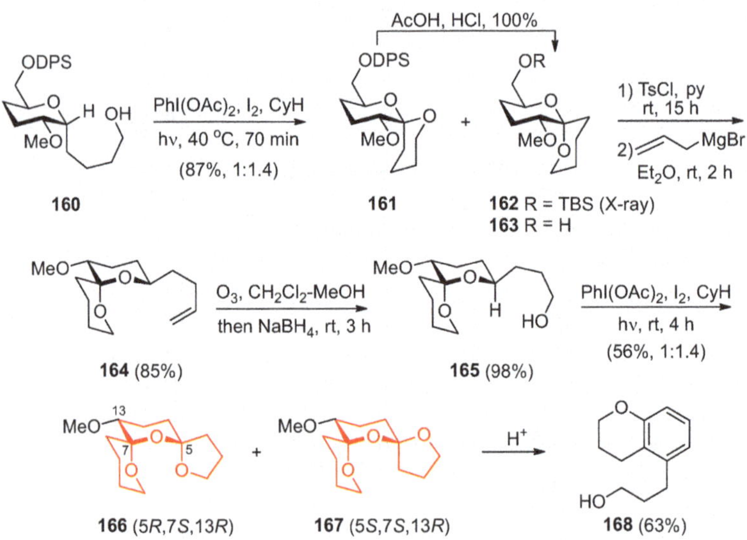

Scheme 38 1,6,8-Trioxadispiro[4.1.5^7.3^5]pentadecanes via 1,6-HAT–SET oxidation–6-*exo*-trig cyclization–1,5-HAT–SET oxidation–5-*exo*-trig cyclization. First approach

162. The stereochemistry of the spirocenter was confirmed by X-ray crystallographic analysis of compound **162**. The kinetic minor isomer **161** can be quantitatively transformed into the thermodynamically more stable **162** by acid-catalyzed isomerization.

The alcohol **163** obtained by deprotection was converted into the alkene **164** by tosylation and treatment with allylmagnesium bromide. Ozonolysis followed by a reductive workup yielded alcohol **165**, which was submitted to a second 1,5-HAT reaction to give a mixture of dispiroacetals **166** and **167**. Their stereochemistries were established by extensive 1D and 2D NMR experiments. Unfortunately, attempts at acid-catalyzed equilibration of **166** and **167** were unsuccessful. The molecules appear to be highly sensitive to these conditions, and 3-(chroman-5-yl) propan-1-ol (**168**), formed by an intramolecular aldol reaction, was the sole isolated product [172].

A second carbohydrate-based synthesis of the 1,6,8-trioxadispiro[4.1.5^7.3^5] pentadecane skeleton has been conveniently accomplished via alcohol **169** (Scheme 39) which, in turn, was prepared from D-glucal [171, 173]. Attempts to elaborate the trioxadispiro system directly from diol **170** by simultaneous double hydrogen abstraction reactions were unsuccessful and gave the dioxaspiro[4.5]decane ring system **171** as the only isolable product. As expected, the tetrahydrofuran ring was formed first, and the 1,3-diaxial interaction with the group at C1 appeared to hinder the abstraction of the C5 hydrogen through a less-favored seven-membered cyclic transition state. Therefore, the two cyclizations were carried out sequentially, the tetrahydropyrane ring constructed first, giving compound **172**, whose stereochemistry was confirmed by X-ray diffraction analysis of the *p*-nitrobenzoyl derivative **174**. After deprotection the alcohol **173** was submitted to the second oxidative radical cyclization to provide a mixture of isomeric dispiroacetals **175** and **176**.

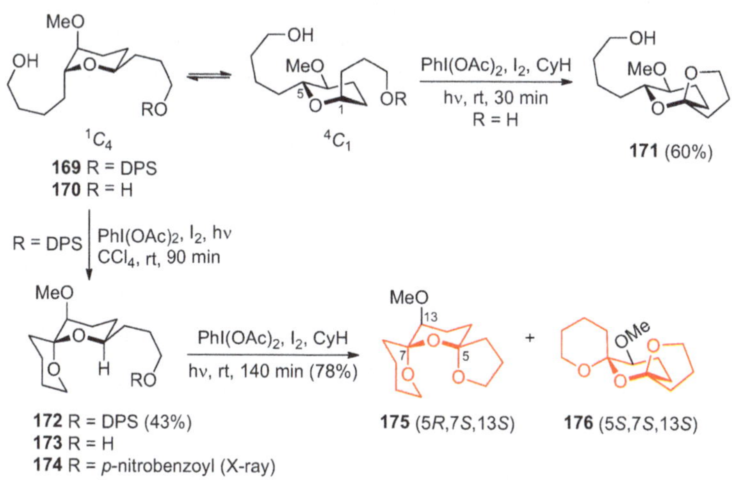

Scheme 39 1,6,8-Trioxadispiro[4.1.5^7.3^5]pentadecanes via 1,6-HAT–SET oxidation–6-*exo*-trig cyclization–1,5-HAT–SET oxidation–5-*exo*-trig cyclization. Second approach

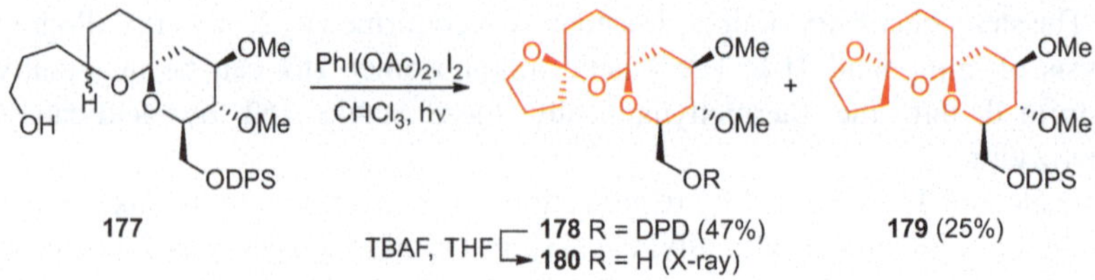

Scheme 40 1,6,8-Trioxadispiro[$4.1.5^7.3^5$]pentadecanes via 1,5-HAT–SET oxidation–5-*exo*-trig cyclization

An alternative approach toward the construction of the 1,6,8-trioxadispiro [$4.1.5^7.3^5$]pentadecane core present in (+)-spirastrellolide A using this methodology has been reported by Hsung and Tang and coworkers (Scheme 40) [174]. The oxidative cyclization of the alcohol **177**, prepared in several steps from D-glucose, under the $PhI(OAc)_2$ and iodine conditions afforded a separable mixture of the isomeric trioxadispiroacetals **178** and **179** in 72% overall yield. The acid-catalyzed equilibrium of pure **178** or **179** gave an isomeric mixture with a 1:1.5 ratio, slightly in favor of the thermodynamic isomer **179**. The isomer **178** was deprotected and the structure of the crystalline alcohol **180** determined by X-ray diffraction analysis.

10 Conclusions

This chapter describes various radical cascade reactions for the preparation of carbohydrate spiro- and dispiro-heterocycles. The fundamental mechanistic steps of the cascade reactions have been classified following the radical chain process: initiation, intermediate sequences, and final radical trap. The sequence 1,*n*-HAT–*n*-*exo*-trig cyclization initiated by *O*- or *N*-radicals via a radical-polar crossover mechanism is especially well suited for the formation of five- ($n = 5$) or six-membered ($n = 6$) rings containing oxygen or nitrogen. The radicals are generated by a mild and convenient reaction of the free alcohols and amides with hypervalent iodine(III) reagents and molecular iodine under irradiation with visible light. Carbocycles are preferentially obtained by intramolecular addition of C(sp^2)- and C(sp^3)-centered radicals to unsaturated partners such as alkenes, alkynes, or aldehydes via kinetically favorable *exo*-trig or *exo*-dig processes or by photoinduced 1,*n*-HAT ($n = 5, 6$) via Norrish type II–Yang cyclization of carbonyl compounds. With the recent discovery of photoredox catalytic methods for the preparation of free radicals under mild conditions, more efficient and selective HAT and cyclization processes are expected in the near future. This will undoubtedly have a very positive effect on the development of new syntheses of heterocycles.

References

1. Pearce AJ, Mallet J-M, Sinaÿ P (2001) Radicals in carbohydrate chemistry. In: Renaud P, Sibi MP (eds) Radicals in organic synthesis, vol 2. Wiley, Weinheim, pp 538–577. https://doi.org/10.1002/9783527618293.ch52
2. Pérez-Martín I, Suárez E (2012) Radicals and carbohydrates. In: Chatgilialoglu C, Studer A (eds) Encyclopedia of radicals in chemistry, biology and materials. Wiley, Chichester, pp 1131–1174. https://doi.org/10.1002/9781119953678.rad031
3. Binkley RW, Binkley ER (2013) Radical reaction of carbohydrates, vols I and II. http://www.carborad.com/default.html
4. McSweeney L, Dénès F, Scanlan EM (2016) Thiyl-radical reactions in carbohydrate chemistry: from thiosugars to glycoconjugate synthesis. Eur J Org Chem 2016:2080–2095. https://doi.org/10.1002/ejoc.201501543
5. Hansen SG, Skrydstrup T (2006) Modification of amino acids, peptides, and carbohydrates through radical chemistry. In: Gansäuer A (ed) Radicals in synthesis II. Topics in current chemistry, vol 264. Springer, Berlin, pp 135–162. https://doi.org/10.1007/128_022
6. Boultadakis-Arapinis M, Lescot C, Micouin L, Lecourt T (2013) Functionalization of the anomeric C–H bond of carbohydrates: old strategies and new opportunities. Synlett 24:2477–2491. https://doi.org/10.1055/s-0033-1339317
7. Frihed TG, Bols M, Pedersen CM (2016) C–H functionalization on carbohydrates. Eur J Org Chem 2016:2740–2756. https://doi.org/10.1002/ejoc.201600121
8. Murakami M, Ishida N (2017) β-Scission of alkoxy radicals in synthetic transformations. Chem Lett 46:1692–1700. https://doi.org/10.1246/cl.170834
9. Salamone M, Bietti M (2014) Reaction pathways of alkoxyl radicals. The role of solvent effects on C–C bond fragmentation and hydrogen atom transfer reactions. Synlett 25:1803–1816. https://doi.org/10.1055/s-0033-1341280
10. Undheim K (2015) Stereoselective reactions in preparation of chiral α-hetera-spiro[m.n] alkanes. Synthesis 47:2497–2522. https://doi.org/10.1055/s-0034-1378863
11. Undheim K (2014) Preparation and structure classification of heteraspiro[m.n]alkanes. Synthesis 46:1957–2006. https://doi.org/10.1055/s-0033-1338640
12. El Ashry ESH (ed) (2007) Heterocycles from carbohydrate precursors. Topics in heterocyclic chemistry, vol 7. Springer, Berlin. https://doi.org/10.1007/978-3-540-72957-0
13. Soengas RG, da Silva G, Estévez JC (2017) Synthesis of spironucleosides: past and future perspectives. Molecules 22:2028. https://doi.org/10.3390/molecules22112028
14. Landais Y (ed) (2018) Free-radical synthesis and functionalization of heterocycles. Topics in heterocyclic chemistry, vol 54. Springer, Cham. https://doi.org/10.1007/978-3-319-89521-5
15. Undheim K (2017) Spirocyclic orthoesters, orthothioesters and orthoaminals in the synthesis and structural modification of natural products. Synthesis 49:705–723. https://doi.org/10.1055/s-0036-1588339
16. Brimble MA, Furkert DP (2003) Chemistry of bis-spiroacetal systems: natural products, synthesis and stereochemistry. Curr Org Chem 7:1461–1484. https://doi.org/10.2174/1385272033486404
17. Brimble MA, Stubbing LA (2014) Synthesis of 5,6- and 6,6-spirocyclic compounds. In: Cossy J (ed) Synthesis of saturated oxygenated heterocycles I. Topics in heterocyclic chemistry, vol 35. Springer, Berlin, pp 189–267. https://doi.org/10.1007/978-3-642-41473-2_5
18. Kiyota H (2006) Synthesis of marine natural products with bicyclic and/or spirocyclic acetals. In: Kiyota H (ed) Marine natural products. Topics in heterocyclic chemistry, vol 5. Springer, Berlin, pp 65–95. https://doi.org/10.1007/7081_029
19. Smith LK, Baxendale IR (2015) Total syntheses of natural products containing spirocarbocycles. Org Biomol Chem 13:9907–9933. https://doi.org/10.1039/C5OB01524C
20. Sherburn MS (2012) Basic concepts on radical chain reactions. In: Chatgilialoglu C, Studer A (eds) Encyclopedia of radicals in chemistry, biology and materials. Wiley, Chichester, pp 57–80. https://doi.org/10.1002/9781119953678.rad017

21. Kittaka A, Asakura T, Kuze T, Tanaka H, Yamada N, Nakamura KT, Miyasaka T (1999) Cyclization reactions of nucleoside anomeric radical with olefin tethered on base: factors that induce anomeric stereochemistry. J Org Chem 64:7081–7093. https://doi.org/10.1021/jo990611d
22. Chatgilialoglu C, Gimisis T, Spada GP (1999) C-1′ radical-based approaches for the synthesis of anomeric spironucleosides. Chem Eur J 5:2866–2876. https://doi.org/10.1002/(SICI)1521-3765(19991001)5:10<2866::AID-CHEM2866>3.0.CO;2-6
23. Taniguchi T (2017) Recent advances in reactions of heteroatom-centered radicals. Synthesis 49:3511–3534. https://doi.org/10.1055/s-0036-1588481
24. Nechab M, Mondal S, Bertrand MP (2014) 1,*n*-hydrogen-atom transfer (HAT) reactions in which $n \neq 5$: an updated inventory. Chem Eur J 20:16034–16059. https://doi.org/10.1002/chem.201403951
25. Sperry J, Liu Y-C, Brimble MA (2010) Synthesis of natural products containing spiroketals *via* intramolecular hydrogen abstraction. Org Biomol Chem 8:29–38. https://doi.org/10.1039/B916041H
26. Dénès F, Beaufils F, Renaud P (2008) Preparation of five-membered rings via the translocation-cyclization of vinyl radicals. Synlett 19:2389–2399. https://doi.org/10.1055/s-2008-1078016
27. Renaud P, Beaufils F, Dénès F, Feray L, Imboden C, Kuznetsov N (2008) Stereoselective radical translocations. CHIMIA Int J Chem 62:510–513. https://doi.org/10.2533/chimia.2008.510
28. Čeković Ž (2005) Reactions of carbon radicals generated by 1,5-transposition of reactive centers. J Serb Chem Soc 70:287–318. https://doi.org/10.2298/JSC0503287C
29. Čeković Ž (2003) Reactions of δ-carbon radicals generated by 1,5-hydrogen transfer to alkoxyl radicals. Tetrahedron 59:8073–8090. https://doi.org/10.1016/S0040-4020(03)01202-X
30. Feray L, Kuznetsov N, Renaud P (2001) Hydrogen atom abstraction. In: Renaud P, Sibi MP (eds) Radicals in organic synthesis, vol 2. Wiley, Weinheim, pp 246–278. https://doi.org/10.1002/9783527618293.ch39
31. Robertson J, Pillai J, Lush RK (2001) Radical translocation reactions in synthesis. Chem Soc Rev 30:94–103. https://doi.org/10.1039/b000705f
32. Majetich G, Wheless K (1995) Remote intramolecular free radical functionalizations: an update. Tetrahedron 51:7095–7129. https://doi.org/10.1016/0040-4020(95)00406-X
33. Mihailović ML, Čeković Ž (1970) Intramolecular oxidative cyclization of alcohols with lead tetraacetate. Synthesis 2:209–224. https://doi.org/10.1055/s-1970-21596
34. Heusler K, Kalvoda J (1964) Intramolecular free-radical reactions. Angew Chem Int Ed 3:525–596. https://doi.org/10.1002/anie.196405251
35. Dorigo AE, McCarrick MA, Loncharich RJ, Houk KN (1990) Transition structures for hydrogen atom transfers to oxygen. Comparisons of intermolecular and intramolecular processes, and open- and closed-shell systems. J Am Chem Soc 112:7508–7514. https://doi.org/10.1021/ja00177a009
36. Dorigo AE, Houk KN (1988) The relationship between proximity and reactivity. An ab initio study of the flexibility of the $OH^{\bullet} + CH_4$ hydrogen abstraction transition state and a force-field model for the transition states of intramolecular hydrogen abstractions. J Org Chem 53:1650–1664. https://doi.org/10.1021/jo00243a011
37. Dorigo AE, Houk KN (1987) Transition structures for intramolecular hydrogen atom transfers: the energetic advantage of seven-membered over six-membered transition structures. J Am Chem Soc 109:2195–2197. https://doi.org/10.1021/ja00241a056
38. Yoshimura A, Zhdankin VV (2016) Advances in synthetic applications of hypervalent iodine compounds. Chem Rev 116:3328–3435. https://doi.org/10.1021/acs.chemrev.5b00547
39. Zhdankin VV (2009) Hypervalent iodine(III) reagents in organic synthesis. ARKIVOC 2009:1–62. https://doi.org/10.3998/ark.5550190.0010.101
40. Togo H, Katohgi M (2001) Synthetic uses of organohypervalent iodine compounds through radical pathways. Synlett 12:565–581. https://doi.org/10.1055/s-2001-13349

41. Hu X-Q, Chen J-R, Xiao W-J (2017) Controllable remote C–H bond functionalization by visible-light photocatalysis. Angew Chem Int Ed 56:1960–1962. https://doi.org/10.1002/anie.201611463
42. Studer A, Curran DP (2016) Catalysis of radical reactions: a radical chemistry perspective. Angew Chem Int Ed 55:58–102. https://doi.org/10.1002/anie.201505090
43. Huang W, Cheng X (2017) Hantzsch esters as multifunctional reagents in visible-light photoredox catalysis. Synlett 28:148–158. https://doi.org/10.1055/s-0036-1588129
44. Shaw MH, Twilton J, MacMillan DWC (2016) Photoredox catalysis in organic chemistry. J Org Chem 81:6898–6926. https://doi.org/10.1021/acs.joc.6b01449
45. Prier CK, Rankic DA, MacMillan DWC (2013) Visible light photoredox catalysis with transition metal complexes: applications in organic synthesis. Chem Rev 113:5322–5363. https://doi.org/10.1021/cr300503r
46. Stateman LM, Nakafuku KM, Nagib DA (2018) Remote C–H functionalization via selective hydrogen atom transfer. Synthesis 50:1569–1586. https://doi.org/10.1055/s-0036-1591930
47. Yan M, Lo JC, Edwards JT, Baran PS (2016) Radicals: reactive intermediates with translational potential. J Am Chem Soc 138:12692–12714. https://doi.org/10.1021/jacs.6b08856
48. Chiba S, Chen H (2014) sp3 C–H oxidation by remote H-radical shift with oxygen- and nitrogen-radicals: a recent update. Org Biomol Chem 12:4051–4060. https://doi.org/10.1039/c4ob00469h
49. Wagner PJ (2005) Abstraction of γ-hydrogens by excited carbonyls. In: Griesbeck AG, Mattay J (eds) Synthetic organic photochemistry. Marcel Dekker, New York, pp 11–39
50. Wessig P, Muhling O (2005) Abstraction of (γ±n)-hydrogen by excited carbonyls. In: Griesbeck A, Mattay J (eds) Synthetic organic photochemistry. Marcel Dekker, New York, pp 41–87
51. Wagner PJ, Klán P (2004) Norrish type II photoelimination of ketones: cleavage of 1,4-biradicals formed by γ-hydrogen abstraction. In: Horspool W, Lenci F (eds) Organic photochemistry and photobiology2nd edn. CRC, Boca Raton, pp 1–31
52. Wagner PJ (2004) Yang photocyclization: coupling of biradicals formed by intramolecular hydrogen abstraction of ketones. In: Horspool W, Lenci F (eds) Organic photochemistry and photobiology2nd edn. CRC, Boca Raton, pp 1–69
53. Arjona O, Gómez AM, López JC, Plumet J (2007) Synthesis and conformational and biological aspects of carbasugars. Chem Rev 107:1919–2036. https://doi.org/10.1021/cr0203701
54. Martínez-Grau A, Marco-Contelles J (1998) Carbocycles from carbohydrates *via* free radical cyclizations: new synthetic approaches to glycomimetics. Chem Soc Rev 27:155–162. https://doi.org/10.1039/a827155z
55. Dalko PI, Sinaÿ P (1999) Recent advances in the conversion of carbohydrate furanosides and pyranosides into carbocycles. Angew Chem Int Ed 38:773–777. https://doi.org/10.1002/(SICI)1521-3773(19990315)38:6<773::AID-ANIE773>3.0.CO;2-N
56. Madsen R (2007) Synthetic strategies for converting carbohydrates into carbocycles by the use of olefin metathesis. Eur J Org Chem 2007:399–415. https://doi.org/10.1002/ejoc.200600614
57. Chen Y-L, Cen J-D (2004) Recent advances in conversion from carbohydrates to functionalized carbocycles. Chin J Org Chem 24:1332–1339
58. Baldwin JE (1976) Rules for ring closure. J Chem Soc Chem Commun:734–736. https://doi.org/10.1039/C39760000734
59. Baldwin JE, Thomas RC, Kruse LI, Silberman L (1977) Rules for ring closure: ring formation by conjugate addition of oxygen nucleophiles. J Org Chem 42:3846–3852. https://doi.org/10.1021/jo00444a011
60. Beckwith ALJ, Easton CJ, Serelis AK (1980) Some guidelines for radical reactions. J Chem Soc Chem Commun:482–483. https://doi.org/10.1039/C39800000482
61. Beckwith ALJ (1981) Regio-selectivity and stereo-selectivity in radical reactions. Tetrahedron 37:3073–3100. https://doi.org/10.1016/S0040-4020(01)98839-8

62. Beckwith ALJ, Schiesser CH (1985) Regio- and stereo-selectivity of alkenyl radical ring closure: a theoretical study. Tetrahedron 41:3925–3941. https://doi.org/10.1016/S0040-4020(01)97174-1
63. Walton JC (2006) Unusual radical cyclisations. In: Gansäuer A (ed) Radicals in synthesis II. Topics in current chemistry, vol 264. Springer, Berlin, pp 163–200. https://doi.org/10.1007/128_021
64. Gilmore K, Alabugin IV (2012) Unusual cyclizations. In: Chatgilialoglu C, Studer A (eds) Encyclopedia of radicals in chemistry, biology and materials. Wiley, Chichester, pp 693–728. https://doi.org/10.1002/9781119953678.rad021
65. Srikrishna A (2001) Unusual cyclizations. In: Renaud P, Sibi MP (eds) Radicals in organic synthesis, vol 2. Wiley, Weinheim, pp 151–187. https://doi.org/10.1002/9783527618293.ch34
66. Ishibashi H, Sato T, Ikeda M (2002) 5-Endo-trig radical cyclizations. Synthesis 34:695–713. https://doi.org/10.1055/s-2002-25759
67. Taniguchi T, Ishibashi H (2013) Synthesis of alkaloids using radical cyclizations. Heterocycles 87:527–545. https://doi.org/10.3987/REV-12-759
68. Ishibashi H (2006) Controlling the regiochemistry of radical cyclizations. Chem Rec 6:23–31. https://doi.org/10.1002/tcr.20069
69. Ishibashi H (2012) Synthesis of alkaloids using radical cyclizations. Yakugaku Zasshi 132:1413–1430. https://doi.org/10.1248/yakushi.12-00241
70. Chatgilialoglu C, Ferreri C, Guerra M, Timokhin V, Froudakis G, Gimisis T (2002) 5-*Endo-trig* radical cyclizations: disfavored or favored processes? J Am Chem Soc 124:10765–10772. https://doi.org/10.1021/ja0261731
71. Hartung J (2001) Stereoselective construction of the tetrahydrofuran nucleus by alkoxyl radical cyclizations. Eur J Org Chem 2001:619–632. https://doi.org/10.1002/1099-0690(200102)2001:4<619::AID-EJOC619>3.0.CO;2-A
72. Herrera AJ, Rondón M, Suárez E (2007) Stereocontrolled photocyclization of 1,2-diketones applied to carbohydrate models: a new entry to *C*-ketosides. Synlett 18:1851–1856. https://doi.org/10.1055/s-2007-984523
73. Herrera AJ, Rondón M, Suárez E (2008) Stereocontrolled photocyclization of 1,2-diketones. Application of a 1,3-acetyl group transfer methodology to carbohydrates. J Org Chem 73:3384–3391. https://doi.org/10.1021/jo702663w
74. Roberts SW, Rainier JD (2005) Substitution and remote protecting group influence on the oxidation/addition of α-substituted 1,2-anhydroglycosides: a novel entry into *C*-ketosides. Org Lett 7:1141–1144. https://doi.org/10.1021/ol0501469
75. Bernasconi C, Cottier L, Descotes G, Rémy G (1979) Dégradations ménagées des sucres V. Synthèse par voie photochimique d'aldonolactones-1,5 à partir d'oxo-2 propylglycosides. Bull Soc Chim Fr 5–6:332–336
76. Descotes G (1982) Photoreactivite acetalique en serie heterocyclique et osidique. Bull Soc Chim Belg 91:973–983
77. Hürzeler M, Bernet B, Vasella A (1993) Glyconothio-*O*-lactones. Part I. Preparation and reactions with nucleophiles. Helv Chim Acta 76:995–1012. https://doi.org/10.1002/hlca.19930760223
78. Binkley RW (1977) A mild process for the oxidation of partially protected carbohydrates. J Org Chem 42:1216–1221. https://doi.org/10.1021/jo00427a025
79. Brunckova J, Crich D (1995) Intramolecular hydrogen atom abstraction: the β-oxygen effect in the Norrish type II photoreaction. Tetrahedron 51:11945–11952. https://doi.org/10.1016/0040-4020(95)00755-W
80. Houlton JS, Motherwell WB, Ross BC, Tozer MJ, Williams DJ, Slawin AMZ (1993) A convenient strategy for replacement of the anomeric hydroxyl group by difluoromethyl functionality in carbohydrate derivatives. Tetrahedron 49:8087–8106. https://doi.org/10.1016/S0040-4020(01)88029-7

81. Herpin TF, Motherwell WB, Tozer MJ (1994) The synthesis of difluoromethylene-linked C-glycosides and C-disaccharides. Tetrahedron Asymmetry 5:2269–2282. https://doi.org/10.1016/S0957-4166(00)86305-3
82. Kittaka A, Tanaka H, Yamada N, Miyasaka T (1996) Nucleoside anomeric radicals *via* 1,5-translocation: Facile access to anomeric spiro nucleosides. Tetrahedron Lett 37:2801–2804. https://doi.org/10.1016/0040-4039(96)00433-9
83. Kittaka A, Tanaka H, Yamada N, Kato H, Miyasaka T (1997) 1,5-translocation strategy for nucleoside anomeric radicals. Nucleosides Nucleotides Nucleic Acids 16:1423–1426. https://doi.org/10.1080/07328319708006197
84. Gimisis T, Chatgilialoglu C (1996) 1,5-radical translocation protocol for the generation of C-1′ radicals in nucleosides. Synthesis of spiro nucleosides through a rare 5-*endo*-trig cyclization. J Org Chem 61:1908–1909. https://doi.org/10.1021/jo952218n
85. Strohmeier J, Nadler A, Heinrich D, Fitzner A, Diederichsen U (2010) Synthesis of 8,1′-etheno and 8,2′-ethano bridged guanosine derivatives using radical cyclization. Heterocycles 82:713–728. https://doi.org/10.3987/COM-10-S(E)51
86. Stella L (1983) Homolytic cyclizations of *N*-chloroalkenylamines. Angew Chem Int Ed Engl 22:337–350. https://doi.org/10.1002/anie.198303373
87. Dénès F (2018) Heteroatom-centred radicals for the synthesis of heterocyclic compounds. In: Landais Y (ed) Free-radical synthesis and functionalization of heterocycles. Topics in heterocyclic chemistry, vol 54. Springer, Cham, pp 151–230. https://doi.org/10.1007/7081_2018_19
88. Freire R, Martın A, Pérez-Martın I, Suárez E (2002) Synthesis of oxa-aza spirobicycles by intramolecular hydrogen abstraction promoted by *N*-radicals in carbohydrate systems. Tetrahedron Lett 43:5113–5116. https://doi.org/10.1016/S0040-4039(02)00983-8
89. Martín A, Pérez-Martín I, Suárez E (2009) Synthesis of oxa-aza spirobicycles by intramolecular hydrogen atom transfer promoted by *N*-radicals in carbohydrate systems. Tetrahedron 65:6147–6155. https://doi.org/10.1016/j.tet.2009.05.049
90. Courtneidge JL, Lusztyk J, Pagé D (1994) Alkoxyl radicals from alcohols. Spectroscopic detection of intermediate alkyl and acyl hypoiodites in the Suárez and Beebe reactions. Tetrahedron Lett 35:1003–1006. https://doi.org/10.1016/S0040-4039(00)79950-3
91. Martín A, Pérez-Martín I, Suárez E (2005) Intramolecular hydrogen abstraction promoted by amidyl radicals. Evidence for electronic factors in the nucleophilic cyclization of ambident amides to oxocarbenium ions. Org Lett 7:2027–2030. https://doi.org/10.1021/ol050526u
92. Kittaka A, Tanaka H, Odanaka Y, Ohnuki K, Yamaguchi K, Miyasaka T (1994) Vinyl radical-based cyclization of 6-substituted 1-(2-deoxy-D-*erythro*-pent-1-enofuranosyl) uracils: synthesis of anomeric spiro nucleosides. J Org Chem 59:3636–3641. https://doi.org/10.1021/jo00092a024
93. Kittaka A, Tsubaki Y, Tanaka H, Nakamura KT, Miyasaka T (1996) Tandem radical cyclization-oxygenation of 6-(2,2-di-bromovinyl)-1-(2-deoxy-D-*erythro*-pent-1-enofuranosyl)-uracil: synthesis of anomeric spiro nucleosides having arabino and ribo configurations. Nucleosides Nucleotides 15:97–107. https://doi.org/10.1080/07328319608002373
94. Kittaka A, Yamada N, Tanaka H, Nakamura KT, Miyasaka T (1996) Radical-mediated cyclization of a 6-chloro-9-(2-deoxy-D-*erythro*-pent-1-enofuranosyl)-8-(2,2-dibromovinyl) purine. Nucleosides Nucleotides 15:1447–1457. https://doi.org/10.1080/07328319608002446
95. Kodama T, Shuto S, Nomura M, Matsuda A (2001) An efficient method for the preparation of 1′α-branched-chain sugar pyrimidine ribonucleosides from uridine: the first conversion of a natural nucleoside into 1′-substituted ribonucleosides. Chem Eur J 7:2332–2340. https://doi.org/10.1002/1521-3765(20010601)7:11<2332::AID-CHEM23320>3.0.CO;2-W
96. Dang SF, Sun JB, Xu XX, Wang P, Wu JC (2008) Synthesis of 4′-spironucleoside via radical translocation cyclization reaction. Chem Res Chin Univ 24:473–476. https://doi.org/10.1016/S1005-9040(08)60099-9
97. Martín A, Salazar JA, Suárez E (1995) Synthesis of chiral spiroacetals from carbohydrates. Tetrahedron Lett 36:4489–4492. https://doi.org/10.1016/0040-4039(95)00766-6

98. Martín A, Salazar JA, Suárez E (1996) Synthesis of chiral spiroacetals from carbohydrates. J Org Chem 61:3999–4006. https://doi.org/10.1021/jo960060g
99. Crich D, Huang X, Newcomb M (1999) Synthesis of tetrahydrofurans by a tandem hydrogen atom abstraction/radical nucleophilic displacement sequence. Org Lett 1:225–228. https://doi.org/10.1021/ol990564b
100. Crich D, Huang X, Newcomb M (2000) Inter- and intramolecular pathways for the formation of tetrahydrofurans from β-(phosphatoxy)alkyl radicals. Evidence for a dissociative mechanism. J Org Chem 65:523–529. https://doi.org/10.1021/jo991570o
101. Sartillo-Piscil F, Vargas M, Anaya de Parrodi C, Quintero L (2003) Diastereoselective synthesis of 1,2-*O*-isopropylidene-1,6-dioxaspiro[4.4]nonane applying the methodology of generation of radical cations under non-oxidizing conditions. Tetrahedron Lett 44:3919–3921. https://doi.org/10.1016/S0040-4039(03)00817-7
102. Cortezano-Arellano O, Quintero L, Sartillo-Piscil F (2015) Total synthesis of cephalosporolide E via a tandem radical/polar crossover reaction. The use of the radical cations under nonoxidative conditions in total synthesis. J Org Chem 80:2601–2608. https://doi.org/10.1021/jo502757c
103. Crich D (2001) Radical rearrangements of esters. In: Renaud P, Sibi MP (eds) Radicals in organic synthesis, vol 2. Wiley, Weinheim, pp 188–206. https://doi.org/10.1002/9783527618293.ch35
104. Sharma GVM, Chander AS, Reddy VG, Krishnudu K, Rao MHVR, Kunwar AC (2000) Radical mediated stereoselective synthesis of chiral spiroacetals from enol-esters. Tetrahedron Lett 41:1997–2000. https://doi.org/10.1016/S0040-4039(00)00083-6
105. Sharma GVM, Rakesh CAS, Reddy VG, Rao MHVR, Kunwar AC (2003) Radical reactions on enol-esters: facile synthesis of 3-ulosonic acid derivatives and chiral spiroacetals. Tetrahedron Asymmetry 14:2991–3004. https://doi.org/10.1016/S0957-4166(03)00448-8
106. Stork G, Sher PM (1986) A catalytic tin system for trapping of radicals from cyclization reactions. Regio- and stereocontrolled formation of two adjacent chiral centers. J Am Chem Soc 108:303–304. https://doi.org/10.1021/ja00262a024
107. Kittaka A, Tanaka H, Kato H, Nonaka Y, Nakamura KT, Miyasaka T (1997) Synthesis of anomeric spiro uracil nucleosides with an orthoester structure: stereoselective cyclization controlled by the C6-substituent. Tetrahedron Lett 38:6421–6424. https://doi.org/10.1016/S0040-4039(97)01486-X
108. Kittaka A, Kato H, Tanaka H, Nonaka Y, Amano M, Nakamura KT, Miyasaka T (1999) Face selective 6,1′-(1-oxo)ethano bridge formation of uracil nucleosides under hypoiodite reaction conditions. Tetrahedron 55:5319–5344. https://doi.org/10.1016/S0040-4020(99)00232-X
109. Gimisis T, Castellari C, Chatgilialoglu C (1997) A new class of anomeric spironucleosides. Chem Commun 21:2089–2090. https://doi.org/10.1039/a705742c
110. El Kouni MH, Naguib FNM, Panzica RP, Otter BA, Chu S-H, Gosselin G, Chu CK, Schinazi RF, Shealy YF, Goudgaon N, Ozerov AA, Ueda T, Iltzsch MH (1996) Effects of modifications in the pentose moiety and conformational changes on the binding of nucleoside ligands to uridine phosphorylase from *Toxoplasma gondii*. Biochem Pharmacol 51:1687–1700. https://doi.org/10.1016/0006-2952(96)00213-4
111. Ogamino J, Mizunuma H, Kumamoto H, Takeda S, Haraguchi K, Nakamura KT, Sugiyama H, Tanaka H (2005) 5-*Exo* versus 6-*endo* cyclization of nucleoside 2-sila-5-hexenyl radicals: reaction of 6-(bromomethyl)dimethylsilyl 1′,2′-unsaturated uridines. J Org Chem 70:1684–1690. https://doi.org/10.1021/jo040260p
112. Sakaguchi N, Hirano S, Matsuda A, Shuto S (2006) Radical reactions with 2-bromobenzylidene group, a protecting/radical-translocating group for the 1,6-radical hydrogen transfer reaction. Org Lett 8:3291–3294. https://doi.org/10.1021/ol061162o
113. Shuto S, Sugimoto I, Abe H, Matsuda A (2000) Mechanistic study of the ring-enlargement reaction of (3-oxa-2-silacyclopentyl)methyl radicals into 4-oxa-3-silacyclohexyl radicals. Evidence for a pentavalent silicon-bridging radical transition state in 1,2-rearrangement reactions of β-silyl radicals. J Am Chem Soc 122:1343–1351. https://doi.org/10.1021/ja993239s

114. Shuto S, Kanazaki M, Ichikawa S, Minakawa N, Matsuda A (1998) Stereo- and regioselective introduction of 1- or 2-hydroxyethyl group via intramolecular radical cyclization reaction with a novel silicon-containing tether. An efficient synthesis of 4'α-branched 2′-deoxyadenosines. J Org Chem 63:746–754. https://doi.org/10.1021/jo971703a
115. Tamao K, Ishida N, Tanaka T, Kumada M (1983) Hydrogen peroxide oxidation of the silicon-carbon bond in organoalkoxysilanes. Organometallics 2:1694–1696. https://doi.org/10.1021/om50005a041
116. Yoshimura Y, Ueda T, Matsuda A (1991) Synthesis of 6,1′-propanouridine, fixed in *syn*-conformation by a spiro-carbon bridge. Tetrahedron Lett 32:4549–4552. https://doi.org/10.1016/0040-4039(91)80036-6
117. Yoshimura Y, Otter BA, Ueda T, Matsuda A (1992) Nucleosides and nucleotides. 108. Synthesis and optical properties of syn-fixed carbon-bridged pyrimidine cyclonucleotides. Chem Pharm Bull 40:1761–1769. https://doi.org/10.1248/cpb.40.1761
118. Rémy G, Cottier L, Descotes G (1979) Photolyse stereospecifique d'(oxo-3 butyl)-tetra-*O*-acetyl-2,3,4,6-β-D-glucopyranoside en spiro-C-1-sucre. Tetrahedron Lett 20:1847–1850. https://doi.org/10.1016/S0040-4039(01)86857-X
119. Rémy G, Cottier L, Descotes G (1980) Photocyclisation des (oxo-3)-butyl glycopyranosides. Can J Chem 58:2660–2665. https://doi.org/10.1139/v80-425
120. Rémy G, Cottier L, Descotes G, Faure R, Loiseleur H, Thomas-David G (1980) Confirmation d'une reaction de photocarbocyclisation stéréospécifique de l'(oxo-3 butyl) tétra-*O*-acétyl-2,3,4,6-β-D-glucopyrannoside par détermination de la structure cristalline du tétra-*O*-acétyl-2,3,4,6-désoxy-1β-D-glucopyrannoside-1-spiro-2′-(méthyl-3′ tétrahydrofurannol-3′)$^{1,4}B$ (*cis* et *trans*). Acta Crystallogr Sect B 36:873–877. https://doi.org/10.1107/S0567740880004736
121. Rémy G, Cottier L, Descotes G (1983) Synthèses et epimérisation de déoxy-spiro C-1 sucres. Can J Chem 61:434–438. https://doi.org/10.1139/v83-077
122. Rémy G, Cottier L, Descotes G (1982) Photolyse des (oxo-3)-butyl-α et β-D-mannopyranosides. J Carbohydr Chem 1:37–47. https://doi.org/10.1080/07328308208085077
123. Rémy G, Descotes G (1983) Synthese et photochimie des (oxo-3)butyl- α et β-L-arabinopyranosides. J Carbohydr Chem 2:159–166. https://doi.org/10.1080/07328308308057864
124. Bron P, Cottier L, Descotes G (1984) Synthèse d'arylhydroxyspirocétals. Influence du groupe aryle sur la photocyclisation d'arylcétoacétals en série hétérocyclique et osidique. J Heterocycl Chem 21:21–25. https://doi.org/10.1002/jhet.5570210106
125. Praly J-P, Descotes G (1981) Photocyclisation de phenyl-glycosides orthocarbonyles: voies d'acces stereo-selectives aux dioxaspirannones aromatiques. Carbohydr Res 95:C1–C4. https://doi.org/10.1016/S0008-6215(00)85306-9
126. Bernasconi C, Cottier L, Descotes G, Praly J-P, Rémy G, Grenier-Loustalot M-F, Metras F (1983) Photocyclisation de phényl-glycosides orthocarbonylés: Voies d'accès stéréosélectives aux dioxaspirannones aromatiques. Carbohydr Res 115:105–116. https://doi.org/10.1016/0008-6215(83)88139-7
127. Saito T, Fujiwara K, Sano Y, Sato T, Kondo Y, Akiba U, Ishigaki Y, Katoono R, Suzuki T (2018) An improved synthesis of the C42–C52 segment of ciguatoxin 3C. Tetrahedron Lett 59:1372–1376. https://doi.org/10.1016/j.tetlet.2018.02.052
128. Kraus GA, Thurston J (1987) Alkoxy radicals in organic synthesis. A novel approach to spiroketals. Tetrahedron Lett 28:4011–4014. https://doi.org/10.1016/S0040-4039(01)83848-X
129. Haudrechy A, Sinaÿ P (1991) A novel type of spiroketalisation on pyranoses. Carbohydr Res 216:375–379. https://doi.org/10.1016/0008-6215(92)84174-Q
130. Ferrier RJ, Hall DW (1992) One-step synthesis of glycosidic spiroketals from 2,3-epoxybutyl glycoside derivatives. J Chem Soc Perkin Trans 1:3029–3034. https://doi.org/10.1039/p19920003029
131. Ösz E, Sós E, Somsák L, Szilágyi L, Dinya Z (1997) A straightforward route to hydantocidin analogues with pyranose ring structure. Tetrahedron 53:5813–5824. https://doi.org/10.1016/S0040-4020(97)00213-5

132. Somsák L, Kovács L, Tóth M, Ösz E, Szilágyi L, Györgydeák Z, Dinya Z, Docsa T, Tóth B, Gergely P (2001) Synthesis of and a comparative study on the inhibition of muscle and liver glycogen phosphorylases by epimeric pairs of D-gluco- and D-xylopyranosylidene-spiro-(thio) hydantoins and *N*-(D-glucopyranosyl) amides. J Med Chem 44:2843–2848. https://doi.org/10.1021/jm010892t
133. Miljković M (2010) Carbohydrate-based antibiotics. In: Carbohydrates. Springer, New York, pp 469–486. https://doi.org/10.1007/978-0-387-92265-2_14
134. Ollis WD, Smith C, Wrigth DE (1979) The orthosomycin family of antibiotics-I: the constitution of flambamycin. Tetrahedron 35:105–127. https://doi.org/10.1016/0040-4020(79)85015-2
135. Wright DE (1979) The orthosomycins, a new family of antibiotics. Tetrahedron 35:1207–1237. https://doi.org/10.1016/0040-4020(79)80046-0
136. Praly J-P, Descotes G (1982) Photocyclisation d'hydroxyalkyl glycosides en orthoesters glycosidiques. Tetrahedron Lett 23:849–852. https://doi.org/10.1016/S0040-4039(00)86965-8
137. Praly J-P, Descotes G, Grenier-Loustalot M-F, Metras F (1984) Nouvelle synthèse de spiro-orthoesters D-glucosidiques par photocyclisation stéréoselective d'hydroxyalkyl-D-glucosides. Carbohydr Res 128:21–35. https://doi.org/10.1016/0008-6215(84)85081-8
138. Praly J-P, Descotes G, Faure R, Loiseleur H (1984) *O*-Isopropylidène-1,2 *O*-méthyl-3 *O*-[tétra-*O*-acétyl-2,3,4,6 désoxy-1 D-glucopyrannosylidène-1-(1S)]-5,6 α-D-glucofurannose, $C_{24}H_{34}O_{15}$. Acta Crystallogr Sect C 40:1206–1207. https://doi.org/10.1107/S0108270184007344
139. Francisco CG, Herrera AJ, Kennedy AR, Melián D, Suárez E (2002) Intramolecular 1,8-hydrogen abstraction between glucopyranose units in a disaccharide model promoted by alkoxy radicals. Angew Chem Int Ed Engl 41:856–858. https://doi.org/10.1002/1521-3773(20020301)41:5<856::AID-ANIE856>3.0.CO;2-F
140. Francisco CG, Herrera AJ, Kennedy AR, Martín A, Melián D, Pérez-Martín I, Quintanal LM, Suárez E (2008) Intramolecular 1,8-hydrogen-atom transfer reactions in (1→4)-*O*-disaccharide systems: conformational and stereochemical requirements. Chem Eur J 14:10369–10381. https://doi.org/10.1002/chem.200801414
141. Crich D, Huang W (2001) Dynamics of alkene radical cation/phosphate anion pair formation from nucleotide C4′ radicals. The DNA/RNA paradox revisited. J Am Chem Soc 123:9239–9245. https://doi.org/10.1021/ja0108411
142. Jahjah R, Gassama A, Bulach V, Suzuki C, Abe M, Hoffmann N, Martinez A, Nuzillard J-M (2010) Stereoselective triplet-sensitised radical reactions of furanone derivatives. Chem Eur J 16:3341–3354. https://doi.org/10.1002/chem.200903045
143. Dorta RL, Francisco CG, Freire R, Suárez E (1988) Intramolecular hydrogen abstraction. The use of organoselenium reagents for the generation of alkoxy radicals. Tetrahedron Lett 29:5429–5432. https://doi.org/10.1016/S0040-4039(00)82887-7
144. Dorta RL, Francisco CG, Suárez E (1994) Organoselenium reagents in the tandem β-fragmentation-cyclization of carbohydrate anomeric alkoxy radicals. Tetrahedron Lett 35:2049–2052. https://doi.org/10.1016/S0040-4039(00)73046-2
145. Dorta RL, Francisco CG, Suárez E (1994) A selenurane derivative promotes β-fragmentation of carbinolamides leading to cyclic imides. Tetrahedron Lett 35:1083–1086. https://doi.org/10.1016/S0040-4039(00)79971-0
146. Martín A, Quintanal LM, Suárez E (2007) Hydrogen atom transfer experiments provide chemical evidence for the conformational differences between C- and O-glycosides. Tetrahedron Lett 48:5507–5511. https://doi.org/10.1016/j.tetlet.2007.05.166
147. León EI, Martín A, Pérez-Martín I, Quintanal LM, Suárez E (2010) Hydrogen atom transfer experiments provide chemical evidence for the conformational differences between *C*- and *O*-disaccharides. Eur J Org Chem 2010:5248–5262. https://doi.org/10.1002/ejoc.201000470
148. Martín A, Pérez-Martín I, Quintanal LM, Suárez E (2007) Intramolecular 1,8- versus 1,6-hydrogen atom transfer between pyranose units in a (1→4)-disaccharide model promoted by alkoxyl radicals. Conformational and stereochemical requirements. Org Lett 9:1785–1788. https://doi.org/10.1021/ol070496q

149. Guyenne S, León EI, Martín A, Pérez-Martín I, Suárez E (2012) Intramolecular 1,8-hydrogen atom transfer reactions in disaccharide systems containing furanose units. J Org Chem 77:7371–7391. https://doi.org/10.1021/jo301153u
150. Alvarez-Dorta D, León EI, Kennedy AR, Martín A, Pérez-Martín I, Suárez E (2016) Radical-mediated C–H functionalization: a strategy for access to modified cyclodextrins. J Org Chem 81:11766–11787. https://doi.org/10.1021/acs.joc.6b02241
151. Foulard G, Brigaud T, Portella C (1997) Radical trifluoromethylation of a D-mannose derived ketene dithioacetal. Synthesis of 2-*C*-trifluoromethyl derivatives of D-*glycero*-D-*galacto*- and D-*glycero*-D-*talo*-heptopyranose. J Org Chem 62:9107–9113. https://doi.org/10.1021/jo971187o
152. Sowa CE, Stark M, Heidelberg T, Thiem J (1996) Novel approach to sugar derived spiroannelated heterocycles. Synlett 7:227–228. https://doi.org/10.1055/s-1996-5370
153. Stark M, Thiem J (2006) Highly functionalized glyco-conjugated hexahydroazepindiones from saccharide imides via the Norrish type II reaction. Carbohydr Res 341:1543–1556. https://doi.org/10.1016/j.carres.2006.03.016. Corrigendum: Stark M, Thiem J (2007). Carbohydr Res 342:772. https://doi.org.10.1016/j.carres.2006.12.001
154. Thiering S, Thiem J, Kopf J (2007) Reactions of glycosan-annelated oxolactams. Heterocycles 74:533–543. https://doi.org/10.3987/COM-07-S(W)31
155. Sowa CE, Kopf J, Thiem J (1995) Syntheses of sugar-derived heterotricyclic lactams. J Chem Soc Chem Commun 2:211–212. https://doi.org/10.1039/c39950000211
156. Sowa CE, Thiem J (1994) Stereoselective intramolecular alkylation of glycosylimides to highly functionalized bicyclic 2,5-azepanediones and heterotricyclic[5.3.1.$0^{2,6}$]undecanamides. Angew Chem Int Ed Engl 33:1979–1981. https://doi.org/10.1002/anie.199419791
157. Cottier L, Descotes G, Grenier MF, Metras F (1981) Synthese photochimique et etude structurale d'alcoxy-spirocetals et de trioxa-bis-spiroacetals. Tetrahedron 37:2515–2524. https://doi.org/10.1016/S0040-4020(01)88911-0
158. Cottier L, Descotes G, Faure R, Loiseleur H (1981) Etude structurale de trioxa-bis-spirocétals: Structure de la configuration *E–E* du diméthyl-4,11 trioxa-1,6,8 dispiro[4.1.4.3] tétradécanediol-4,11. Acta Crystallogr Sect B 37:1155–1157. https://doi.org/10.1107/S0567740881005359
159. Cottier L, Descotes G (1985) Syntheses stereoselectives des isomeres cinetiques et thermodynamiques d'alcoxy-spirocetals et de trioxa-bis-spirocetals. Tetrahedron 41:409–412. https://doi.org/10.1016/S0040-4020(01)96433-6
160. Hu T, Curtis JM, Oshima Y, Quilliam MA, Walter JA, Watson-Wright WM, Wright JLC (1995) Spirolides B and D, two novel macrocycles isolated from the digestive glands of shellfish. J Chem Soc Chem Commun 20:2159–2161. https://doi.org/10.1039/C39950002159
161. Meilert K, Brimble MA (2006) Synthesis of the bis-spiroacetal moiety of the shellfish toxins spirolides B and D using an iterative oxidative radical cyclization strategy. Org Biomol Chem 4:2184–2192. https://doi.org/10.1039/b604334h
162. Meilert K, Brimble MA (2005) Synthesis of the bis-spiroacetal moiety of spirolides B and D. Org Lett 7:3497–3500. https://doi.org/10.1021/ol051260u
163. Furkert DP, Brimble MA (2002) Synthesis of the C10-C22 bis-spiroacetal domain of spirolides B and D via iterative oxidative radical cyclization. Org Lett 4:3655–3658. https://doi.org/10.1021/ol026605c
164. Berg DH, Hamill RL (1978) The isolation and characterization of narasin, a new polyether antibiotic. J Antibiot 31:1–6. https://doi.org/10.7164/antibiotics.31.1
165. Dutton CJ, Banks BJ, Cooper CB (1995) Polyether lonophores. Nat Prod Rep 12:165–181. https://doi.org/10.1039/NP9951200165
166. Keller-Juslén C, King HD, Kuhn M, Loosli H-R, Von Wartburg A (1978) Noboritomycins A and B, new polyether antibiotics. J Antibiot 31:820–828. https://doi.org/10.7164/antibiotics.31.820
167. Allen P, Brimble MA, Turner P (2001) Polyether antibiotic CP44,161. Acta Crystallogr Sect C 57:95–96. https://doi.org/10.1107/S0108270100013950

168. Westley JW, Evans RH, Sello LH, Troupe N, Liu C-M, Blount JF, Pitcher RG, Williams TH, Miller PA (1981) Isolation and characterization of the first halogen containing polyether antibiotic X-14766A, a product of *Streptomyces malachitofuscus* subsp. *Downeyi*. J Antibiot 34:139–147. https://doi.org/10.7164/antibiotics.34.139
169. Warabi K, Williams DE, Patrick BO, Roberge M, Andersen RJ (2007) Spirastrellolide B reveals the absolute configuration of the spirastrellolide macrolide core. J Am Chem Soc 129:508–509. https://doi.org/10.1021/ja068271i
170. Dorta RL, Martín A, Salazar JA, Suárez E, Prangé T (1996) Synthesis of dispiroacetals from carbohydrates by intramolecular hydrogen abstractions. Tetrahedron Lett 37:6021–6024. https://doi.org/10.1016/0040-4039(96)01265-8
171. Dorta RL, Martín A, Salazar JA, Suárez E, Prangé T (1998) Syntheses of chiral dispiroacetals from carbohydrates. J Org Chem 63:2251–2261. https://doi.org/10.1021/jo972023a
172. Dorta RL, Martín A, Betancor C, Suárez E (1997) Serendipitous acid-catalyzed rearrangement of 13-methoxy-1,6,8-trioxadispiro[4.1.5.3]pentadecane to 3-chroman-5-yl-propan-1-ol. J Org Chem 62:2273–2274. https://doi.org/10.1021/jo962065j
173. Dorta RL, Martín A, Suárez E, Prangé T (1996) Synthesis of (5*R*,7*S*,13*S*)-13-methoxy-1,6,8-trioxadispiro[4.1.5.3]pentadecane. Tetrahedron Asymmetry 7:1907–1910. https://doi.org/10.1016/0957-4166(96)00227-3
174. Wu Y-B, Tang Y, Luo G-Y, Chen Y, Hsung RP (2014) An approach toward constructing the trioxadispiroketal core in the DEF-ring of (+)-spirastrellolide A. Org Lett 16:4550–4553. https://doi.org/10.1021/ol502103b

Top Heterocycl Chem (2019) 57: 105–136
DOI: 10.1007/7081_2019_32

Published online: 3 April 2019

Carbohydrate-Derived Spiroketals and Spirocyclic Lactones

Perali Ramu Sridhar

Contents

Abstract Various methods for the synthesis of carbohydrate-derived spiroketals and spirocyclic lactones starting from *endo*- and *exo*-glycals are discussed. Further conversion of the spiroketals and spirolactones to the natural products is also emphasized wherever applicable.

P. R. Sridhar (✉)
School of Chemistry, University of Hyderabad, Hyderabad, Telangana, India
e-mail: p_ramu_sridhar@uohyd.ac.in

Keywords Bicycles · bis-*C*,*C*-glycosides · *Endo*-glycals · *Exo*-glycals · Natural products

Abbreviations

Ac	Acetyl
Bn	Benzyl
Bu	Butyl
Cp	Cyclopentadienyl
CSA	Camphor-10-sulfonic acid
DBU	1,8-Diazabicyclo[5.4.0]undec-7-ene
DDQ	2,3-Dichloro-5,6-dicyano-1,4-benzoquinone
DMDO	3,3-Dimethyldioxirane
DMF	*N*,*N*-dimethylformamide
DMP	Dess-Martin periodinane
DMSO	Dimethyl sulfoxide
EDA	Ethyl diazoacetate
IBX	2-Iodoxybenzoic acid
i-Pr	Isopropyl
LAH	Lithium aluminum hydride
LiDBB	Lithium 4,4′-di-*tert*-butylbiphenylide
LiHMDS	Lithium hexamethyldisilazide
mCPBA	*m*-chloroperoxybenzoic acid
MDA	Methyl diazoacetate
MS	Molecular sieves
NBS	*N*-bromosuccinimide
NIS	*N*-iodosuccinimide
PMB	*p*-methoxybenzyl
PPTS	Pyridinium *p*-toluenesulfonate
SIBX	Stabilized 2-iodoxybenzoic acid
TBDPS	*tert*-butyldiphenylsilyl
TBS	*tert*-butyldimethylsilyl
t-Bu	*tert*-butyl
Tf	Trifluoromethanesulfonyl
TFA	Trifluoroacetic acid
THF	Tetrahydrofuran
THP	Tetrahydropyran-2-yl
TIPS	Triisopropylsilyl
TMEDA	*N*,*N*,*N*′,*N*′-tetramethylethylenediamine
TMS	Trimethylsilyl
TMSOTf	Trimethylsilyl trifluoromethanesulfonate
Ts	*p*-toluenesulfonyl

1 Introduction

Carbohydrates have been extensively used as raw materials in the total synthesis of natural products because of their high abundance and inherent chiral centers [1]. Spirocyclic compounds in which the quaternary carbon center of the spiro system is attached to two oxygen atoms of different cycles, frequently called as spirocyclic ketals, ubiquitously exist in nature as subunits of a number of bioactive natural products (Reviews: [2–6]). However, despite of their wide occurrence, methods for the predicted and straightforward synthesis of enantiomerically pure spirocyclic systems are rather scarce [2, 7]. A major challenge in the construction of these spirocycles is the stereoselective formation of the quaternary ketal carbon center [8–10]. Interestingly, the biosynthetic origin of most of the spirocyclic ketals is derived from carbohydrates, and a major portion of these spirocyclic compounds exist in highly oxygenated form. The structural similarity of spiroketals with highly oxygenated carbohydrate scaffolds opens a unique opportunity to use abundant sugars as chiral building blocks for the construction of spiroketal units.

Many biologically active spirocyclic systems existing in nature are thermodynamic products in which there is a high preference for the carbon-oxygen bonds to be in a bis-diaxial C-O conformation at the spirocentre. This consistency especially dominates in the case of 1,7-dioxaspiro[5.5]undecane ring systems. For example, in avermectin B_{1a} and B_{2a} aglycons [11], norhalichondrin A [12], okadaic acid (C29–C38) [13], and calcimycin [14], the spiroketal systems reside in the bis-diaxial C-O conformation favored by anomeric effects. A similar observation was found even in the case of spiro[4.5]ketal systems that are present in polyether antibiotics like calyculin A [15], okadaic acid (C16–C26), and A204A [16] (Fig. 1). As a result, the thermodynamically favored acid-catalyzed spirocyclization of dihydroxy ketones is one of the universal methods for the construction of these kinds of spiroketals (Scheme 1) [17, 18].

Although a major number of spiroketal units in nature exists in a thermodynamically stable conformation, few naturally occurring compounds do exist that do not

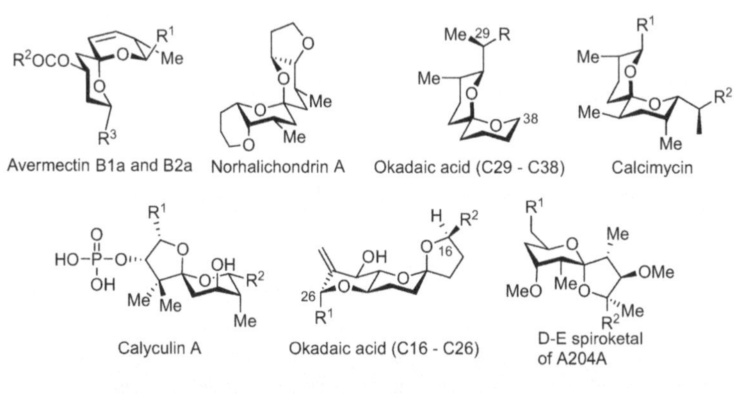

Fig. 1 Conformations of spiroketal substructures of various natural products

Scheme 1 Acid-catalyzed spirocyclization of dihydroxy ketones

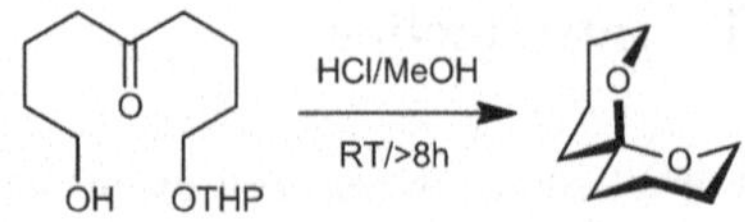

1 2

Fig. 2 Natural products that do not reside in the bis-diaxial C-O conformation

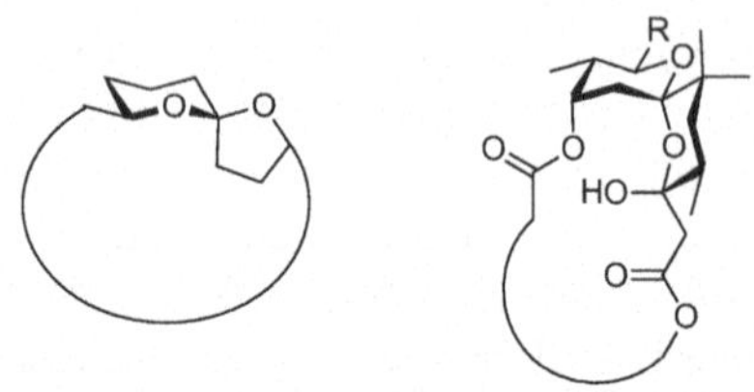

Pectenotoxin 1 Aplysiatoxin-Oscillatoxin spiroketal conformation

reside in a bis-diaxial C-O conformation. This arrangement of the spirocentre makes their total synthesis much more difficult. For example, in aplysiatoxin-oscillatoxin [19] and pectenotoxin 1 [20, 21] spiroketals, one spiro C-O bond is oriented equatorially (Fig. 2). In these cases, the steric factors influences the conformation of the spiroketal. However, there is no much information in the literature regarding predictability and consistency in the formation of spiro [4.5] and spiro [4.4] ring systems.

This chapter highlights various existing synthetic strategies for the construction of carbohydrate-derived spiroketals and spirolactones and their significant application in the total synthesis of natural products. Apart from that, the efforts are also devoted to cover the synthesis of spirocyclic bis-*C*,*C*-glycosyl derivatives, carbohydrate-derived spirocycles possessing only carbon atoms at α and β orientations at the spirocentre, and their utility. The purpose of this chapter is to summarize the syntheses of glycal-derived spiroketals and their versatile application in the synthesis of natural or unnatural building blocks.

2 Synthetic Methods for Spiroketals Based on Glycals

There are a large number of reports on the spiroketal synthesis which involve the acid-catalyzed spirocyclization of appropriately positioned dihydroxy ketones or related synthons. In general, these reactions produce the thermodynamically stabilized spiroketal center. In recent years, considerable efforts have been made toward the formation of spiroketals from suitably functionalized glycals involving an intramolecular glycosylation reaction. Because glycals play a pivotal role in the synthesis of spiroketals, this chapter highlights the synthetic methods that involve the use of *exo*-glycals or *endo*-glycals in spiroketalization reactions.

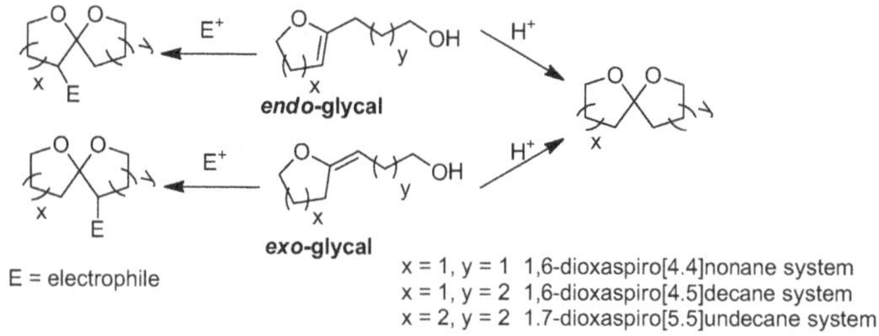

Fig. 3 General routes in the syntheses of various spiroketal frameworks

2.1 *Synthesis of Spiroketals from* Endo*-Glycals*

Endo-glycals, the 1,2-unsaturated cyclic sugar derivatives, have been one of the major sources for the generation of oxocarbenium ion which could help in glycosylation reaction. A number of complex carbohydrates have been synthesized using *endo*-glycals as the glycosyl acceptors. An intramolecular version of the glycosylation reaction using 1-*C*-branched *endo*-glycals possessing a hydroxyl group on the 1-*C*-branch has been used as a substrate for the synthesis of spiroketals at the anomeric position. This technique was very often utilized for the synthesis of noncarbohydrate-derived spiroketals possessing cyclic vinyl ether functionality (Fig. 3).

2.2 *Synthesis of Spiroketals from* Exo*-Glycals*

Exo-glycals are unsaturated sugars possessing an exocyclic C=C bond involving the former C-1 center of the sugar ring. Methylene *exo*-glycals are the simplest version of *exo*-glycals. Although a couple of methods for the synthesis of *exo*-glycals have been reported in the literature [22, 23], the application of these electron-rich olefins in the synthesis of spirocyclic architectures is not extensively studied. One of the major reasons for this is the migratory aptitude of the *exo*-olefin to the more stable *endo*-olefin. When the *exo*-glycal olefin is a trisubstituted one, having a hydroxyl group on *exo*-methylene substituent, it could serve as a precursor for the spiroketalization reaction. A schematic diagram of the use of *endo*- and *exo*-glycals in the spiroketalization reaction is depicted in Fig. 3.

3 1,6-Dioxaspiro[4.4]nonane Systems

Spiroketals possessing 1,6-dioxaspiro[4.4]nonane systems have been less studied than the 1,6-dioxaspiro[4.5]decane and 1,7-dioxaspiro[5.5]undecane counterparts. The anomeric effects are also difficult to predict in these systems. Few

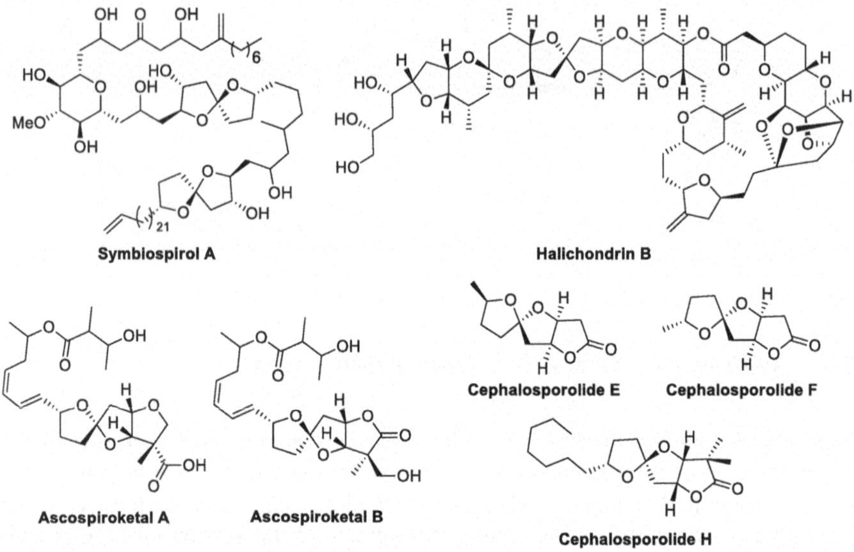

Fig. 4 Natural products possessing 1,6-dioxaspiro[4.4]nonane framework

marine-derived natural products like symbiospirol A and halichondrin B and fungal metabolites such as cephalosporolides E, F, and H and ascospiroketal A and B are some of the examples which possess 1,6-dioxaspiro[4.4]nonane framework (Fig. 4).

3.1 *Synthesis of 1,6-Dioxaspiro[4.4]nonane Systems from Unsaturated Compounds Related to Furanoid* Endo-*Glycals*

The synthesis of 1,6-dioxaspiro[4.4]nonane systems from the corresponding furanoid glycals is one of the relatively unexplored areas in the synthesis of carbohydrate spiroketals. Although several methods have been reported for the synthesis of various furanoid *endo*-glycals [24], their conversion to 1-*C*-branched furanoid glycals, which could be the synthons for the spiroketalization, is very scarce [25, 26]. To the best of our knowledge, there were no reports on the conversion of carbohydrate-derived 1-*C*-branched furanoid *endo*-glycals to the corresponding spiro systems in the literature. The only report that is similar to this kind of transformation is the conversion of 2-*C*-branched 4,5-dihydrofuran to the corresponding spiroketal system. Boyce et al. reported rhenium(VII)-oxide-mediated oxidative spirocyclization of 3-(4,5-dihydrofuran-2-yl)propan-1-ol **3** [27]. The reaction proceeded through *syn*-oxidation of the olefin providing a single diastereomer **4** in which the newly formed C-O spiro bond and the adjacent C-OH bond were in *cis* relationship. In 2012, Čorić et al. [28] reported an asymmetric

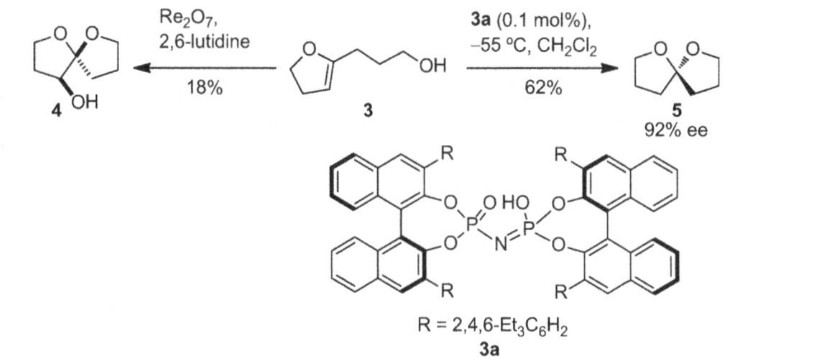

Scheme 2 Synthesis of 1,6-dioxaspiro[4.4]nonane systems from 2-*C*-substituted 4,5-dihydrofuran

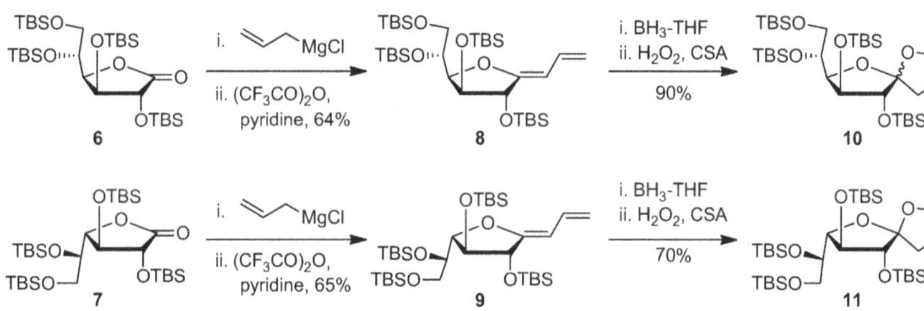

Scheme 3 Synthesis of 1,6-dioxaspiro[4.4]nonane systems from furanoid *exo*-glycals

spiroketalization of **3** catalyzed by a confined Brønsted acid **3a** to provide the spiroketal **5** (Scheme 2).

3.2 Synthesis of 1,6-Dioxaspiro[4.4]nonane Systems from Furanoid Exo-Glycals

There are a very few synthetic procedures involving furanoid *exo*-glycals as the precursors for the preparation of 1,6-dioxaspiro[4.4]nonane systems. Yang et al. [29] reported an efficient stereoselective method in which furanoid lactones **6** and **7** were reacted with allylmagnesium chloride which on dehydration provided the sugar dienes **8** and **9**, respectively. Hydroboration-oxidation of these dienes in the presence of CSA provided the spiroketals **10** and **11** via the formation of the corresponding *exo*-glycal intermediate [30] (Scheme 3). The lack of selectivity in the formation of spiroketal **10** may be due to the developing steric hindrance between the substituents on C-3 and C-4 and the incoming hydroxyl nucleophile while approaching the oxocarbenium ion toward the formation of spiroketal.

Ramakrishna and Sridhar [31] reported the synthesis of 1,6-dioxaspiro[4.4]nonan-2-one systems **20–23** by ring opening and spiroketalization of

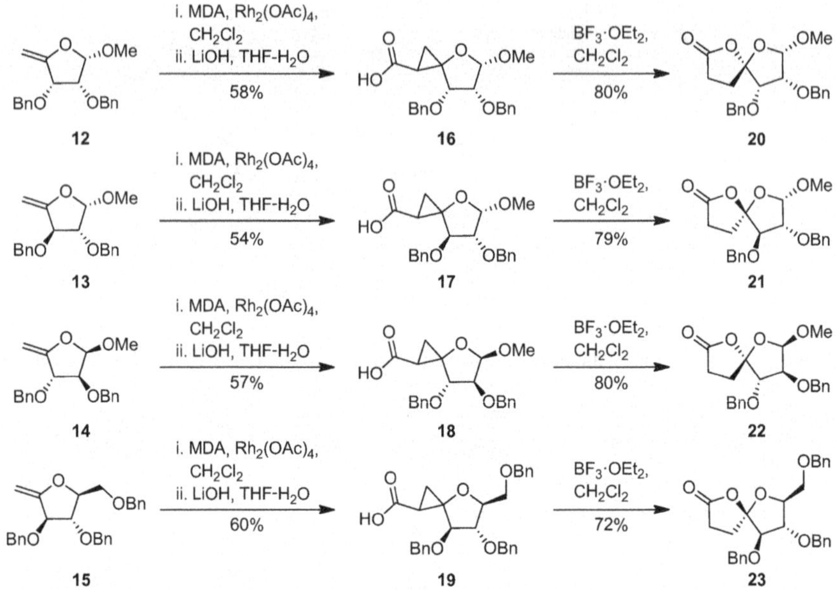

Scheme 4 Synthesis of 1,6-dioxaspiro[4.4]nonane-2-one systems from 4-*exo*-methylene furanosides or furanoid *exo*-glycals. *MDA* methyl diazoacetate

spiro-cyclopropanecarboxylic acids **16–19** which were synthesized from *exo*-methylene derivatives of various pentoses **12–15**, respectively (Scheme 4) [32].

In these reactions, an interesting observation was that all spirolactones **20–23** exhibited a *cis* relative configuration between the spirolactone C-O bond and the adjacent C-O bond of the furanose sugar unit. This similarity relates to the stabilization of the chair-like oxocarbenium ion intermediate by the C2 stereocenter in hexoses that influences the stereochemistry of the newly generated spirocentre.

The cyclopropane ring-opening and spirolactonization reactions were further extended to the stereoselective total synthesis of (+)-2,3-dihydro pyranolide D [31]. Accordingly, fused bicyclic lactone **24** upon olefination with Tebbe's reagent provided the *exo*-methylene derivative **25**. Cyclopropanation of this olefin with MDA gave spiro-cyclopropane carboxylate **26** which upon ester hydrolysis resulted in acid **27**. Lewis acid-mediated spirolactonization of **27** led to the formation of spirolactones **28a** and **28b** in 1:1 ratio. Hydrogenolysis of these lactones resulted into the formation of (+)-2,3-dihydro pyranolide D **29** and (+)-2,3-dihydro 4-*epi*-pyranolide D **30** (Scheme 5).

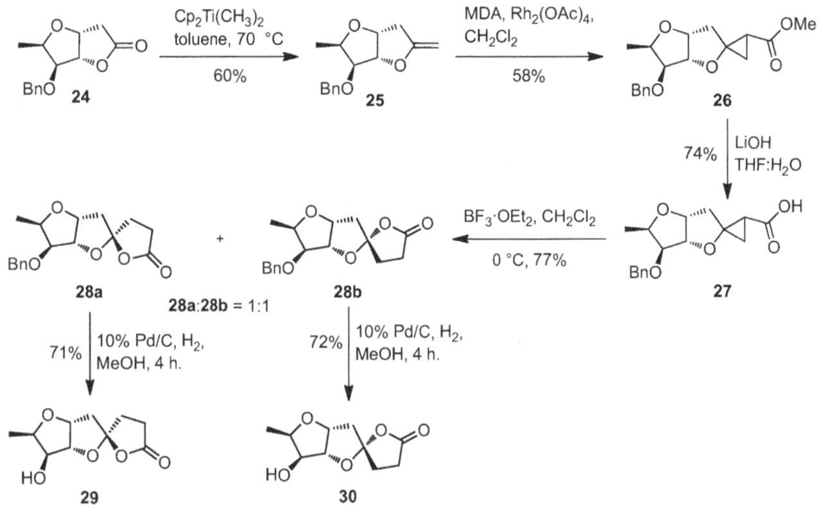

Scheme 5 Stereoselective total synthesis of (+)-2,3-dihydro pyranolide D **29** and (+)-2,3-dihydro 4-*epi*-pyranolide D **30**

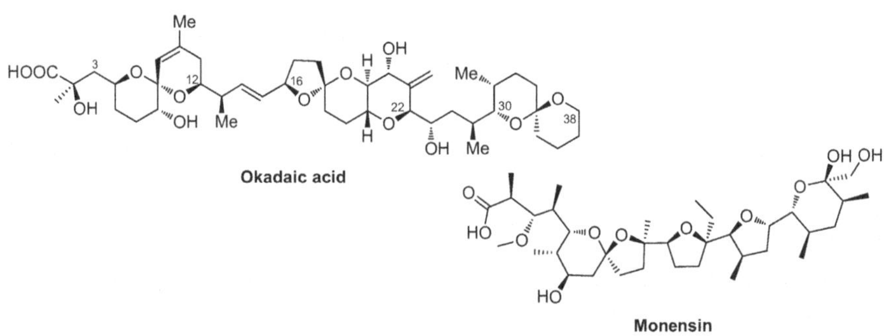

Fig. 5 Structures of some of natural products possessing 1,6-dioxaspiro[4.5]decane framework

4 1,6-Dioxaspiro[4.5]decane Systems

Several marine natural products possess 1,6-dioxaspiro[4.5]decane framework. For example, monensin [33, 34], a polyether antibiotic isolated from *Streptomyces cinnamonensis*, has 1,6-dioxaspiro[4.5]decane as the only spirocyclic motif, whereas, in okadaic acid [13], this framework is part of a tricyclic architecture (Fig. 5). A major number of these spiroketals exist in the thermodynamically more stable conformation in which the anomeric oxygen is placed in the axial position. Carbohydrate-derived 1,6-dioxaspiro[4.5]decane systems could be synthesized by either using appropriately functionalized 1-*C*-branched furanoid or pyranoid glycals. Due to the availability of quite a few protocols for the synthesis of pyranoid *exo*- and *endo*-glycals, they have been used extensively in the synthesis of spiroketals.

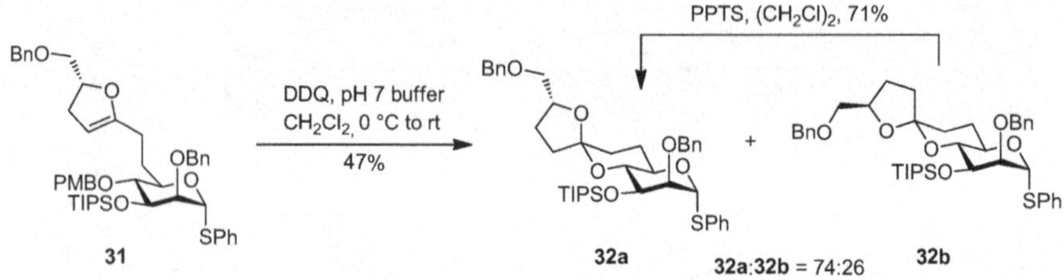

Scheme 6 Synthesis of the tricyclic core of okadaic acid. *PPTS* pyridinium *p*-toluenesulfonate

In contrast, fully carbohydrate-derived 1-*C*-branched furanoid *endo*- or *exo*-glycals for the synthesis of 1,6-dioxaspiro[4.5]decane systems have not been reported. Instead, only a very few substituted deoxy-furanoid *endo*-glycals were used for the spirocyclization reactions in the total synthesis of natural products. On the other hand, methodologies have been reported on the synthesis of 1-*C*-branched pyranoid *endo*- and *exo*-glycals which were subsequently converted to 1,6-dioxaspiro[4.5]decane systems.

4.1 *Synthesis of 1,6-Dioxaspiro[4.5]decane Systems from Furanoid* Endo-*Glycals*

Fuwa et al. [35], in their attempts toward the synthesis of the C15-C38 fragment of okadaic acid, reported a spontaneous spiroketalization of furanoid *endo*-glycal **31–32a** and its diastereomer **32b** in 74:26 ratio, respectively, upon removal of *p*-methoxybenzyl (PMB) protecting group with DDQ (Scheme 6). An internal epimerization of **32b** to the more stable **32a** was achieved by exposing **32b** to pyridinium *p*-toluenesulfonate (PPTS).

4.2 *Synthesis of 1,6-Dioxaspiro[4.5]decane Systems from Pyranoid* Endo-*Glycals*

It has been already established that the spiroketal subunit is an important structural motif in a variety of natural products. However, until 2012, the stereoselective synthesis of this moiety has not been explored much. The very first report [36] on the chiral catalyst-controlled diastereoselective spiroketalization of 1-*C*-branched pyranoid *endo*-glycal derivatives was reported by using chiral phosphoric acid **33a**, which was an excellent chiral catalyst for enantioselective spiroketalization of achiral 3-(1*H*-2-benzopyran-3-yl)-1,1-diphenylpropan-1-ol to give 1,6-dioxaspiro [4.5]decane-type product in 89% yield and 90% ee [36]. The use of **33a** in a diastereoselective approach using pyranoid *endo*-glucals **33–35** led to spiroketals

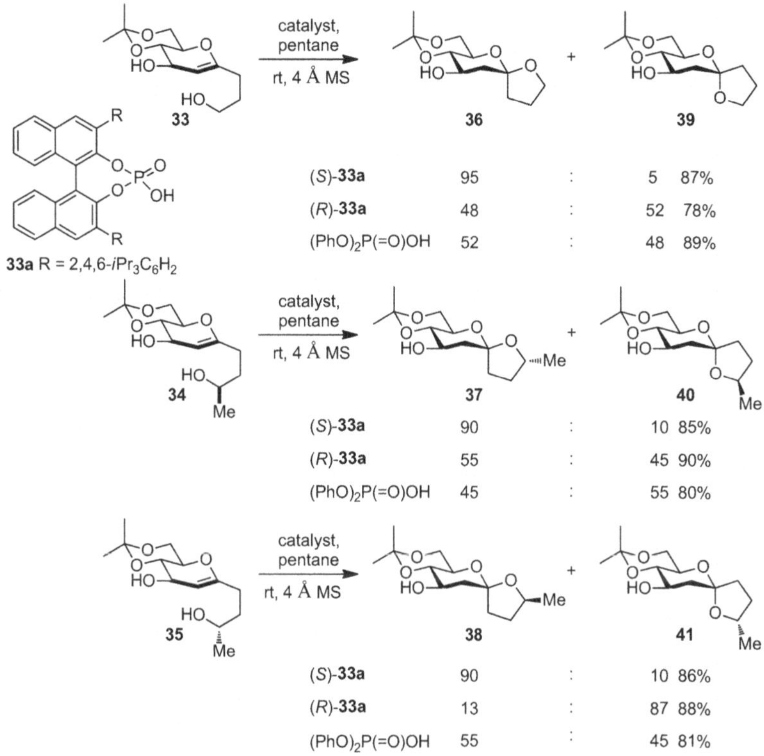

Scheme 7 Chiral phosphoric acid-catalyzed diastereoselective spiroketalization of *endo*-glycals

as follows: treatment of glucal derivative **33** with (*S*)-**33a** (5 mol%) provided the non-thermodynamic spiroketal **36** in a highly diastereoselective fashion with 95:5 ratio of **36** and **39**, respectively, whereas the reaction of **33** with a catalytic amount of (*R*)-**33a** or achiral $(PhO)_2PO_2H$ provided a 1:1 mixture of the diastereomeric spiroketals **36** and **39** (Scheme 7). It has been shown that (*S*)-**33a** catalyzed reactions were not sensitive to other substituents on the tethered alcohol. This was proved by cyclization of *endo*-glycals **34** and **35** with (*S*)-**33a** to give the non-thermodynamic spiroketals **37** and **38** (dr = 90:10, respectively) along with the minor thermodynamic spiroketals **40** and **41**. On the other hand, (*R*)-**33a**-catalyzed reaction is sensitive to α-methyl group on the tethered alcohol. Potuzak et al. [37] have developed stereoselective spiroketalization protocol involving an initial asymmetric epoxidation reaction of 1-*C*-branched glycals. The D-*threo*-glycals **42–44** upon anti-epoxidation with DMDO at −78°C gave the 1,2-anhydrosugars **45–47**. In situ addition of MeOH at −63°C to the reaction mixture resulted in the formation of kinetically controlled spiroketals **48a–50a** as the major products (Scheme 8). It was presumed that the stereoselective ring opening of the epoxide involving C1-inversion is due to MeOH-induced hydrogen-bonding catalysis [38]. This observation was further confirmed by an equilibration of intermediate epoxides **45–47**

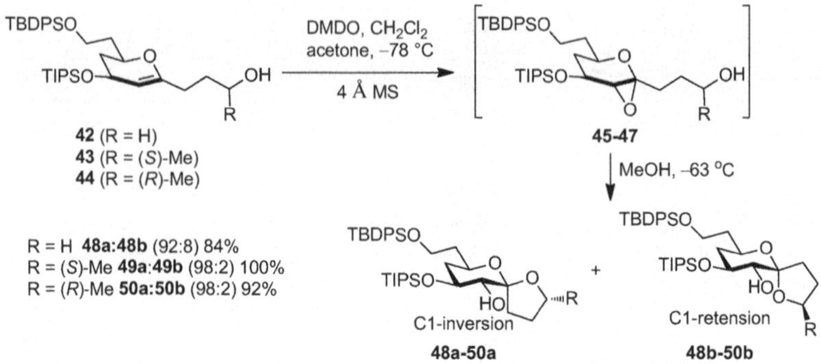

Scheme 8 MeOH-induced hydrogen-bonding catalyzed stereoselective synthesis of kinetic spiroketals from D-*threo*-glycals

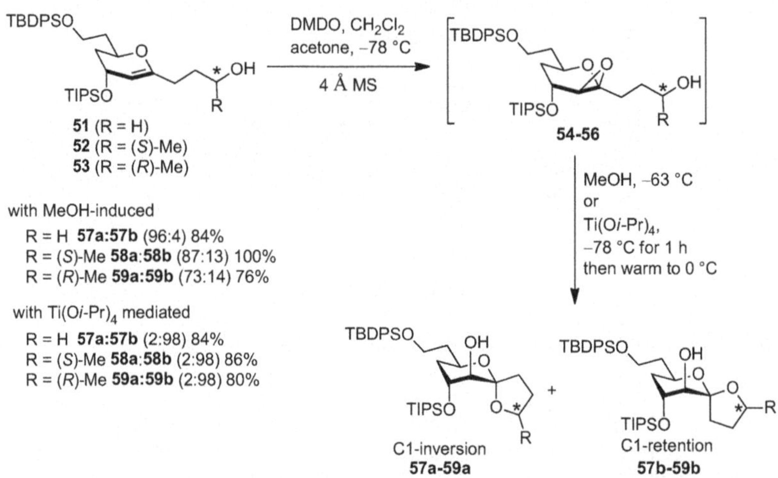

Scheme 9 MeOH-induced hydrogen-bonding or titanium(IV) isopropoxide catalyzed stereoselective synthesis of kinetic and thermodynamic spiroketals from D-*erythro*-glycals

with TsOH, which resulted in the exclusive formation of the thermodynamically favored products **48b–50b**. Performing similar reactions with D-*erythro*-glycal series **51–53**, in which epoxidation gave **54–56**, followed by the MeOH-induced spirocyclization reactions provided inverted spiroketals **57a–59a** as the major products and very minor percentage of kinetic products **57b–59b**, possibly through the retained configuration at C1 (Scheme 9). In order to obtain the retention spiroketals in D-*erythro*-glycal series, a Lewis acid, in particular, Ti(O*i*-Pr)$_4$, was proved to be very efficient. In this case, 1,2-anhydrosugar derivatives provided the kinetic spiroketals **57b–59b** as the major products along with very minor amounts of (<2%) the thermodynamic spiroketals **57a–59a**. A tethered mechanism has been proposed for this kind of kinetic spirocyclization reaction (Scheme 9) [39]. These protocols are certainly very useful for the synthesis of thermodynamic as well as the kinetic spiroketals depending on the application of the appropriate reagents.

Takaoka et al. [40] reported the synthesis of contra-thermodynamic (or kinetic) spiroketals involving reductive cyclization of a dialkoxylithium intermediate which could undergo an intramolecular alkylation reaction. The ketene monothioacetal **60** upon reaction with a diol in the presence of CSA produced the spirocyclic orthoester **61** as a mixture of diastereomers at the orthoester center. Cleavage of **61** by treating with TMSCN and $BF_3{\cdot}OEt_2$ followed by deprotection of the generated TMS ether and reprotection with TBS provided cyano acetal **62**. Reductive cyclization of **62** upon treatment with lithium 4,4′-di-*tert*-butylbiphenylide (LiDBB also called as Freeman's reagent) led to the formation of the axial alkyllithium intermediate **62b** via alkyl radical intermediate **62a** which further undergoes alkylation with retention of configuration to form contra-thermodynamic spiroketal **63** (Scheme 10).

Peczuh et al. reported an interesting rearrangement of enol-ether-spiroketal **64** to the bis-spiroketal **65** triggered under epoxidation reaction conditions [41] (Scheme 11).

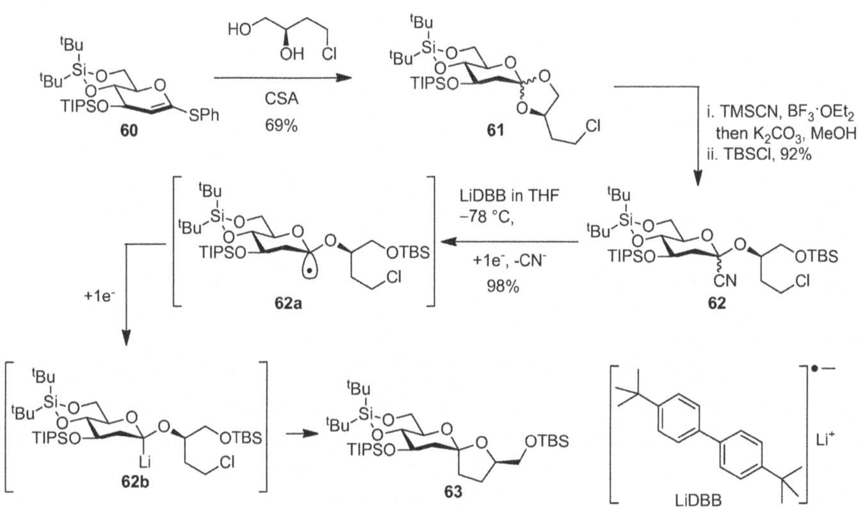

Scheme 10 Stereoselective reductive lithiation and cyclization in the spiroketalization reaction

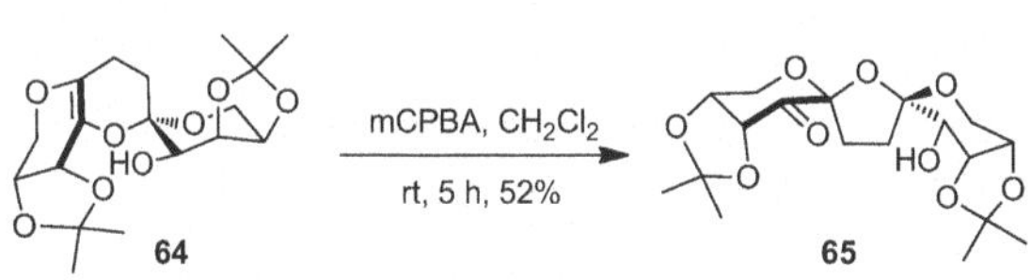

Scheme 11 Inside-out rearrangement of fused spiroketal to bis-spiroketal

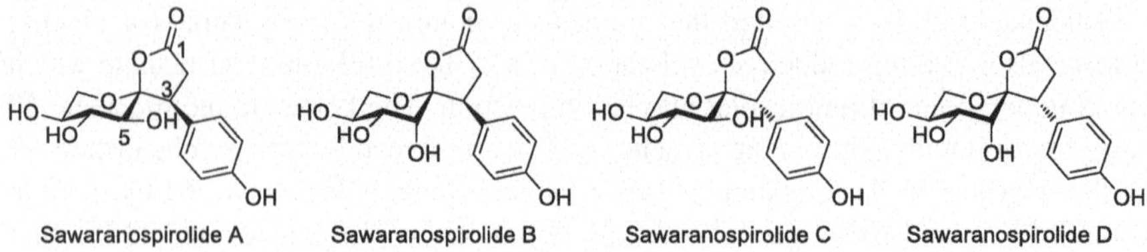

Fig. 6 The molecular structures of sawaranospirolides A-D

4.3 *The Application of 1,6-Dioxaspiro[4.5]decane System in Natural Product Synthesis*

A 1-*C*-branched *endo*-glycal has been used as the starting material for the stereoselective total synthesis of natural product sawaranospirolide enantiomers. These are spiroacetal γ-butyrolactones, isolated from the heartwood of the Japanese tree "Sawara" (*Chamaecyparis pisifera*). A total of the four diastereomers were isolated which differ merely in the configurations at C-3 and C-5 (Fig. 6) [42]. Robertson et al. [43] initiated the total synthesis of sawaranospirolides starting from the *endo*-glycal **66**. The initial introduction of the formyl group was achieved by reacting in situ generated glycal anion with *N,N*-dimethylformamide to form the aldehyde **67**. Horner-Wadsworth-Emmons reaction of **67** with phosphonate delivered diphenyloxazoline **68** which upon reaction with *p*-(benzyloxy)phenyllithium resulted in the formation of a single diastereomer **69**. Sequential activation of the oxazoline with methyl triflate followed by partial hydrolysis to an amide with LiOH, and complete hydrolysis with KOH, provided the acid **70** without compromising the stereochemical integrity.

*m*CPBA-mediated oxidative spirocyclization of **70** provided spirolactone **71**. TIPS deprotection and hydrogenolysis of the phenolic *O*-benzyl group led to the formation of *ent*-sawaranospirolide C **72** (Scheme 12). On the other hand, pyridinium *p*-toluenesulfonate (PPTS)-mediated spirocyclization of **70** to give lactone **73** followed by dihydroxylation, TIPS deprotection, and hydrogenolysis provided the *ent*-sawaranospirolide D **74**.

4.4 *Synthesis of 1,6-Dioxaspiro[4.5]decane Systems from Pyranoid* Exo-*Glycals*

Methods for the syntheses of 1,6-dioxaspiro[4.5]decane systems using *exo*-glycals are very limited. Alcaraz et al. [44] synthesized the *exo*-glycal **76** via the Ramberg-Bäcklund rearrangement of **75**. Subjecting **76** to CSA-mediated spiroketalization reaction gave a mixture of spiroketals **77a** and **77b** in 70:30 ratio, respectively (Scheme 13).

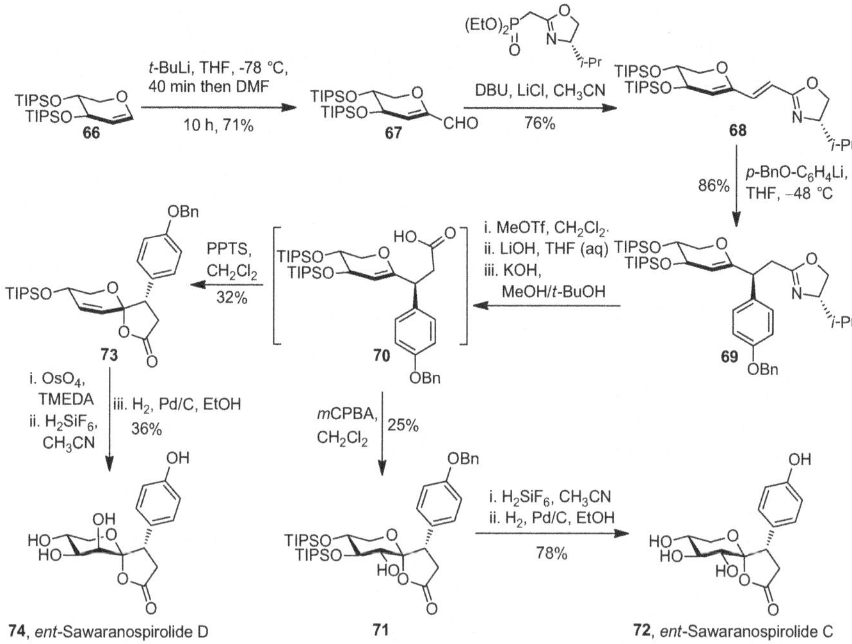

Scheme 12 Total synthesis of *ent*-sawaranospirolide C **72** and *ent*-sawaranospirolide D **74**. *PPTS* pyridinium *p*-toluenesulfonate

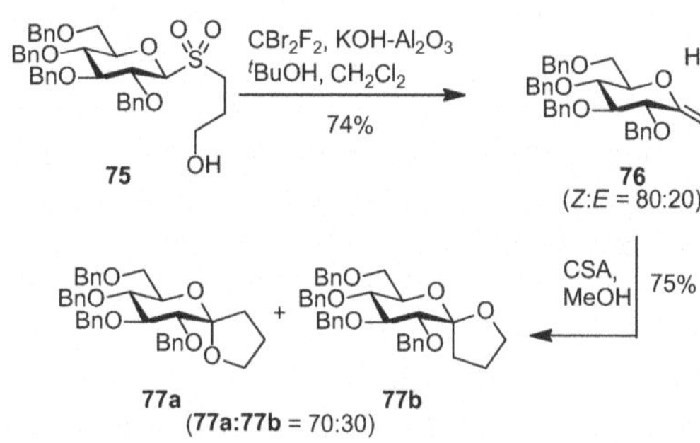

Scheme 13 CSA-mediated spirocyclization of an *exo*-glycal synthesized by Ramberg-Bäcklund rearrangement

Brand et al. [45] reported an oxidative ring expansion reaction of carbohydrate-derived spiro-cyclopropanes to yield various spiroketal frameworks. Glucose-derived *exo*-glycal **78** upon cyclopropanation with ethyl diazoacetate (EDA) followed by LAH reduction provided alcohol **79** as the diastereomeric mixture. An oxidative ring enlargement of **79**, facilitated by 2-iodoxybenzoic acid (IBX) and $Yb(OTf)_3$, provided a mixture of thermodynamic and kinetic spiroketals **80a** and **80b**, respectively, in 5:1 ratio. A similar kind of reaction sequence was observed

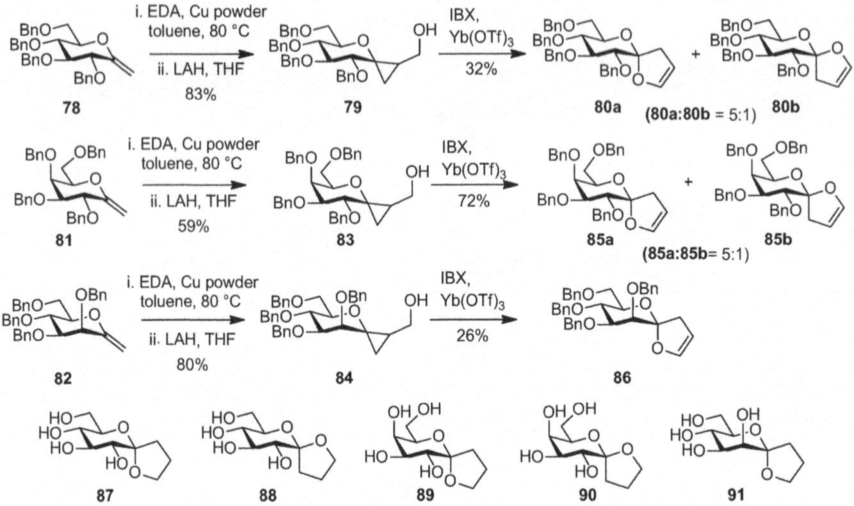

Scheme 14 Oxidative ring expansion of spiro-cyclopropanated sugar derivatives for the synthesis of 1,6-dioxaspiro[4.5]decane system

with spiro-cyclopropane **83**, derived from galactose-based *exo*-glycal **81**, producing spiroketals **85a** and **85b**, while mannose-derived spiro-cyclopropane **84**, synthesized from *exo*-glycal **82**, provided the thermodynamic product **86** as a single diastereomer. The spiroketals **80a**, **80b**, **85a**, **85b**, and **86** synthesized in this route were subjected to hydrogenation to give fully saturated and globally deprotected spiroketals **87–91** (Scheme 14).

Ramakrishna et al. [31] reported the stereoselective synthesis of 1,6-dioxaspiro [4.5]decane-2-one frameworks from *exo*-glycals involving spiro-cyclopropanecarboxylic acids as intermediates. In this route, the D-fructose-derived *exo*-glycal **92** was cyclopropanated with methyl diazoacetate (MDA) to provide a diastereomeric mixture of spiro-cyclopropane carboxylate **93**. Hydrolysis of the ester to give acid **94**, followed by treatment with $BF_3{\cdot}OEt_2$, led to the formation of 1,6-dioxaspiro[4.5]decane framework **95** as a single diastereomer (Scheme 15). Utilizing this methodology, various sugar-derived *exo*-glycals **96–98** were converted to the spiro-cyclopropanecarboxylic acids **99–101** which on exposure to $BF_3{\cdot}OEt_2$ provided the spirolactones **102–104** as single diastereomers favoring the thermodynamic criteria (Scheme 15).

Brand et al. [45] revealed that spiro-cyclopropanecarboxylated sugar derivative **106**, obtained from *exo*-glycal **105**, upon reduction to alcohol **109** followed by oxidative ring expansion with DMP/$Yb(OTf)_3$ provided the kinetic product **110** as a single diastereomer (Scheme 16). On the other hand, in Sridhar's approach [31] hydrolysis of **106–107** followed by $BF_3{\cdot}OEt_2$-mediated ring-opening cyclization reaction led to the exclusive formation of the thermodynamic product **108**. Thus, these methodologies are very advantageous to synthesize spirocyclic systems possessing bis-diaxial C-O conformation as well as spiroketals that do not reside in bis-diaxial C-O conformation.

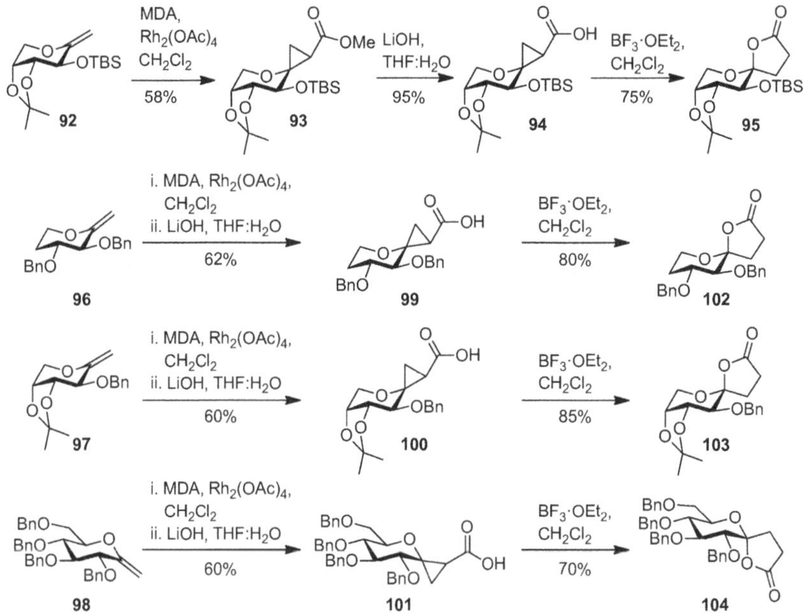

Scheme 15 Lewis acid-mediated ring expansion of sugar-derived spiro-cyclopropanecarboxylic acids

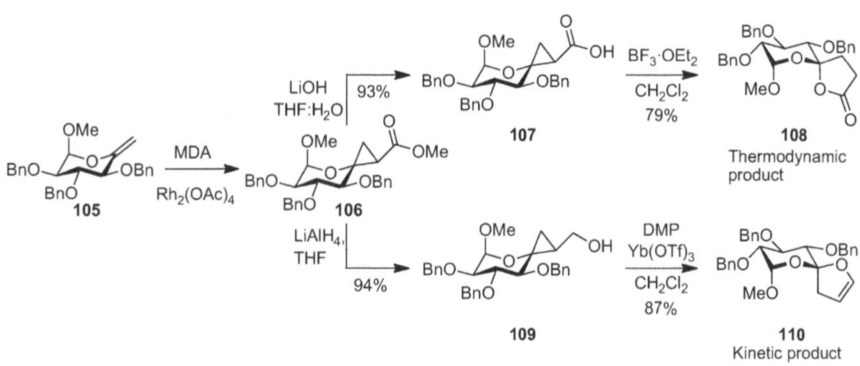

Scheme 16 Synthesis of kinetic and thermodynamic spiroketals possessing 1,6-dioxaspiro[4.5] decane framework

5 1,7-Dioxaspiro[5.5]undecane Systems

1,7-Dioxaspiro[5.5]undecane systems are the most studied frameworks in the spiroketal family of natural products. Many complex marine macrolides, antibiotics, and pheromones possess this skeleton, which is often crucial for their biological activity [46]. The stereochemistry at the spirocentre plays a major role in spatial orientation and conformation of the spirocyclic system. For instance, talaromycins are a class of spiroketal mycotoxins which are produced by the fungus *Talaromyces stipitatus* [47–49]. There are a total of six talaromycin isomers (talaromycin A–F) all possessing a 1,7-dioxaspiro[5.5]undecane system differing only at stereocenters (Fig. 7) and exhibiting variable biological profile.

On the other hand, spongistatins, which are isolated from the marine sponges belonging to genus *Spongia*, possess two sets of highly functionalized and oxygenated 1,7-dioxaspiro[5.5]undecane subunits [50]. Interestingly, the spiroketal fragment present in the spongistatins is also shown to inhibit the microtubule assembly [51], a property of the parent molecule. Similarly, a structurally related spiroketal motif present in okadaic acid and tautomycin is shown to induce apoptosis [52] (Fig. 8). Due to the presence of 1,7-dioxaspiro[5.5]undecane system in a number of bioactive natural products, their stereoselective synthesis has been an attractive target to a number of synthetic chemists worldwide.

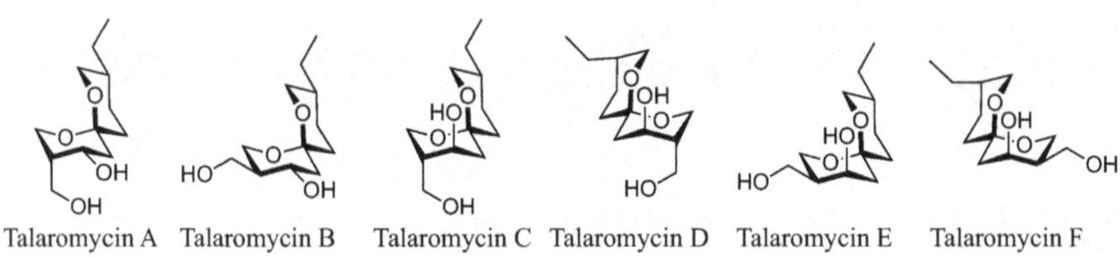

Fig. 7 Structures of talaromycins A–F

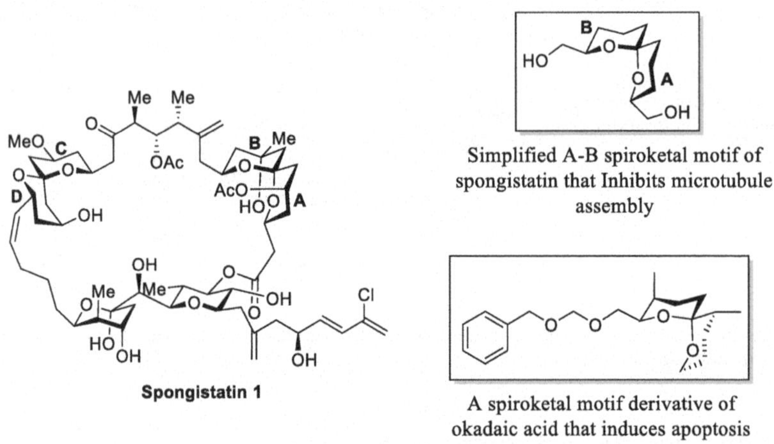

Fig. 8 Structure of spongistatin 1 and the corresponding biologically active spiroketal fragments

5.1 *Synthesis of 1,7-Dioxaspiro[5.5]undecane Systems from Pyranoid* Endo-*Glycals*

In general, protocols applied for the synthesis of 1,6-dioxaspiro[4.5]decane systems also could be adopted for the synthesis of 1,7-dioxaspiro[5.5]undecane systems. An acid-catalyzed ketalization of 1-*C*-branched glycals are one of the attractive approaches for the synthesis of carbohydrate-derived 1,7-dioxaspiro[5.5]undecane systems. Conway et al. [53] subjected the *endo*-glycal **111** to CSA in CH_2Cl_2 to access the thermodynamically favored spiroketal **112** as the major product along with a minor component **113**. Inducing the spirocyclization with PhSeCl and pyridine in CH_2Cl_2 at −78°C afforded a mixture of spiroketals **114**, **115,** and **116** (1.5:1:3 ratio, respectively) in 43% yield (Scheme 17). It is interesting to note that the kinetic spiroketal **116** as the major product in the phenylselenyl ion induced spiroketalization reaction.

Fuwa and Sasaki et al. [54] reported the spirocyclization of 1-*C*-branched glycals **117** and **118** upon treatment with CSA in CH_2Cl_2 resulting in the exclusive formation of thermodynamically favored spiroketals **119** and **120**, respectively, in high yield (Scheme 18).

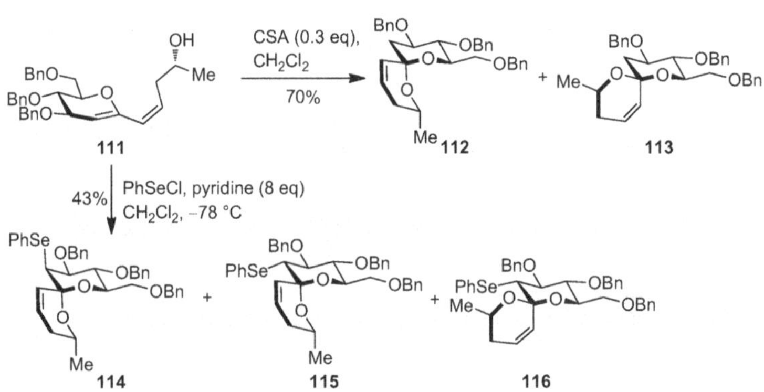

Scheme 17 Spiroketalization of 1-*C*-branched pyranoid *endo*-glycal

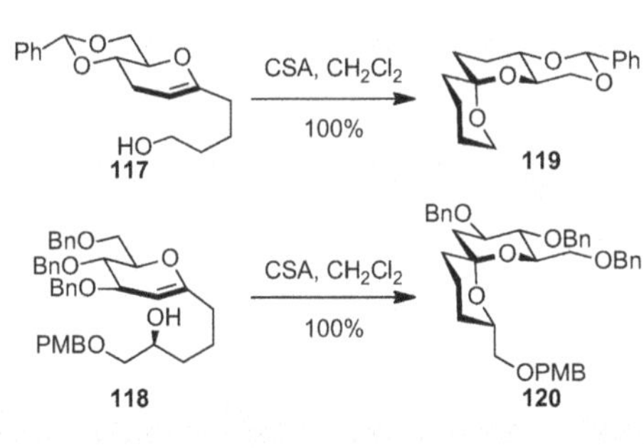

Scheme 18 Camphorsulfonic acid-mediated spirocyclization of 1-*C*-branched *endo*-glycals

On the other hand, glucal **121** under similar reaction conditions produced a mixture of kinetic spiroketals **122a** and **122b**. The formation of these nonanomeric or contra-thermodynamic spiroketals was probably due to the unfavorable 1,3-diaxial interactions in the corresponding doubly anomeric isomer. Alternatively, kinetically controlled spirocyclization of 1-*C*-branched glycal **121** using NIS at −90°C in CH_2Cl_2 provided the corresponding spiroketals **123a** and **123b** (Scheme 19).

CSA-catalyzed spirocyclization of D-*erythro* glycals **124** and **125** resulted in the formation of doubly anomeric spiroketals **126** and **127**, respectively, as the only products, however, in poor yield. The poor yield, in this case, was attributed to the competitive Ferrier cyclization reaction. In contrast, iodocyclization of **124** and **125** provided the doubly anomeric spiroketal **128** and nonanomeric spiroketal **129**, respectively, in excellent yield with high stereoselectivity (Scheme 20). These methods suggest that kinetically controlled spiroketal formation of *endo*-glycals

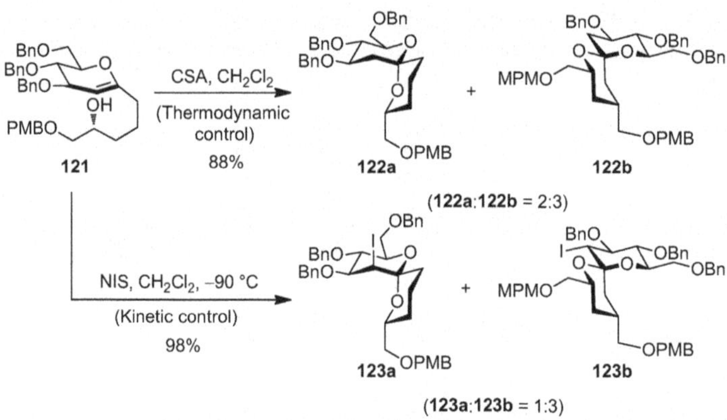

Scheme 19 Thermodynamic vs kinetic control in spiroketalization of 1-*C*-branched *endo*-glycals

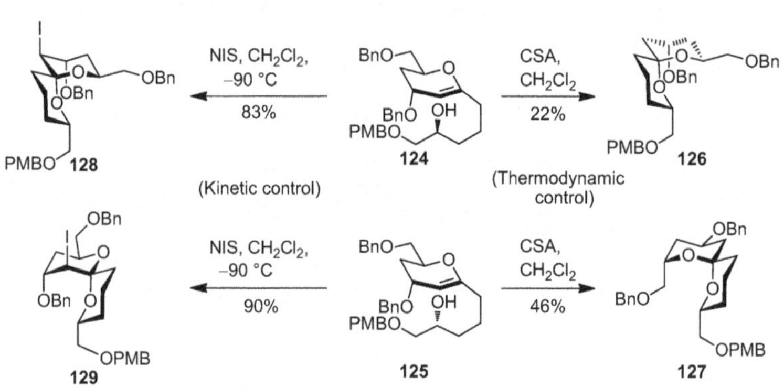

Scheme 20 Thermodynamic vs kinetic control in spiroketalization of 1-*C*-branched D-*erythro* *endo*-glycals

could be a powerful method for the synthesis of both doubly anomeric and nonanomeric spiroketals.

Roush et al. used the kinetically controlled iodo-spirocyclization reaction of 1-*C*-branched glycal **130** to obtain spiroketal subunits C–D [55] as well as A–B [56] in spongistatin 1 (Fig. 8). Accordingly, **130** was treated with NIS in dichloromethane at −90°C to obtain a diastereomeric mixture of **131** and **132** in 8:1 ratio. Methylation of the free OH in **131** gave **133**. Reductive deiodination of **133** provided the spiroketal **134**, which is the C–D spiroketal unit in spongistatin 1. On the other hand, inversion of both hydroxyls in **130** to form **136**, via diacetate **135**, followed by NIS-mediated spiroketalization provided the iodospiroketal **137**. Reductive dehalogenation of **137** and oxidation of the secondary hydroxyl with Dess-Martin periodinane (DMP) provided the ketone **138**. Subjecting **138** to Grignard reaction with MeMgBr resulted in the formation of **139** which resembles the A–B spiroketal subunit of spongistatin 1 (Scheme 21).

The methanol-induced ring opening [37] of epoxides **143–145** derived from D-*threo*-glycals **140–142** provided kinetic spiroketals **146–148**, respectively, as

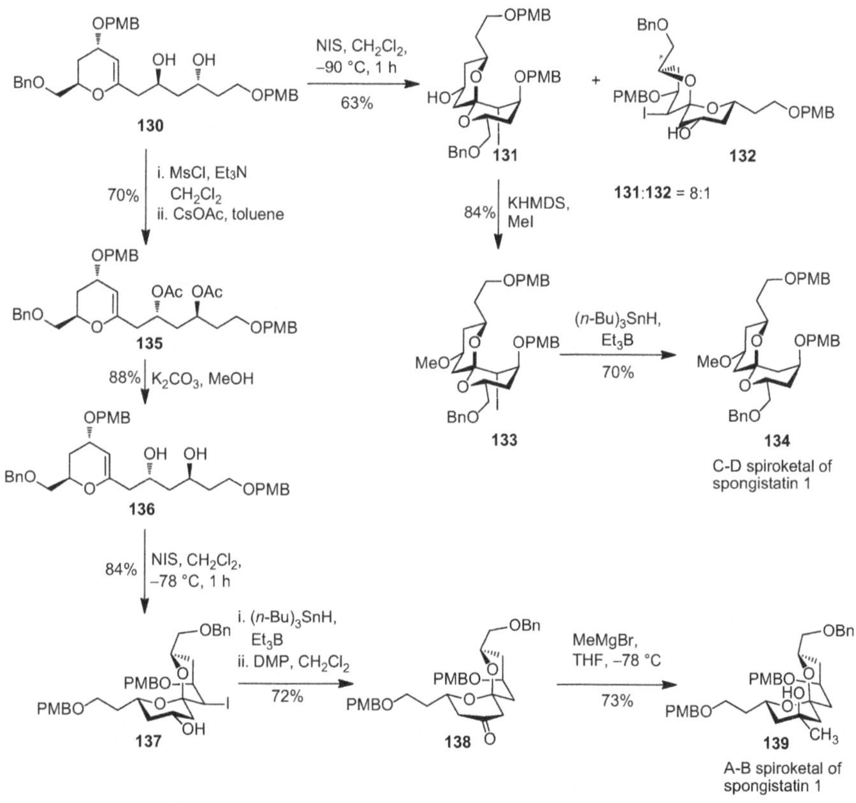

Scheme 21 Synthesis of A–B and C–D spiroketal subunits of spongistatin 1

the only isolable products with very high stereoselectivity and in an excellent yield (Scheme 22).

On the other hand, the Ti(O*i*-Pr)$_4$-mediated kinetic spirocyclization [39] of epoxides **152–154**, which are derived from D-*erythro* glycals **149–151**, provided 1,7-dioxaspiro[5.5]undecane systems **155–157** with >98% diastereoselectivity (Scheme 23). However, application of these highly stereoselective methods for the synthesis of kinetic spiroketals in the total synthesis of natural products has not been reported yet.

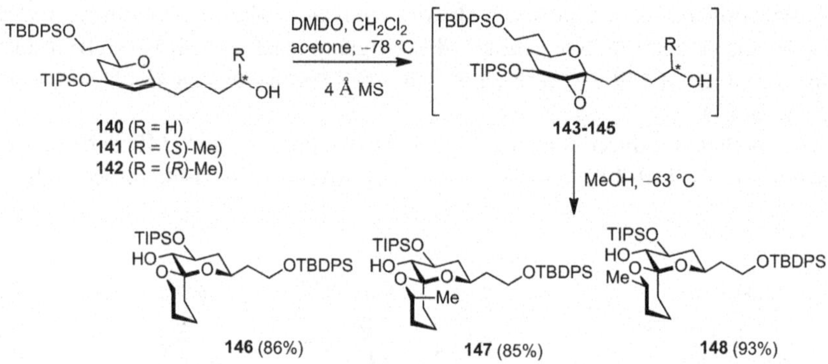

Scheme 22 Methanol-induced spiroketalization for the synthesis of 1,7-dioxaspiro[5.5]undecane systems

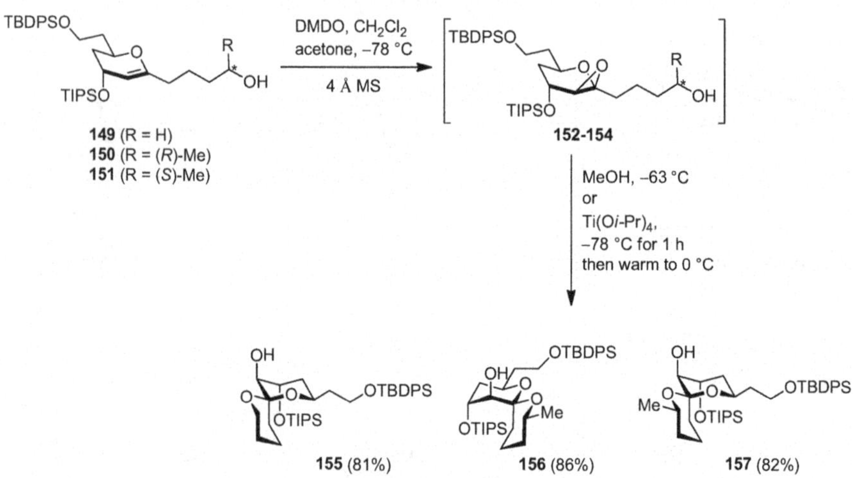

Scheme 23 Synthesis of kinetic spiroketals by methanol-induced or Ti(O*i*-Pr)$_4$-mediated spiroketalization reactions

5.2 *Synthesis of 1,7-Dioxaspiro[5.5]undecane Systems from Pyranoid* Exo-*Glycals*

Exo-glycals have been another type of the useful starting materials for the synthesis of 1,7-dioxaspiro[5.5]undecane systems. A few methods have been reported in the literature for the preparation of the trisubstituted *exo*-glycal precursors which could be used for the subsequent spiroketalization reaction. For example, Ramberg-Bäcklund olefination [44], palladium-catalyzed coupling reactions [53], addition-elimination reactions, Wittig reactions [57], and Julia olefination [58] reaction could be used for the preparation of trisubstituted *exo*-glycals from the carbohydrate-derived lactones. Gueyrard and Goekjian [58] reported that the trisubstituted *exo*-glycal **160**, synthesized by Julia olefination of **158** with **159**, in the presence of *p*-TsOH undergoes spiroketalization to give the thermodynamically stable doubly anomeric spiroketal **161** as the only product. Similarly synthesized *exo*-glycals **162** and **163**, when subjected to mild acidic conditions, provided the spiroketals **164**, **165** and **166**, **167**, respectively (Scheme 24).

The *exo*-glycal **168** has been converted to benzene-fused 1,7-dioxaspiro[5.5] undecane systems **169–175** by reacting with *para*-substituted phenols in the presence of a catalytic amount of $BF_3{\cdot}OEt_2$ [59] (Scheme 25). The methodology was successfully applied to the synthesis of 1-oxa-7-thiaspiro[5.5]undecane systems **176–180** as well. In all the products, the glycosidic bond was found to be in axial orientation.

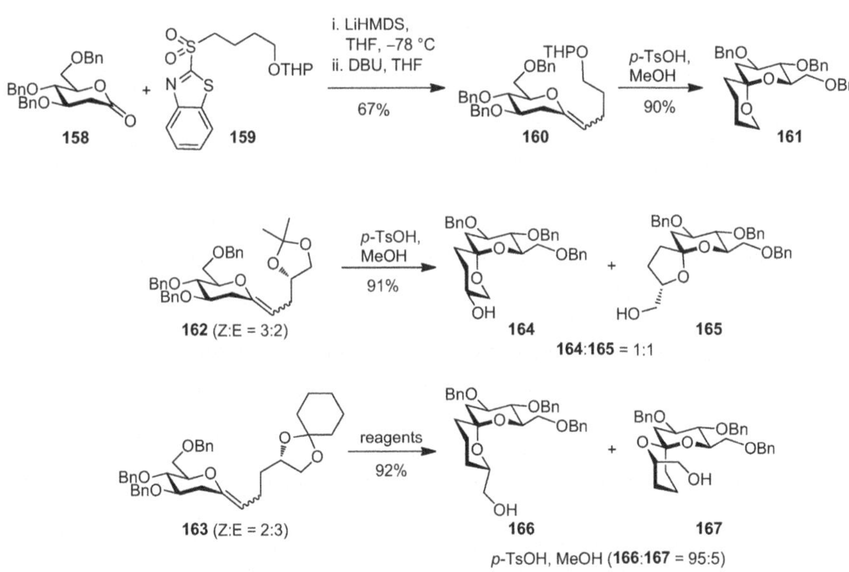

Scheme 24 Synthesis of 1,7-dioxaspiro[5.5]undecane systems from *exo*-glycals

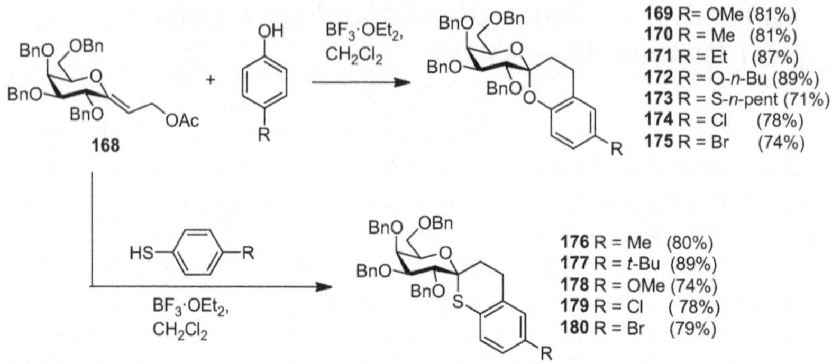

Scheme 25 A one-pot *exo*-Ferrier rearrangement and an intramolecular Friedel-Crafts alkylation reaction to generate 1,7-dioxaspiro[5.5]undecane systems

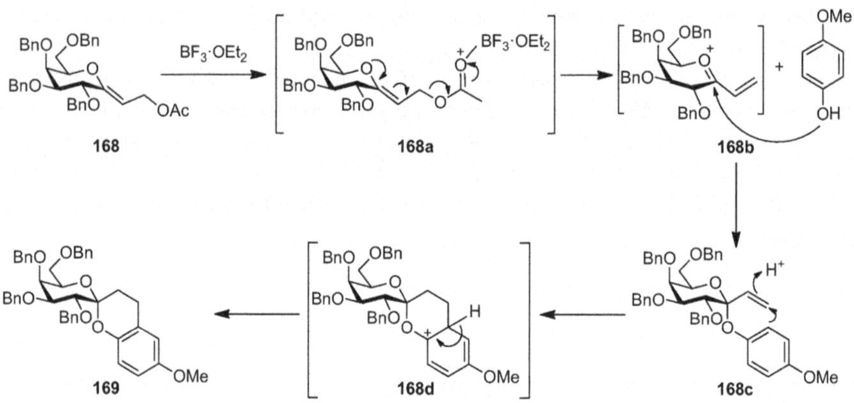

Scheme 26 Proposed mechanism for the formation of 1,7-dioxaspiro[5.5]undecane framework

The possible mechanism involves $BF_3 \cdot OEt_2$-mediated activation of the allylic acetate **168** to form the intermediate **168a**, which led to the formation of α,β-conjugated oxonium ion **168b** involving an *exo*-Ferrier type rearrangement. Glycosylation of the oxonium ion intermediate **168b** with *p*-methoxyphenol would provide the adduct **168c**. An intramolecular Friedel-Crafts alkylation in **168c** would give the carbocation intermediate **168d** which eventually will lead to the formation of benzene-fused 1,7-dioxaspiro[5.5]undecane system **169** (Scheme 26).

Matsuda et al. [60] reported a novel approach for the synthesis of *exo*-glycals which were further utilized in the synthesis of 1,7-dioxaspiro[5.5]undecane systems. 1-*C*-vinyl sugar derivative **181** upon reaction with silyl enolate **182** in the presence of TMSOTf resulted in the formation of *exo*-glycal **183** [61]. Conversion of the aldehyde **183** to the corresponding alcohol **184** with $NaBH_4$ followed by an intramolecular glycosylation reaction in the presence of $BF_3 \cdot OEt_2$ provided thermodynamically stable bis-diaxial spiroketal **185** (Scheme 27).

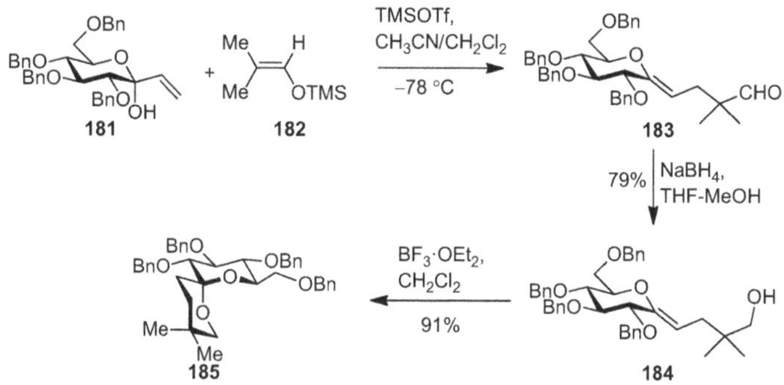

Scheme 27 $BF_3{\cdot}OEt_2$-catalyzed spiroketalization of a trisubstituted *exo*-glycal

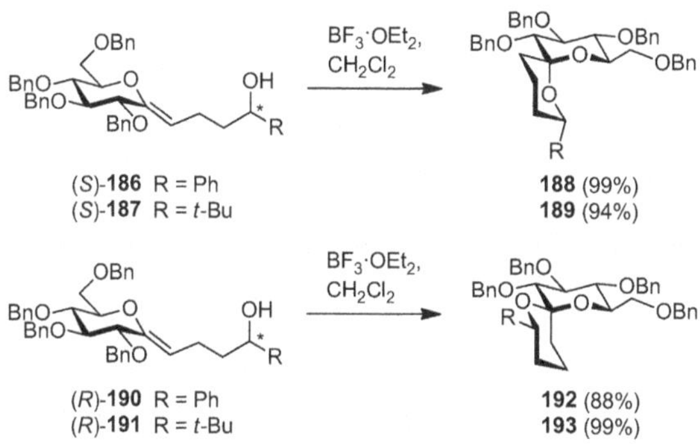

Scheme 28 Substituent stereochemistry-directed thermodynamic vs kinetic spiroketalization of *exo*-glycals

Similarly, carrying out the spirocyclization reaction of *exo*-glycals **186** and **187** in which configuration of the alcohol is (*S*), the formation of bis-diaxial products **188** and **189** was observed. When the configuration of the alcohol in *exo*-glycals **190** and **191** is (*R*), spiroketals **192** and **193**, where double anomeric effect is absent, were formed as a single diastereomer (Scheme 28). The formation of the kinetic products **192** and **193** was explained based on the relative stabilities of the substituted spiroketals ([62], for a book: [63]).

6 Spirocyclic Bis-*C*,*C*-Glycosyl Derivatives

Compounds formally derived from an *O*-glycoside by replacement of both the exocyclic oxygen and hydrogen atoms on the anomeric center by carbon atoms are bis-*C*,*C*-glycosides. If the two carbon atoms (α and β carbons) of a bis-*C*,*C*-glycoside form a cyclic structure, then they are called spirocyclic bis-*C*,*C*-glycosides [64]. One such kind of framework is the 1,7-dioxaspiro[4.4]nonane system. These *C*-spirocycles are present in a number of bioactive natural products like secosyrins

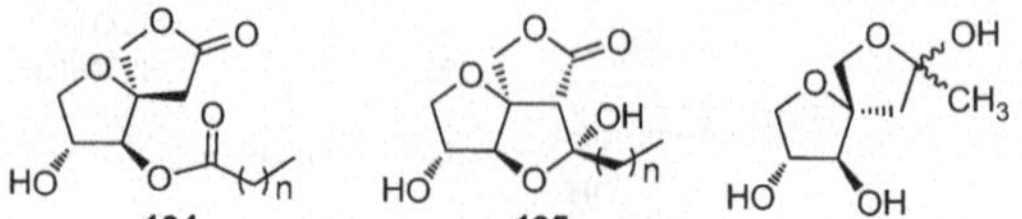

194
n = 4 Secosyrin 1
n = 6 Secosyrin 2

195
n = 4 Syringolide 1
n = 6 Syringolide 2

196
Sphydrofuran

Fig. 9 Structures of some of the bioactive natural products possessing 1,7-dioxaspiro[4.4]nonane framework

1 and 2 [65, 66] (**194**), syringolide 1 and 2 (**195**), sphydrofuran **196** [67, 68], etc. (Fig. 9). The stereoselective formation of the quaternary spirocentre is one of the major challenges in the construction of these sugar-derived spirocyclic architectures.

6.1 Synthesis of 1,7-Dioxaspiro[4.4]nonane Systems from Endo-*Glycals*

Sridhar et al. [69] reported a one-pot ring contraction of 2-*C*-branched sugars by a sequential dehydrogenation, stereoselective intramolecular hetero-Michael addition, and ester hydrolysis to produce sugar-derived *C*-spirocyclic-lactols. 2-*C*-Branched sugars were synthesized from the corresponding 1,2-cyclopropanecarboxylated sugars which could be obtained from the appropriate *endo*-glycal precursors. Based on this strategy, *endo*-glycal **197** was cyclopropanated with MDA in the presence of catalytic amount of $Rh_2(OAc)_4$ to produce 1,2-cyclopropanecarboxylate **198**. NBS-mediated electrophilic ring opening of **198** led to the formation of bromohydrin **199**. Treatment of **199** with K_2CO_3 in MeOH gave a diastereomeric mixture of spirocyclic lactols **202** in 92% yield. The formation of **202** was explained by a stereoselective dehydrohalogenation of **199** to form *E*-olefin **199a**, which exists in equilibrium with the acyclic aldehyde **199b**. Intramolecular hetero-Michael addition (For a few unsuccessful IHMA reactions for spirolactones: [70, 71]) reaction of intermediate **199b** through the transition state **200** led to the formation of the aldehyde-ester **201**. Hydrolysis of the ester **201** produced the spirocyclic lactol **202** which was further converted to the *C*-spiroglycoside **203** by dehydroxylation reaction using Et_3SiH/TFA (Scheme 29).

The methodology was successfully applied to synthesize a series of *C*-spirocyclic lactones **204–208** using the appropriate 1,2-cyclopropanecarboxylated sugar derivatives (Fig. 10).

6.2 The 2,6-Dioxaspiro[4.5]decane Framework

The formation of bis-*C,C*-glycosides was also investigated for the synthesis of 2,6-dioxaspiro[4.5]decane framework [69]. In this process, the seven-membered *endo*-glycal derivative **209** was cyclopropanated to give **210** and then subjected to NBS-mediated ring opening followed by treatment with K_2CO_3 in MeOH provided the spirolactone **211** (Scheme 30).

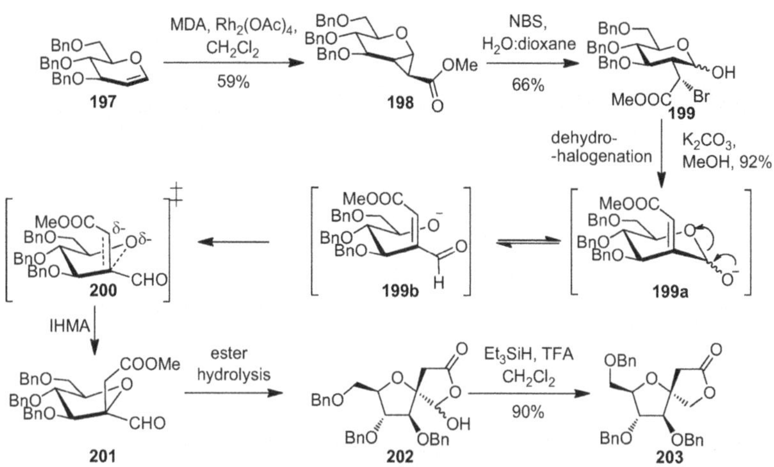

Scheme 29 Synthesis of a 1,7-dioxaspiro[4.4]nonane framework from a glycal-derived 1,2-cyclopropanecarboxylate

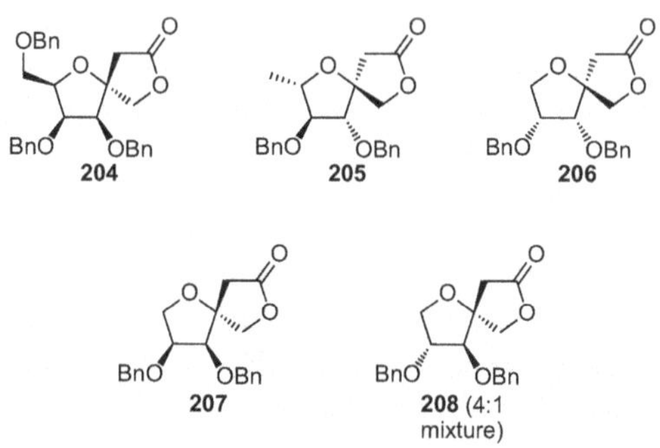

Fig. 10 Structures of various bis-*C*,*C*-glycosides synthesized by electrophilic ring opening of glycal-derived 1,2-cyclopropanecarboxylates

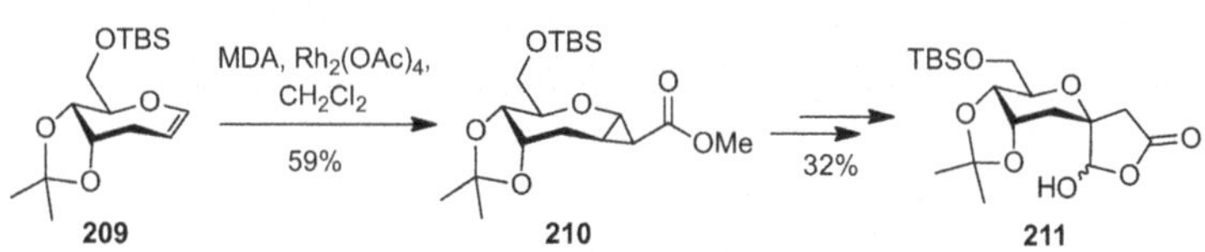

Scheme 30 Synthesis of 2,6-dioxaspiro[4.5]decane framework form septanose-derived *endo*-glycal

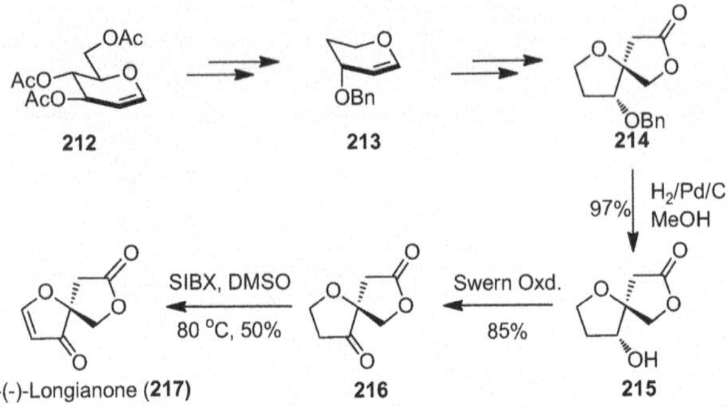

Scheme 31 Stereoselective total synthesis of (*S*)-(−)-longianone

6.3 Application of Bis-C,C-Glycoside Preparation in the Total Synthesis of the Natural Product (S)-(−)-Longianone

One of the very simple natural products that possesses 1,7-dioxaspiro[4.4]nonane framework is (*S*)-(−)-longianone **217** [72], a mycotoxin metabolite of fungus *Xylaria longiana*. A methodology developed for the synthesis of bis-*C*,*C*-glycosides has been successfully applied to the synthesis of this natural product [73]. The triacetyl glucal **212** was converted to deoxy-glycal **213** which was further transformed into the spirocyclic lactone **214** as a single diastereomer by adopting the above methodology. Hydrogenolysis of **214** to give **215** and Swern oxidation of the secondary alcohol provided keto-lactone **216**. Dehydrogenation of compound **216** using stabilized 2-iodoxybenzoic acid (SIBX) [74] provided enantiomerically pure (*S*)-(−)-longianone **217** (Scheme 31).

7 Conclusion

As demonstrated above, although highly stereoselective methods are available for the synthesis of spiroketals from carbohydrate derivatives, their direct application in the total synthesis of natural products from glycals is still not extensively studied. The use of pentose-derived *endo*- and *exo*-glycals for the synthesis of 1,6-dioxaspiro [4.4]nonane framework is still in the preliminary stage of development. Apart from obtaining the required stereochemistry at the spirocentre, accessing the precursors possessing the appropriate chiral centers at nonanomeric positions from carbohydrate sources requires multiple transformations. Discovery of novel synthetic transformations to convert the easily accessible glycals to deoxy-glycals would help in wide application of *endo*- and *exo*-glycals in the synthesis of spiroketal natural products. This certainly requires further research in the stereoselective synthesis of carbohydrate-derived spiroketals.

References

1. Nicolaou KC, Mitchell HJ (2001) Adventures in carbohydrate chemistry: new synthetic technologies, chemical synthesis, molecular design, and chemical biology. Angew Chem Int Ed 40:1576–1624
2. Perron F, Albizati KF (1989) Chemistry of spiroketals. Chem Rev 89:1617–1661
3. Jacobs MF, Kitching W (1998) Spiroacetals of marine origin. Curr Org Chem 2:395–436
4. Mead KT, Brewer BN (2003) Strategies in spiroketal synthesis revisited: recent applications and advances. Curr Org Chem 7:227–256
5. Brimble MA, Furkert DP (2003) Chemistry of bis-spiroacetal systems: natural products, synthesis and stereochemistry. Curr Org Chem 7:1461–1484
6. Ley SV, Milroy LG, Myers RM (2007) Product class 9: spiroketals. Sci Synth 29:613–690
7. Verano AL, Tan DS (2017) Stereocontrolled synthesis of spiroketals: an engine for chemical and biological discovery. Isr J Chem 57:279–291
8. Macial FA, Galindo JLG, Varela RM, Torres A, Molinillo JMG, Fronczek FR (2006) Heliespirones B and C: two new plant heliespiranes with a novel spiro heterocyclic sesquiterpene skeleton. Org Lett 8:4513–4516
9. Noguchi N, Nakada M (2006) Synthetic studies on (+)-ophiobolin A: asymmetric synthesis of the spirocyclic CD-ring moiety. Org Lett 8:2039–2042
10. Entzeroth M, Blackman AJ, Myndersel JS, Moore RE (1985) Structures and stereochemistries of oscillatoxin B, 31-noroscillatoxin B, oscillatoxin D, and 30-methyloscillatoxin D. J Org Chem 50:1255–1259
11. Springer JP, Arison BH, Hirshfield JM, Hoogsteen K (1981) The absolute stereochemistry and conformation of avermectin B2a aglycone and avermectin B1a. J Am Chem Soc 103:4221–4224
12. Uemura D, Takahashi K, Yamamoto T, Katayama C, Tanaka J, Okumura Y, Hirata Y (1985) Norhalichondrin A: an antitumor polyether macrolide from a marine sponge. J Am Chem Soc 107:4796–4798
13. Tachibana K, Scheuer PJ, Tsukitani Y, Kikuchi H, van Engen D, Clardy J, Gopichand Y, Schmitz FJ (1981) Okadaic acid, a cytotoxic polyether from two marine sponges of the genus Halichondria. J Am Chem Soc 103:2469–2471
14. Chaney MO, Demarco PV, Jones ND, Occolowitz JL (1974) Structure of A23187, a divalent cation ionophore. J Am Chem Soc 96:1932–1933
15. Kato Y, Fusetani N, Matsunaga S, Hashimoto K, Fujita S, Furuya T (1986) Bioactive marine metabolites. Part 16. Calyculin A. A novel antitumor metabolite from the marine sponge *Discodermia calyx*. J Am Chem Soc 108:2780–2781
16. Jones ND, Chaney MO, Chamberlin JW, Hamill RL, Shen S (1973) Structure of A204A, a new polyether antibiotic. J Am Chem Soc 95:3399–3400
17. Deslongchamps P, Rowan DD, Pothier N, Saunders JK (1981) 1,7-Dithia and 1-oxa-7-thiaspiro [5.5]undecanes. Excellent systems for the study of stereoelectronic effects (anomeric and exo-anomeric effects) in the monothio and the dithioacetal functions. Can J Chem 59:1122–1131
18. Pothier N, Rowan DD, Deslongchamps P, Saunders JK (1981) 13C chemical shift data for 1,7-dioxaspiro[S.S]undecanes and related compounds. Can J Chem 59:1132–1139
19. Moore RE, Blackman AJ, Cheuk CE, Mynderse JS (1984) Absolute stereochemistries of the aplysiatoxins and oscillatoxin A. J Org Chem 49:2484–2489
20. Yasumoto T, Murata M, Oshima Y, Sano M, Matsumoto GK, Clardy J (1985) Diarrhetic shellfish toxins. Tetrahedron 41:1019–1025
21. Jung JH, Sim CJ, Lee CO (1995) Cytotoxic compounds from a two-sponge association. J Nat Prod 58:1722–1726
22. Taillefumier C, Chapleur Y (2004) Synthesis and uses of *exo*-glycals. Chem Rev 104:263–292
23. Frederic CJM, Vincent SP (2018) Synthesis of *exo*-glycals and their biochemical applications. Tetrahedron 74:6512–6519

24. Pal P, Shaw AK (2017) The evolution of comprehensive strategies for furanoid glycal synthesis and their applications. RSC Adv 7:25897–25963
25. Haraguchi K, Konno K, Yamada K, Kitagawa Y, Nakamura KT, Tanaka H (2010) Electrophilic glycosidation employing 3,5-*O*-(di-*tert*-butylsilylene)-erythro-furanoid glycal leads to exclusive formation of the β-anomer: synthesis of 2′-deoxynucleosides and its 1′-branched analogues. Tetrahedron 66:4587–4600
26. Paquette LA, Brand S, Behrens C (1999) An enantioselective ring expansion route leading to furanose and pyranose nucleosides featuring spirodiketopiperazines at the anomeric position. J Org Chem 64:2010–2025
27. Boyce RS, Kennedy RM (1994) The oxidative spirocyclization of 2-(ω-(OH)-alkyl)cyclic enol ethers by rhenium (VII)-oxide. Tetrahedron Lett 35:5133–5136
28. Čorić I, List B (2012) Asymmetric spiroacetalization catalysed by confined Brønsted acids. Nature 483:315–319
29. Yang WB, Chang CF, Wang SH, Teo CF, Lin CH (2001) Expeditious synthesis of *C*-glycosyl conjugated dienes and aldehydes from sugar lactones. Tetrahedron Lett 42:4657–4660
30. Chang CF, Yang WB, Chang CC, Lin CH (2002) Inter- and intramolecular alcohol additions to *exo*-glycals. Tetrahedron Lett 43:6515–6519
31. Ramakrishna B, Sridhar PR (2015) Stereoselective synthesis of 1,6-dioxaspirolactones from spiro-cyclopropanecarboxylated sugars: total synthesis of dihydro-pyrenolide D. RSC Adv 5:8142–8145
32. Schneider TF, Kaschel J, Werz DB (2014) A new golden age for donor–acceptor cyclopropanes. Angew Chem Int Ed 53:5504–5523
33. Agtarap A, Chamberlin JW, Pinkerton M, Steinrauf L (1967) Structure of monensic acid, a new biologically active compound. J Am Chem Soc 89:5737–5739
34. Haney Jr ME, Hoehn MM (1967) Monensin, a new biologically active compound. I. Discovery and isolation. Antimicrob Agents Chemother 7:349–352
35. Fuwa H, Sakamoto K, Muto T, Sasaki M (2015) Concise synthesis of the C15-C38 fragment of okadaic acid, a specific inhibitor of protein phosphatases 1 and 2A. Tetrahedron 71:6369–6383
36. Zhankui S, Grace AW, Alina B, Pavel N (2012) Chiral phosphoric acid-catalyzed enantioselective and diastereoselective spiroketalizations. J Am Chem Soc 134:8074–8077
37. Potuzak JS, Moilanen SB, Tan DS (2005) Stereocontrolled synthesis of spiroketals via a remarkable methanol-induced kinetic spirocyclization reaction. J Am Chem Soc 127:13796–13797
38. Wurst JM, Liu G, Tan DS (2011) Hydrogen-bonding catalysis and inhibition by simple solvents in the stereoselective kinetic epoxide-opening spirocyclization of glycal epoxides to form spiroketals. J Am Chem Soc 133:7916–7925
39. Moilanen SB, Potuzak JS, Tan DS (2006) Stereocontrolled synthesis of spiroketals via Ti(O*i*-Pr)$_4$-mediated kinetic spirocyclization of glycal epoxides with retention of configuration. J Am Chem Soc 128:1792–1793
40. Takaoka LR, Buckmelter AJ, LaCruz TE, Rychnovsky SD (2005) Rational synthesis of contra-thermodynamic spiroacetals by reductive cyclizations. J Am Chem Soc 127:528–529
41. Lorenc C, Saur J, Moser A, Buevich AV, Williams AJ, Williamson RT, Martin GE, Peczuh MW (2015) Turning spiroketals inside out: a rearrangement triggered by an enol ether epoxidation. ChemistryOpen 4:577–580
42. Hasegawa S, Koyanagi H, Hirose Y (1990) Decarboxylated ascorbigens in the heartwood of *Chamaecyparis pisifera*. Phytochemistry 29:261–266
43. Robertson J, Chovatia PT, Fowler TG, Withey JM, Woollaston DJ (2010) Oxidative spirocyclisation routes towards the sawaranospirolides. Synthesis of *ent*-sawaranospirolides C and D. Org Biomol Chem 8:226–233
44. Alcaraz ML, Griffin FK, Paterson DE, Taylor RJK (1998) Synthetic applications of Ramberg-Bäcklund derived *exo*-glycals. Tetrahedron Lett 39:8183–8186

45. Brand C, Rauch G, Zanoni M, Dittrich B, Werz DB (2009) Synthesis of [*n*,5]-spiroketals by ring enlargement of donor-acceptor-substituted cyclopropane derivatives. J Org Chem 74:8779–8786
46. Kulkarni BA, Roth GP, Lobkovsky E, Porco JA (2002) Combinatorial synthesis of natural product-like molecules using a first-generation spiroketal scaffold. J Comb Chem 4:56–72
47. Lynn DG, Phillips NJ, Hutton WC, Shabanowitz J, Fennell DI, Cole RJ (1983) Talaromycins: application of homonuclear spin correlation maps to structure assignment. J Am Chem Soc 104:7319–7322
48. Hutton WC, Phillips NJ, Graden DW, Lynn DG (1983) The application of two-dimensional N.M.R. Cross relaxation spectroscopy to natural product structure determination: talaromycin B. J Chem Soc Chem Commun 16:864–866
49. Phillips NJ, Cole RJ, Lynn DG (1987) Talaromycins C,D,E, and F. Tetrahedron Lett 28:1619–1621
50. Petit GR, Cichacs ZA, Gao F, Herald CL, Boyd MR, Schmidt JM, Hooper JNA (1993) Antineoplastic agents. 257. Isolation and structure of spongistatin 1. J Org Chem 58:1302–1304
51. Uckun FM, Mao C, Vassilev AO, Huang H, Jan ST (2009) Structure-based design of a novel synthetic spiroketal pyran as a pharmacophore for the marine natural product spongistatin 1. Bioorg Med Chem Lett 10:541–545
52. Mitsuhashi S, Shima H, Kawamura T, Kikuchi K, Oikawa M, Ichihara A, Oikawa H (1999) The spiroketals containing a benzyloxymethyl moiety at C8 position showed the most potent apoptosis-inducing activity. Bioorg Med Chem Lett 9:2007–2012
53. Conway JC, Urch CJ, Quayle P, Xu J (2006) Spiroketalization reactions on a carbohydrate template. Synlett 5:776–780
54. Fuwa H, Sasaki M (2008) An efficient strategy for the synthesis of endocyclic enol ethers and its application to the synthesis of spiroacetals. Org Lett 10:2549–2552
55. Holson EB, Roush WR (2002) Diastereoselective synthesis of the C(17)−C(28) fragment (the C−D spiroketal unit) of spongistatin 1 (altohyrtin A) via a kinetically controlled iodo-spiroketalization reaction. Org Lett 4:3719–3722
56. Holson EB, Roush WR (2002) Synthesis of the C(2)−C(13) fragment (the A−B spiroketal unit) of spongistatin 1 (altohyrtin A): use of a common intermediate for the synthesis of both spongistatin spiroketals. Org Lett 4:3723–3725
57. Mori K, Ikunaka M (1987) Synthesis of (-)-talaromycins a and b. Tetrahedron 43:45–58
58. Corbet M, Bourdon B, Gueyrard D, Goekjian PGA (2008) Julia olefination approach to the synthesis of functionalized enol ethers and their transformation into carbohydrate-derived spiroketals. Tetrahedron Lett 49:750–754
59. Lin HC, Chen YB, Lin ZP, Wong FF, Lin CH, Lin SK (2008) Synthesis of 1,7-dioxaspiro[5.5] undecanes and 1-oxa-7-thiaspiro[5.5]undecanes from *exo*-glycal. Tetrahedron 66:5229–5234
60. Matsuda S, Yoshida A, Nakagawa J, Watanabe M, Oda Y, Yamanoi T (2014) Stereocontrolled spirocyclization of exo-glucal derivatives for stereodivergent synthesis of spiro[5.5]ketals. Tetrahedron Lett 55:6394–6398
61. Yamanoi T, Nara Y, Matsuda S, Oda Y, Yoshida A, Katsuraya K, Watanabe M (2007) Synthetic approach to *exo*-glycals from 1-*C*-vinyl-*D*-glycopyranose derivatives via an S_N1'-substitution mechanism. Synlett 5:785–789
62. Deslongchamps P, Rowan DD, Pothier N, Sauvé G, Saunders JK (1981) 1,7-Dioxaspiro[5.5] undecanes. An excellent system for the study of stereoelectronic effects (anomeric and exo-anomeric effects) in acetals. Can J Chem 59:1105–1121
63. Deslongchamps P (1983) Stereoelectronic effects in organic chemistry. Pergamon, Oxford
64. Paquette LA, Kinney MJ, Dullweber U (1997) Practical synthesis of spirocyclic bis-C,C-glycosides. Mechanistic models in explanation of rearrangement stereoselectivity and the bifurcation of reaction pathways. J Org Chem 62:1713–1722
65. Smith MJ, Mazzola EP, Sims JJ, Midland SL, Keen NT, Burton V, Stayton MM (1993) The syringolides: bacterial C-glycosyl lipids that trigger plant disease resistance. Tetrahedron Lett 34:223–226

66. Midland SL, Keen NT, Sims JJ, Midland MM, Stayton MM, Burton V, Smith MJ, Mazzola EP, Graham KJ, Clardy J (1993) The structures of syringolides 1 and 2, novel C-glycosidic elicitors from *Pseudomonas syringae* pv. Tomato. J Org Chem 58:2940–2945
67. Umezawa S, Usui T, Umezawa H, Tsuchiya T, Takeuchi T, Hamada M (1971) A new microbial metabolite, sphydrofuran. I. J Antibiot 24:85–92
68. Umezawa S, Tsuchiya T, Naganawa H, Takeuchi T, Umezawa H (1971) A new microbial metabolite, sphydrofuran. II. J Antibiot 24:93–106
69. Sridhar PR, Seshadri K, Reddy GD (2012) Stereoselective synthesis of sugar fused β-disubstituted γ-butyro-lactones: *C*-spiro-glycosides from 1,2-cyclopropanecarboxylated sugars. Chem Commun 48:756–758
70. Honda T, Mizutani H, Kanai K (1996) Enantioselective syntheses of syributin 1 and novel *C*-glycosidic elicitors syringolides 1 and 2. J Org Chem 61:9374–9378
71. Mukai C, Moharram SM, Azukizawa S, Hanaoka M (1997) Total syntheses of (+)-secosyrins 1 and 2 and (+)-syributins 1 and 2. J Org Chem 62:8095–8103
72. Edwards RL, Maitland DJ, Oliver CL, Pacey MS, Shields L, Whalley AJS (1999) Metabolites of the higher fungi. Part 31. Longianone, a $C_7H_6O_4$ spiro bicyclic lactone from the fungus *Xylaria longiana* (Rehm.). J Chem Soc Perkin Trans 1:715–720
73. Sridhar PR, Seshadri K (2012) First enantioselective total synthesis of (*S*)-(−)-longianone. Tetrahedron 68:3725–3728
74. Lykakis IN, Zaravinos IP, Raptis C, Stratakis M (2009) Divergent synthesis of the Co-isolated mycotoxins longianone, isopatulin, and (*Z*)-ascladiol via furan oxidation. J Org Chem 74:6339–6342

Top Heterocycl Chem (2019) 57: 137–170
DOI: 10.1007/7081_2019_35

Published online: 9 May 2019

Cyanohydrins and Aminocyanides as Key Intermediates to Various Spiroheterocyclic Sugars

Solen Josse and Denis Postel

Contents

Abstract Derivatives with a double functionalization attract great interest in organic synthesis. The association on the same carbon atom of a nitrile group and a hydroxyl or amine function allows access to promising heterocyclic compounds of particular interest resulting from reactions taking advantage of the electrophilic character of the cyano group and the nucleophilic character of hydroxyl and amino groups. Thus, α-hydroxynitriles (cyanohydrins) or α-aminonitriles represent important classes of organic intermediates. The development of these families in glycochemistry has allowed syntheses of compounds with chain elongations (Kiliani-Fischer synthesis, Strecker synthesis) or even the preparation of chiral

S. Josse (✉) and D. Postel
Laboratoire de Glycochimie, des Antimicrobiens et des Agroressources (UMR7378), Amiens, France

Institut de Chimie de Picardie (FRE 3085), UFR des Sciences-UPJV, Amiens, France
e-mail: solen.josse@u-picardie.fr; denis.postel@u-picardie.fr

building blocks versatile in asymmetric syntheses of biologically active compounds or their intermediates. In this chapter, we summarize the synthesis of quaternized glycoderivatives such as cyanohydrins or glycoaminonitriles and their uses as intermediates to access spiro-heterocycles. Beyond the great structural diversity, stereochemical aspects will also be identified.

Keywords Aminocyanides · Carbohydrates · Cyanohydrins · Spiro-heterocycles

1 Synthesis of Cyanohydrins

The synthesis of sugar-derived cyanohydrins is widely reported in the literature. It is relatively simple and carried out by the addition of cyanide ions to a carbonyl derivative. Thus, the cyanohydrin can be formed and isolated or used as intermediate using the reactions of Kiliani-Fischer, Strecker, or Bucherer-Bergs from aldoses or uloses. However, depending on the conditions used, the stereochemistry of the quaternized center may be oriented.

Bourgeois et al. [1–3] studied the addition of cyanide ions on ulose **1** to access aminonitrile **2** (Scheme 1). Using Strecker conditions, only decomposition of the 1,2:5,6-di-*O*-isopropylidene-α-D-*ribo*-hexofuranos-3-ulose (**1**) was observed. The cyanohydrins **3** and **4** were synthetized in order to be used as precursor of **2**. Treated by an alkaline cyanide, the conversion rate of **1**–**3** remained low (10–20%). A control of the pH by adding buffer (HCO_3^-, CO_3^{2-}) allowed isolating **4** with an efficiency of over 90% yield. In addition, the 3-*R* cyanohydrin **3** can be easily converted into the 3-*S* epimer **4** as on standing in solution cyanohydrins reversed to ulose.

Addition of methyl nitroacetate (MNA) [4] contributed to improve the stereoselectivity of the access to cyanhydrins with D-*allo* (**4**) and D-*gluco* (**3**) configuration, respectively, using Kiliani-Fischer conditions especially by controlling the order of addition of the reagents. Thus, the addition of MNA after the reaction of ulose **1** with the cyanogen agent led to the formation of the preferred D-*gluco* configuration (57%), while adding **1** to the mixture of NaCN and MNA later gave the D-*allo* product with a high stereoselectivity (85%).

It has been confirmed more recently that it was possible to control access to the D-*gluco* or D-*allo* epimers of cyanohydrin resulting from the addition of cyanide

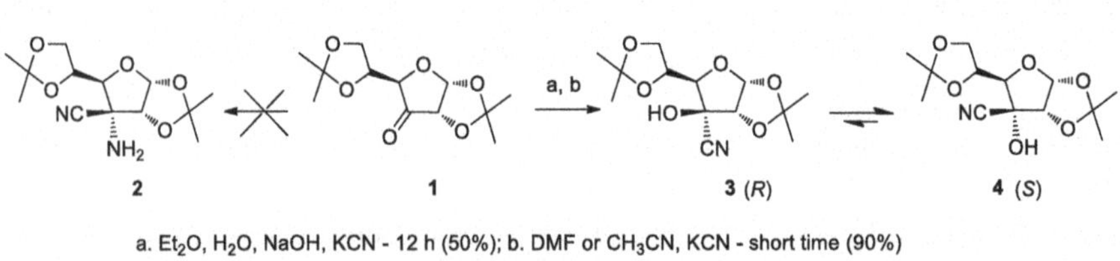

Scheme 1 Cyanohydrins formation on C3 position

ions onto **1**. Thus, while the use of the classically applied mixture (Et_2O-H_2O, $NaHCO_3$) led to the D-*allo* compound [5], the use of NaCN in MeOH led to the D-*gluco* counterpart [6].

This cyanation reaction in C3 position of pentuloses has been widely used to obtain D-*ribo* and D-*xylo* epimers as precursors of [2′,5′-bis-*O*-(*tert*-butyldimethylsilyl)-D-ribofuranosyl]-3-spiro-5″-(4″-amino-1″,2″-oxathiole-2″,2″-dioxide) (TSAO) nucleoside compounds containing the sugar-embedded 3′-spiro-5″-(4″-amino-1″,2″-oxathiole-2″,2″-dioxide) ring system. The authors have studied the synthesis of the TSAO from nucleoside scaffolds or by total synthesis starting from a pentose. Thus, the cyanation of **7**, obtained from 1-(β-D-ribofuranosyl)thymine (**5**) [7, 8], previously protected in the 2′- and 5′-positions by a TBDMS group followed by oxidation of **6** at C3′ to **7**, was carried out in a biphasic system (Et_2O-H_2O) by addition of NaCN in the presence of sodium bicarbonate (Scheme 2). The two cyanohydrin epimers (D-*xylo* **8**; D-*ribo* **9**) were obtained (85%) in a 12:1 mixture and mesylated without being isolated to give compounds **10** and **11** in 53% and 10% yield, respectively.

From pentulose **12**, the cyanation achieved under the same conditions provided the kinetic derivative **13** stereoselectively with a D-*ribo* configuration (94%) (Fig. 1) [9]. Compound **13** was easily epimerized into the D-*xylo*-configured cyanohydrin **14** (89%) by treatment with DBU in acetonitrile. This difference in stereoselectivity

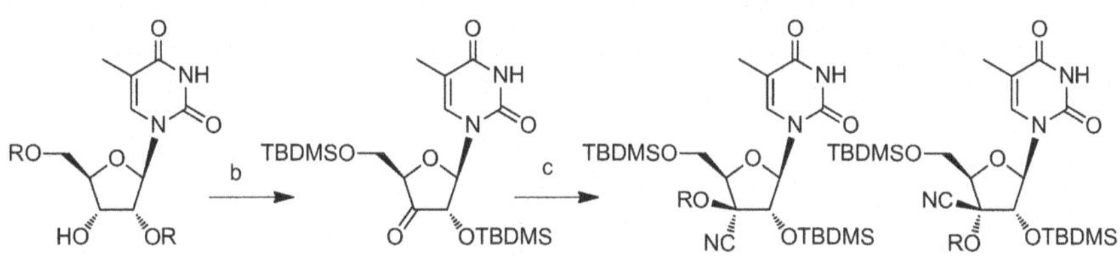

a (**5** R = H; **6** R = TBDMS (68%)

7 (73%)

8 R = H; **10** R = Ms (53%)

9 R = H; **11** R = Ms (10%)

a. TBDMSCl, pyridine; b. CrO_3/Pyridine/Ac_2O; c. NaCN, Et_2O/H_2O, $NaHCO_3$

Scheme 2 Cyanohydrins formation on C3' position

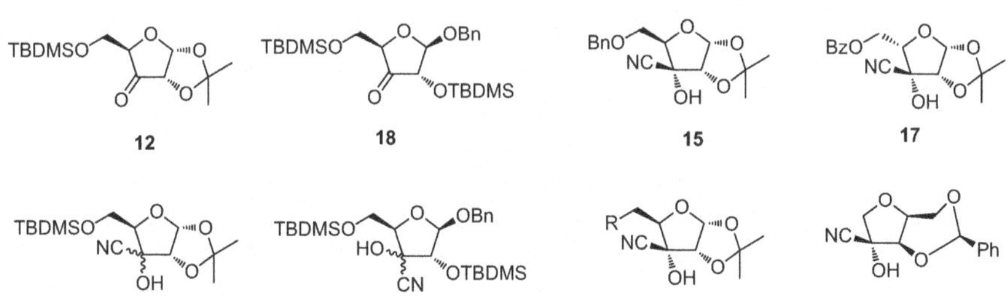

12 **18** **15** **17**

13 *(D-ribo)*
14 *(D-xylo)*

19 *(D-xylo) major*
20 *(D-ribo) minor*

16a R = H
16b R = $CH_2CH(CH_3)_2$
16c R = cyclohexyl

21

Fig. 1 Cyanohydrins derivatives obtained from pentulose

observed between the reactivity of a nucleoside and that of a pentulose highlighted the importance of steric constraints in the approach of the cyanide ion, including thymine (or another bulky group at C1) [10] or the isopropylidene group in the 1,2-position. Thus, starting from pentulose **18**, the D-*xylo*-configured cyanohydrin **19** was obtained as the main compound. The same scaffold with various groups on C5′ led stereoselectively to the D-*ribo*-cyanohydrins **15–16a–c**. On performing the reaction on an L-*lyxo* pentulose with a benzoyl group on C5′ **17** was obtained [11, 12]. The cyanohydrin **21** resulted from addition of the cyano group from the upper side onto C2 of the corresponding ulose in 66% yield [13].

The stereoselectivity observed during formation of cyanohydrin **17**, resulting from the addition of cyano group from opposite side of the benzoyloxymethyl group, was confirmed by Zhang during the cyanation of 5-*O*-benzoyl-1,2-*O*-isopropylidene-α-D-*erythro*-pentos-3-ulose (**22**) to synthesize **23** in 90% yield (Scheme 3) [14].

Introduction of a cyano group was also applied to hexopyranosid-3- and 2-uloses **24**, **26**, and **28** (Scheme 4) [9]. Even if the corresponding cyanohydrins were not isolated, the stereochemistry of the resulting cyanomesylates obtained in the next step illustrated that an *O*-benzoate in position 2 (compound **24**) led to the axial CN (**25**) under thermodynamic control, while a bulky group such as TBDMS in compound **26** led to the equatorial CN (**27**) under kinetic control. However, the authors noted that by using the conditions to get CN in the equatorial position from 3-ulose **26** to give **27**, the nitrile group was obtained at the axial position from 2-ulose **28** to yield compound **29**.

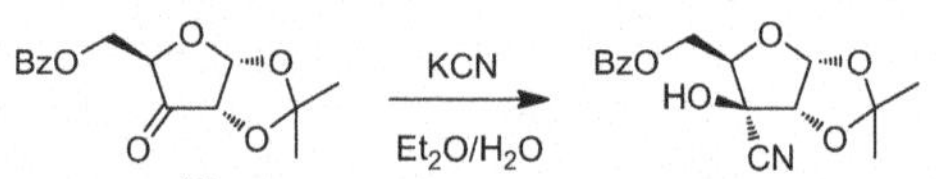

Scheme 3 Cyanohydrin formation on benzoylated derivative

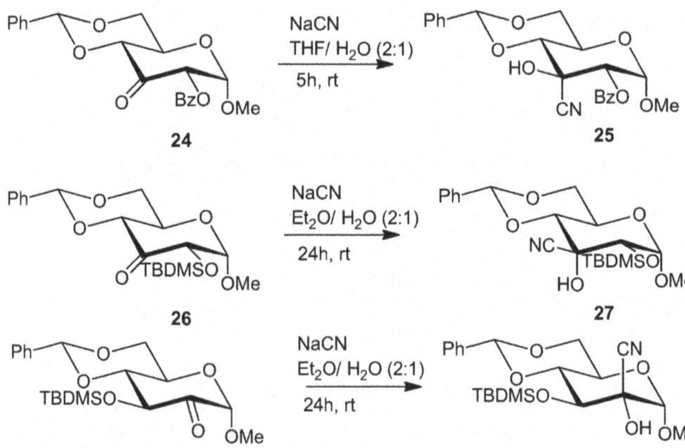

Scheme 4 Cyanohydrins formation on hexopyranosid-3 and 2-uloses

The cyanation was also developed on disaccharidic skeletons similar to trehalose in order to obtain compounds that may have interesting biological activities. Starting from the corresponding uloses, the reactions yielded all the epimeric forms for cyanohydrins without observing stereoselectivity (compounds **30–34**) (Fig. 2) [15].

More recently, a very interesting study was carried out by Koos et al. [16, 17] on the formation of cyanohydrin on a 6-deoxy-2,3-*O*-isopropylidene-α-L-*lyxo*-hexofuranosid-4-ulose derivative (**35** or **36**) through Bucherer-Bergs, Strecker, and Kiliani reactions, respectively (Scheme 5). They reported the relationship between reaction conditions and the final distribution of cyanohydrins resulting from the addition of the cyanide on the uloses **35** and **36**. Thus, the conditions of Bucherer-Bergs (KCN, $(NH_4)_2CO_3$, EtOH-H_2O 1:1) and the conditions of Strecker (KCN, NH_4Cl, NH_3 gas, MeOH anh.) led in a preferred way to the formation of the cyanohydrins **37** and **38** in a 1:2 ratio. In this case, the addition of cyanide ions on the ulose was faster than the base-catalyzed isomerization of **35–36**. This same responsiveness was observed from **36**, allowing the formation of the

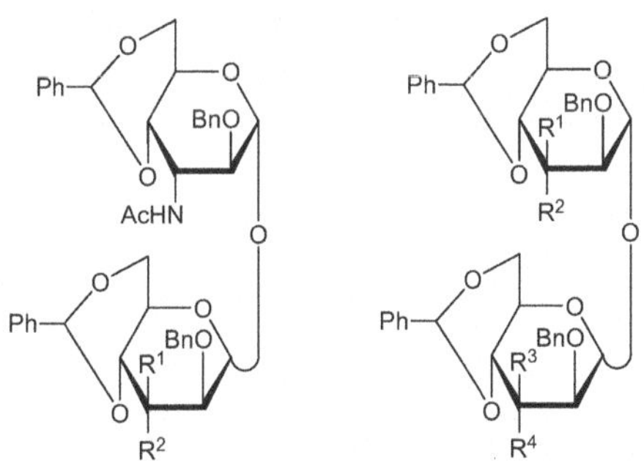

30 R^1 = CN, R^2 = OH
31 R^1 = OH, R^2 = CN

32 R^1 = R^3 = CN, R^2 = R^4 = OH
33 R^1 = R^4 = CN, R^2 = R^3 = OH
34 R^1 = R^3 = OH, R^2 = R^4 = CN

Fig. 2 Cyanohydrins obtained from disaccharides

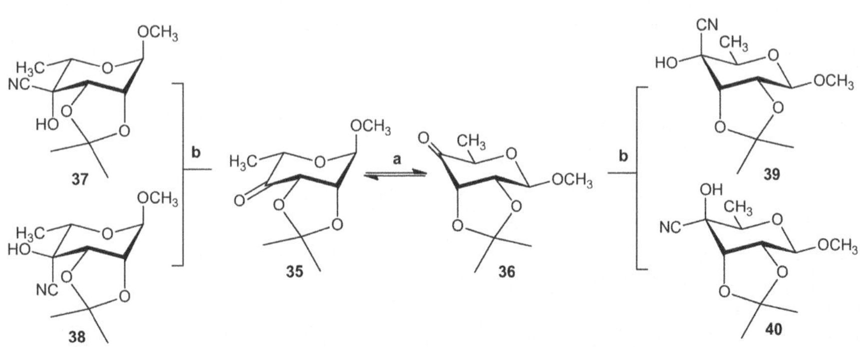

a. base catalyzed epimerisation, b. CN- addition on carbonyl through the Bucherer-Bergs, Strecker and Kiliani reactions

Scheme 5 Stereoselectivity obtained depending of the reaction conditions cyanohydrins **39** and **40** in a 2:1 ratio. The use of Kiliani conditions (KCN, $NaHCO_3$, H_2O) at low temperature could increase the observed stereoselectivity in getting from the uloses **35** and **36** the cyanohydrins **37** and **39**, respectively, as major compounds in a 6:1 ratio compared to their respective epimers. The preferential formation of **38** (α-L-*manno*) over **37** (α-L-*talo*) was surprising because the CN group must access the C4 carbonyl group from the more hindered concave face of the pyranose ring.

A wide range of cyanohydrins (**41–53**) has been synthesized by our group to study the reactivity of alkylidene carbenes on hexopyranose derivatives (Fig. 3) [18, 19]. On all the structures studied, the reaction of Kiliani was used without isolating the cyanohydrins before in situ mesylation. These compounds were intended for use as precursors to the corresponding exomethylidene carbenes; therefore, identification of the stereoisomers and proportion of each of them were not carried out.

The use of a biphasic medium associated with phase-transfer catalyst was reported by Yu et al. [20] in order to prepare oxadiazolyl derivatives of β-D-psicopyranose. The addition of NaCN onto 1,2:4,5-di-*O*-isopropylidene-β-D-*erythro*-2-hexulopyranose-3-ulose in the presence of TBAB in a mixture of CH_2Cl_2:H_2O 1:1 led thus to cyanohydrin **54** (Fig. 3) with a total stereoselectivity and high yield (95–97%).

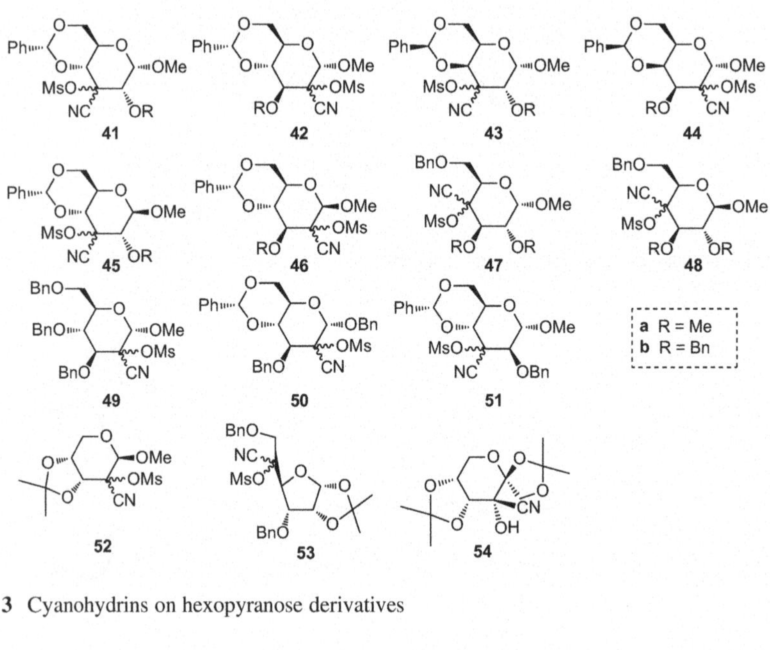

Fig. 3 Cyanohydrins on hexopyranose derivatives

2 Synthesis of Glyco-α-Aminonitriles

The application of the Strecker reaction to carbohydrate substrates has been known for more than hundred years. It was used to prepare many aminosaccharides, by homologation reaction, taking advantage of the aldehyde group. However, very few investigations were conducted on other sites in order to get the carbohydrate derivatives with α-aminonitrile functions. Many attempts of application of Strecker reactions to uloses resulted in obtaining only cyanhydrins as in the work of Bourgeois, mentioned previously. In the same way, Czernecki et al. [21, 22] applied the Strecker conditions (KCN, NH_4Cl, NH_3 gas, MeOH anh.) to the 1,2:3,4-di-*O*-isopropylidene-α-D-*galacto*-hexodialdo-1,5-pyranose (**55**) to obtain only the mixture of the two cyanohydrin epimers. Compound **55** was then subjected to Knoevenagel-Bucherer conditions ($NaHSO_3$, NaCN, rt) to lead to a mixture of α-aminonitriles **56** and **57**. Regardless of the nature of the amine, the α-aminonitriles were obtained with **56** of *R* configuration as the major diastereomer (Scheme 6).

Steiner et al. [23] reported the result of Strecker reaction on a methyl 6-deoxy-2,3-*O*-isopropylidene-α-L-*lyxo*-hexopyranosid-4-ulose (**35**). In this study, the authors pointed out the importance of the temperature on the α-aminocyanation and α-hydroxycyanation competition. Thus, the treatment of **35** by gaseous ammonia in methanol in the presence of NaCN and NH_4Cl led after 4 days at 40°C to a mixture of α-aminonitriles (**58** and **59** in 54% and 12% yield, respectively) (Route A) and hydroxynitrile (**60**, 3%), whereas at 20°C, using the same reagents (Route B), the final mixture was mostly made up of hydroxynitrile **60** (49%) (Scheme 7).

The presence of the methyl 4-amino-4-cyano-4,6-dideoxy-2,3-*O*-isopropylidene-β-D-allopyranoside (**59**) can be explained by base-catalyzed epimerization of **35** into the D-*ribo* uloside (**36**). Furthermore, the use of ammonium carbonate exclusively led to the formation of the hydroxynitrile **60** (73%) (Route C). The latter, placed in conditions similar to Route A, gave, after 4 days, the α-aminonitrile **58** as the major compound (Route D). The authors did not specify if in those circumstances, the reactive intermediate was hydroxynitrile (Tiemann reaction) or imine (Strecker reaction).

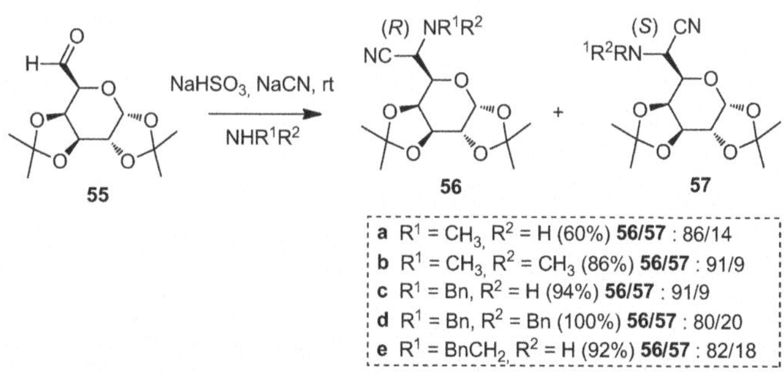

Scheme 6 Aminonitriles obtained from Knoevenagel-Bucherer conditions

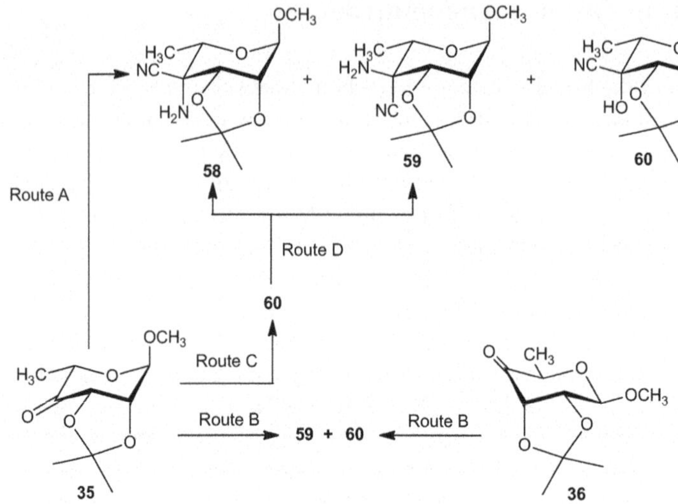

Route A : NH_3 (g), NaCN, NH_4Cl, MeOH, 40°C, 4 days (**58** : 54% ; **59** : 12%; **60** : 3%).
Route B : same reagents as for route A, 25°C, 20h (from **35** : **60** : 49%, from **36** : **58** : 54% ; **59** : 2%;)
Route C : $(NH_4)_2CO_3$ (2 éq ;), NaCN, MeOH, 25°C, 24h (**60** : 73%)
Route D : NH_3 (g), NH_4Cl, MeOH, 40°C, 4 days (**58** : 73% ; **59** : 4%)

Scheme 7 Controlled formation of cyanohydrins or aminonitriles

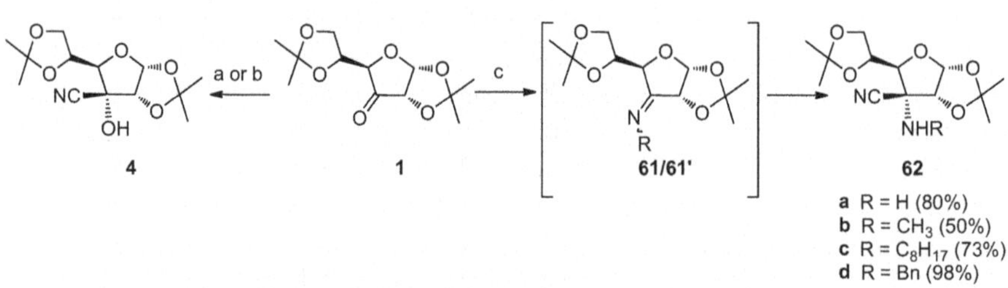

a. Octylamine or benzylamine, KCN, NH_4Cl, MeOH ; b. Benzylamine, molecular sieves (4 Å), iPrOH ;
c. RNH_2, $Ti(OiPr)_4$ then TMSCN

Scheme 8 Aminonitriles formation using $Ti(OiPr)_4$

Our group widely studied the conditions for efficient synthesis of quaternary glyco-α-aminonitriles [24]. Treatment of the ulose derivative **1** using classical Strecker synthesis (KCN, NH_4Cl) and either octylamine or benzylamine gave exclusively the cyanohydrin **4** (Scheme 8).

In order to increase the electrophilic character of the ulose carbonyl, the aminocyanation was performed in the presence of $Ti(OiPr)_4$ as a Lewis acid catalyst. Starting with NH_3, methylamine, octylamine, and benzylamine, respectively, the (*Z/E*) imine intermediates **61/61′** were not isolated, and the cyanating agent TMSCN was added after 12 h. Under these conditions, the α-aminonitrile derivatives **62a–d** were obtained in 50–98% yield.

Similarly, ulose derivatives **63–65** and **66** stereoselectively led to the α-aminonitriles **67–69** and **70**, respectively, in 60–98% yields (Fig. 4) [24].

Probably $Ti(OiPr)_4$ acts as both a dehydrating agent and a Lewis acid to give the imine intermediate and induce stereoselectivity by chelating the N-atom of the imine group and the O-atom at C-2; such a complex would hinder the α-face of the sugar ring and give preference for the addition of CN^- at the β-face (Fig. 5).

The Strecker reaction was also studied using amino acids instead of alkylamines [25]. A series of L-amino acid esters was added onto hexulose **1** and pentuloses **63** and **64** [1.2 eq. $Ti(OiPr)_4$, $R_2OOC\text{-}CHR_1\text{-}NH_2.HCl$, TEA, MeOH (24 h); (ii) TMSCN (1 night)]. The glyco-α-aminonitriles **71–86** were obtained in 15–78% yield (Fig. 6).

The same reaction applied to D-amino acids led to a significant increase in the rate of aminocyanation [25]. This difference in reactivity could be explained by steric congestion linked to the R^1 group, involved in the transition state during the addition of cyanide ions (Fig. 7). The geometry would be conditioned by chelation involving the titanium atom, the carbonyl group as well as the imine nitrogen atom. This phenomenon would increase the steric congestion of the β-face of the carbohydrate.

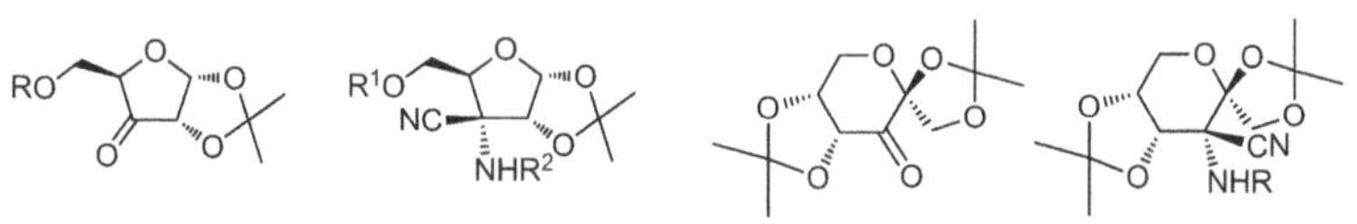

63 R = Tr
64 R = Bn
65 R = Bz

67a R^1 = Tr, R^2 = H (98%)
67b R^1 = Tr, R^2 = CH_3 (60%)
67c R^1 = Tr, R^2 = C_8H_{17} (83%)
67d R^1 = Tr, R^2 = Bn (86%)
68 R^1 = Bn, R^2 = H (70%)
69 R^1 = Bz, R^2 = H (82%)

66

70a R = H (98%)
70b R = C_8H_{17} (80%)

Fig. 4 Structural variety of aminonitriles obtained

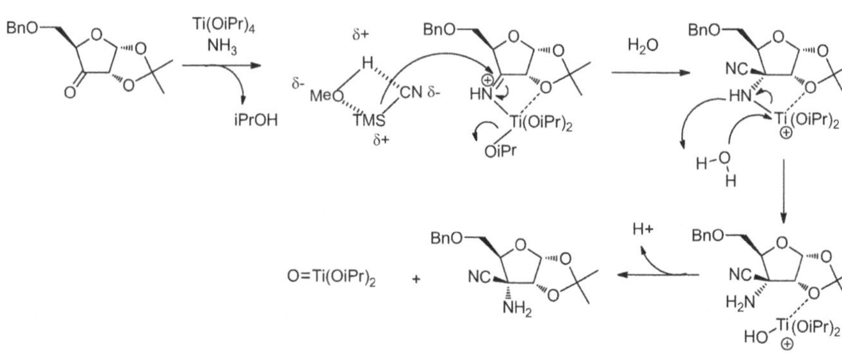

Fig. 5 Proposed role of $Ti(OiPr)_4$ to explain stereoselectivity of the reaction

Fig. 6 Aminonitriles obtained from amino acids

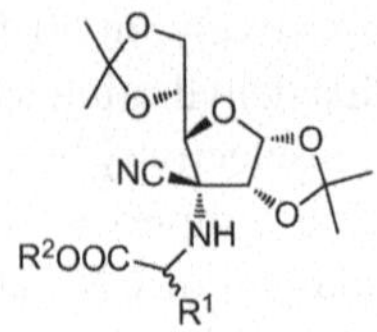

71-79

L series

71 R^1 = H, R^2 = CH_3 (56%)
72 R^1 = CH_3, R^2 = CH_3 (18%)
73 R^1 = $CH_2CH(CH_3)_2$, R^2 = CH_3 (20%)
74 R^1 = $CH_2CH_2SCH_3$, R^2 = CH_3 (16%)
75 R^1 = $CH_2CH_2COOCH_3$, R^2 = CH_3 (15%)
76 R^1 = CH_2Ph, R^2 = CH_3 (30%)
77 R^1 = H, R^2 = tBu (35%)

D series

78 R^1= CH_3, R^2 = CH_3 (30%)
79 R^1= CH_2Ph, R^2 = CH_3 (40%)

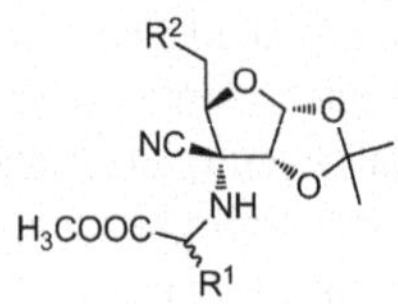

80-86

L series

80 R^1 = H, R^2 = Bn (63%)
81 R^1 = $CH_2CH(CH_3)_2$, R^2 = Bn (20%)
82 R^1 = H, R^2 = Tr (78%)
83 R^1 = CH_3, R^2 = Tr (31%)
84 R^1 = CH_2Ph, R^2 = Tr (57%)

D series

85 R^1 = CH_3, R^2 = Tr (58%)
86 R^1 = CH_2Ph, R^2 = Tr (62%)

Fig. 7 Steric congestion during cyanide ions addition

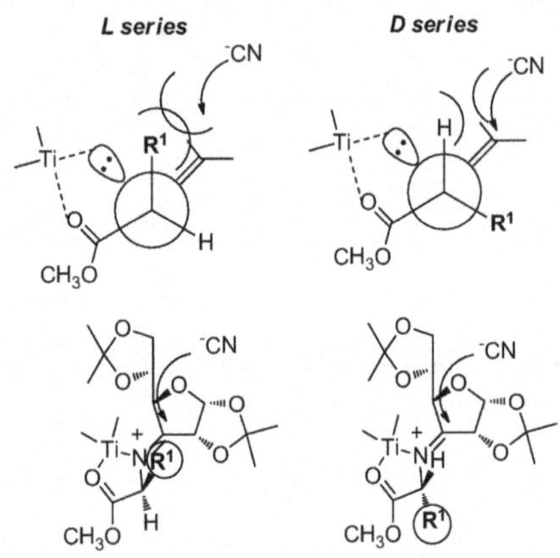

3 Spiro-heterocycles from Cyanohydrins

Cyanohydrins and their *O*-substituted derivatives are useful synthetic intermediates, which have been used to access a variety of organic functions. Among the cycles easily accessible, we can note cycles fused with the saccharidic cycle or even integrating an atom of this scaffold and forming a spiro-cycle [26].

3.1 Oxathiole

The introduction of a sulfonyl group on the cyanohydrins resulting in the corresponding alkanesulfonates may allow cyclization reactions through an α-sulfonyl carbanion. It has been shown that a large range of bases, from n-BuLi to DBU, were able to abstract protons in α positions to sulfur atoms in alkanesulfonates to give reactive carbanion species that reacted with various electrophiles such as nitriles.

Our group has named and classified these transformations as CSIC reactions, taken from the initials of the keywords that describe and define the process for the intermolecular [carbanion-mediated sulfonate (or sulfonamide) intermolecular coupling] and intramolecular [carbanion-mediated sulfonate intramolecular cyclization] conversions [27].

De las Heras and co-workers [28] reported the first example of a CSIC reaction on a sugar cyanomesylate. Mesylation of **13** and **14** with mesyl chloride in pyridine yielded α-mesyloxynitriles **87** (80%) and **88** (71%), respectively, as syrups (Scheme 9). Treatment of **87** and **88** with DBU in acetonitrile afforded 3-spiro-D-*ribo*-oxathiole **89** and the 3-spiro-D-*xylo*-oxathiole isomer **90**.

The work initiated by De las Heras has been widely developed by Camarasa and her team to prepare more than 700 spiro-nucleoside derivatives which represent a particular type of specific HIV-1 reverse transcriptase (RT) inhibitor known as TSAO derivatives ([2′,5′-bis-*O*-(*tert*-butyldimethylsilyl)-D-ribofuranosyl]-3-spiro-5″-(4″-amino-1″,2″-oxathiole-2″,2″-dioxide)) (Fig. 8).

Two synthetic routes have been developed by Camarasa to access the TSAO. The first was to operate from a nucleoside and build on it the oxathiole heterocycle [29, 30].

The key intermediate was obtained by a base-catalyzed addition of cyanide ions onto the ketonucleoside **I** in a biphasic medium (Et_2O/H_2O, $NaHCO_3$). The hydroxynitrile was obtained as a mixture of two epimers (**II**, Scheme 10).

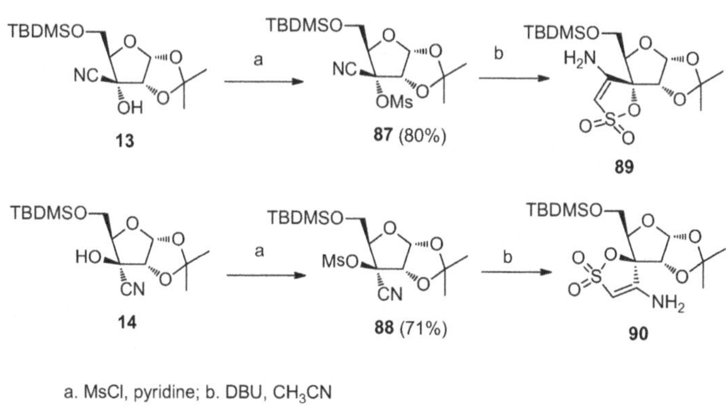

Scheme 9 Synthesis of spiro-oxathiole on C3 position

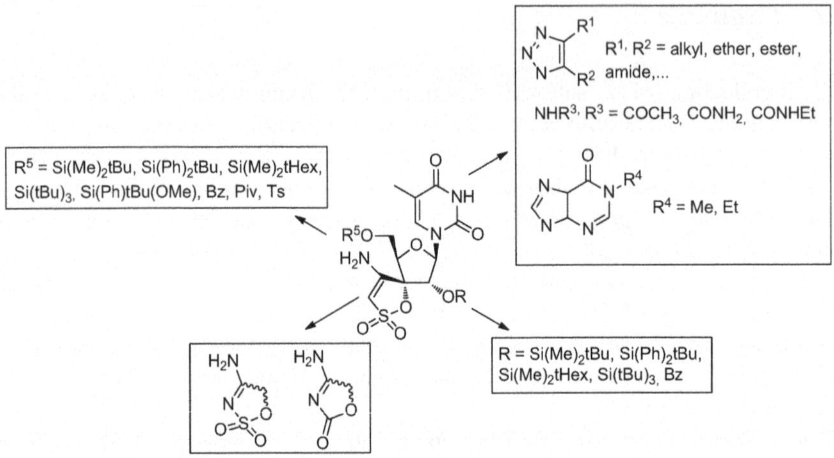

Fig. 8 Structural variety of TSAO derivatives

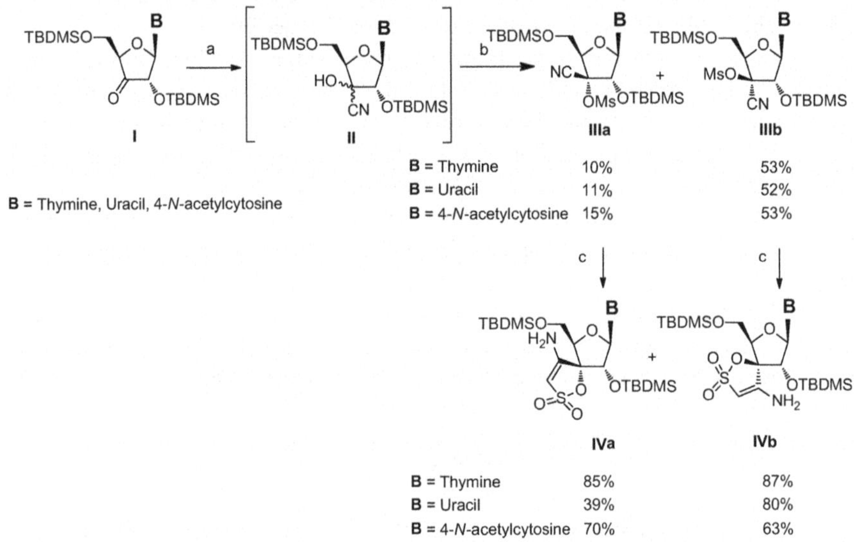

a. NaCN, Et_2O/H_2O, $NaHCO_3$; b. MsCl, pyridine; c. Cs_2CO_3 or DBU

Scheme 10 TSAO synthesis starting from a nucleoside

These were not isolated due to instability but treated by mesyl chloride in pyridine to give the D-*ribo*- (**IIIa** minor) and D-*xylo*- (**IIIb** major) configured 3′-*C*-cyano-3′-*O*-mesyl derivatives with an overall yield of approximatively 60% [29].

The cyclization was conducted according to a carbanionic heterocyclization (CSIC reaction), the carbanion being generated at the α-carbon of the mesylate group. Treatment of the cyanomesylate derivatives **IIIa** and **IIIb** with a base

(Cs_2CO_3 or DBU) gave the spiro-heterocyclic derivatives **IVa** and **IVb** with yields ranging between 69% and 87% for the CSIC reaction. This strategy, although involving a reduced number of steps, has the major disadvantage of leading to a mixture of epimers. In addition, the use of a natural nucleoside limits the number of TSAO that can be prepared in this way.

For these reasons, a second method was developed by Camarasa consisting in functionalization on the C-3 of a non-nucleosidic substrate followed by introduction of the nucleobase to the anomeric carbon using a classical Vorbrüggen glycosylation (Scheme 11) [30].

This strategy involved eight steps from the protected and oxidized carbohydrate derivative **22**. The D-*ribo*-cyanohydrin **91** obtained with a total control of stereoselectivity was mesylated to **92** which was deprotected with aqueous TFA to give **93**, and subsequent acetylation led to the 1,2-di-*O*-acetylated compound **94** (95%). Silylated thymine (thymine, HMDS, $(NH_4)_2SO_4$) was applied in the presence of **94** and TMSOTf to give the 3′-*C*-cyano-3′-*O*-mesyl nucleoside **95** (77%). Treatment of the obtained nucleoside with Cs_2CO_3 yielded the spiro-oxathiole derivative **96** (65%). This heterocyclic derivative was debenzoylated and deacetylated with a saturated methanolic ammonia solution to furnish 3′-spiro-nucleoside **97** in 66% yield. TSAO derivative **98** was finally obtained by silylation (74%). The TSAO-m^{3}T (abbreviation for TSAO alkylated in *N*-3 position with a methyl) derivative (**99**) was obtained by alkylation, (MeI, K_2CO_3) with a yield of 55%.

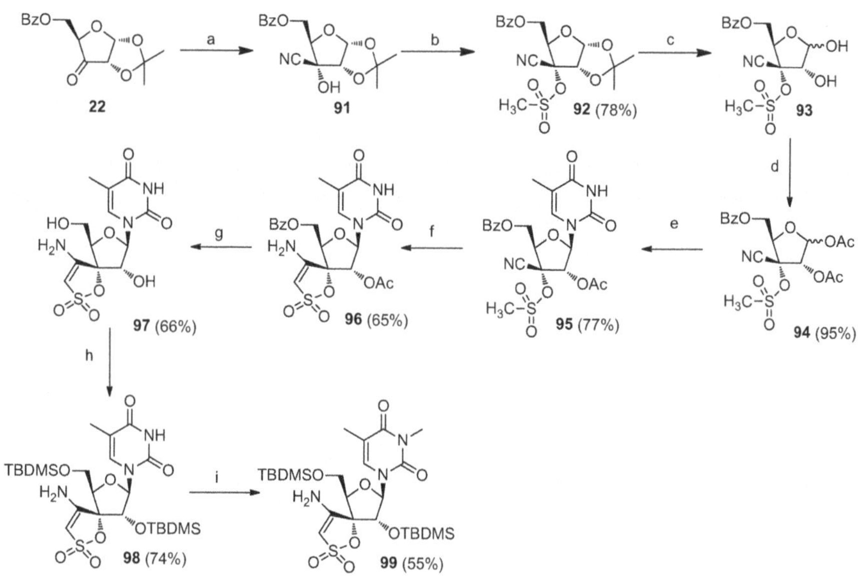

a. NaCN, $NaHCO_3$; b. MsCl, pyr., 10°C, 48h ; c. TFA-H_2O ; d. Ac_2O, pyr. ; e. silylated thymine, TMSOTf ; f. Cs_2CO_3, CH_3CN ; g. NH_3, MeOH ; h. TBDMSCl-DMAP ; i. K_2CO_3, MeI

Scheme 11 TSAO synthesis from a carbohydrate derivative

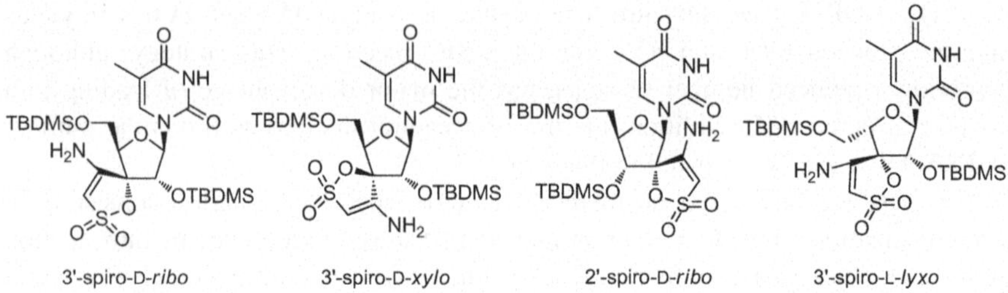

Fig. 9 TSAO analogs

The biological study of the TSAO structures with the spiro ring on C-3′ (D-*ribo* and D-*xylo* configuration) or C-2′ (D-*ribo* configuration) (Fig. 9) was carried out [11, 29, 31]. An activity as anti-HIV-1 inhibitor was observed only for the derivative 3′-spiro-D-*ribo* (EC_{50} = 0.06 μM). Other derivatives, 3′-spiro-D-*xylo*, 2′-spiro-D-*ribo*, and 3′-spiro-L-*lyxo* showed no activity [12].

Among all the structural variations described by Camarasa, a large number corresponded to pentofuranose derivatives. Other studies also reported the introduction of the oxathiole cycle on the C3 and C2 positions of hexopyranoses. Treatment of the benzoate **103** obtained from **25** after mesylation gave a mixture of spiro compounds **104** (12%) and its *N*-benzoyl derivative **105** in 20% yield (the formation of **105** can be explained by the migration of the benzoyl group in position 2) [9]. Treatment of the cyanomesylates **106** and **108**, easily obtained by in situ mesylation of the corresponding cyanohydrins **27** and **29**, with DBU in acetonitrile, afforded spiro derivatives **107** and **109** in 77% and 73% yields, respectively (Scheme 12).

Kiritsis et al. reported the synthesis of pyranosyl nucleosides starting from 2′-*C*-cyano pyranonucleosides of uracil and fluorouracil protected as acetal derivatives. Mesylation of the 2′-*C*-cyanohydrins **110a,b** and **111a,b** led to **112a,b** and **113a,b**, respectively (Scheme 13). The desired 2′-spiro pyrimidine pyranonucleosides were obtained through a subsequent treatment with DBU in acetonitrile in fair to good yields (**114a** 60%, **114b** 74%, **115a** 70%) [32].

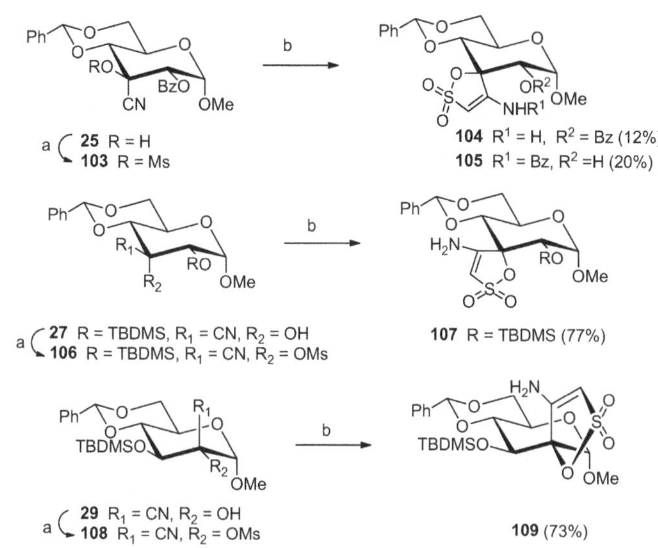

a. MsCl, pyridine; b. DBU, CH_3CN

Scheme 12 Synthesis of spiro-oxathiole on C3 and C2 positions of hexopyranoses

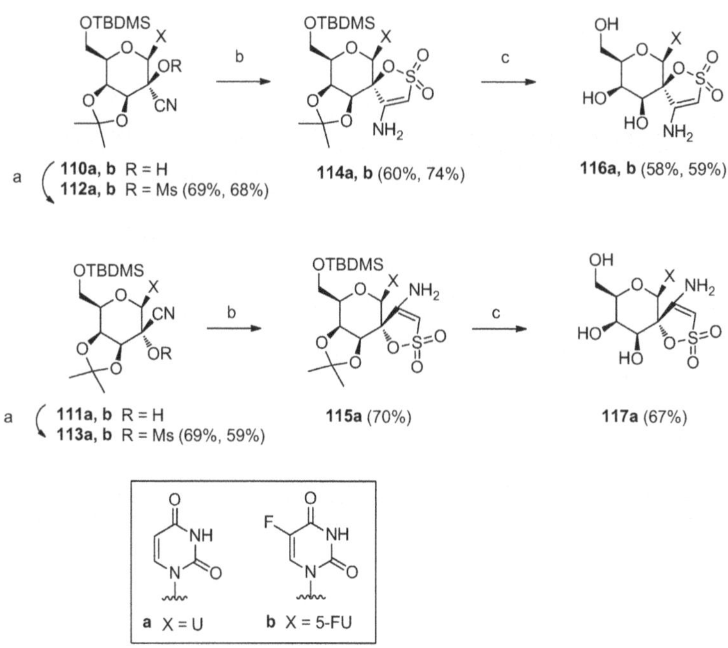

a. MsCl, pyridine; b. DBU, CH_3CN; c. HCl/MeOH, CH_2Cl_2

Scheme 13 Synthesis of 2' spiro-oxathiole pyrimidine pyranonucleosides

3.2 Oxathiole Fused with Another Cycle

Chamorro et al. [33] highlighted the formation of compounds resulting from an intramolecular cyclization from 5′-*O*-tosylate via a S_N2 reaction involving the enamino group of the spiro ring and the 5′-leaving group. These early results led Cordeiro et al. [34] to study this cyclization starting from the D-ribofuranose derivative **118** (Scheme 14). It was demonstrated that the reactivity of this substrate in the presence of a non-nucleophilic base (DBU) yielded **119** in 76% via an intramolecular cyclization in only 15 min at 80°C. After 1 h reaction, intermolecular addition of **119** gave the dimer **120** in 67% yield.

The authors focused on exploring the reactivity of **119** toward various nucleophiles including amino acids. When methanol, ethanol, and ethanethiol were used, the corresponding *O*- and *S*-substituted tricyclic sugars **121** were obtained in 60–70% yields (Scheme 15). When cyanide was used as the nucleophile, nitrile derivative **122** was isolated in 60% yield. All of these additions were fully regio- and stereoselective [34].

3.3 Oxazolone and Oxathiazole

The influence of the nature of the spiro ring has also been studied by comparing the activity of TSAO-T, 4-amino-2-oxazolone derivatives, and 4-amino-1,2,3-

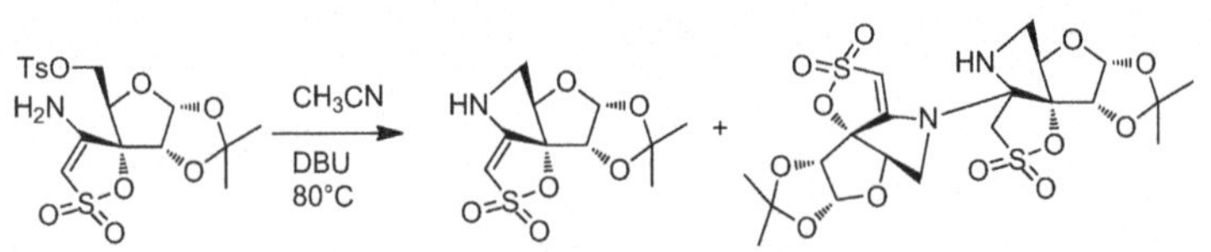

118 **119** **120**

t = 15min (76%) (19%)
t = 1h (31%) (67%)

Scheme 14 Fused cycles obtained from oxathiole derivative

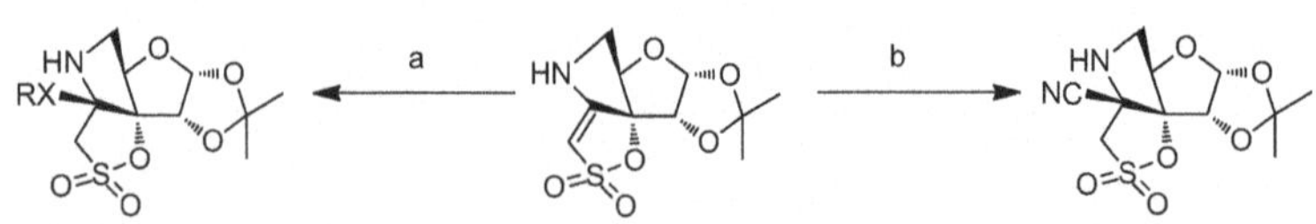

121a RX = OCH_3 (70%)
121b RX = OC_2H_5 (68%)
121c RX = SC_2H_5 (60%)

119

122 (60%)

a. ROH or RSH, 80°C; b. TMSCN, $BF_3.OEt_2$, 80°C

Scheme 15 Reactivity of fused spiro-oxathiole cycles with nucleophiles

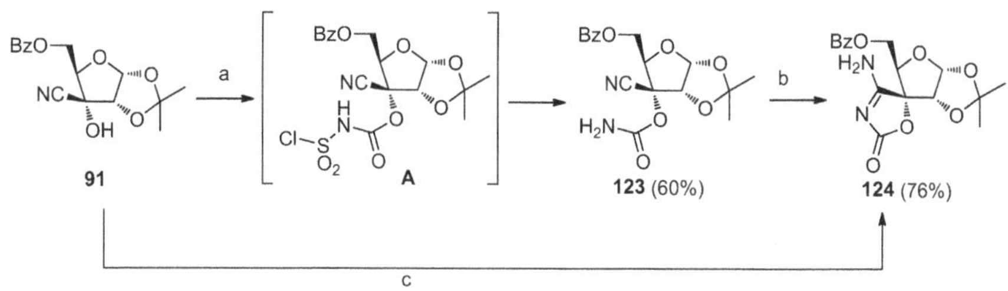

a. CSI, H_2O; b. $NaHCO_3$; c. (i) CSI, H_2O, (ii) $NaHCO_3$

Scheme 16 Spiro-oxazolone synthesis

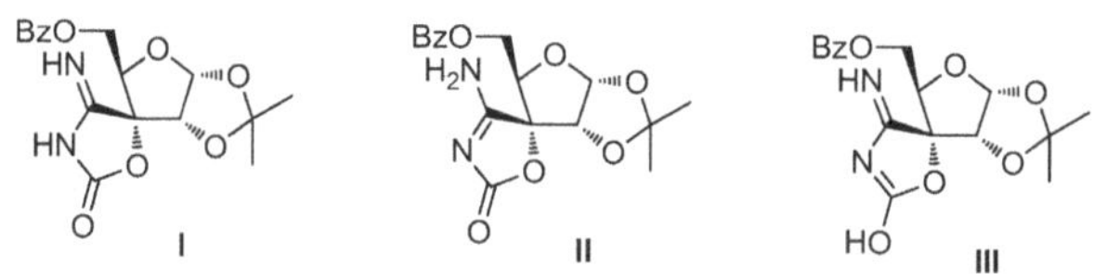

Fig. 10 Tautomeric forms of oxazolone

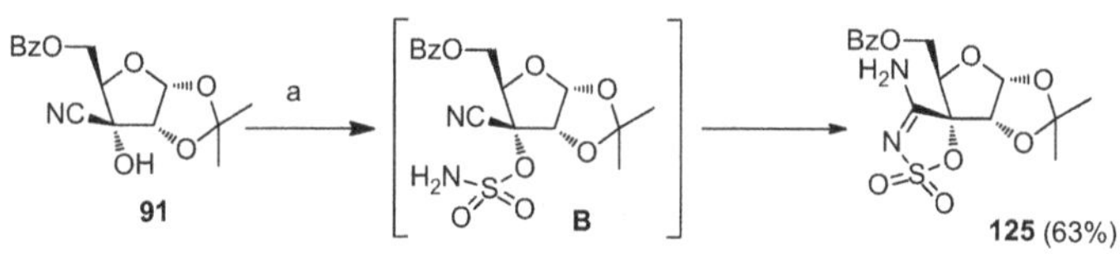

a. sulfamoyl chloride, DMAP, dioxane

Scheme 17 Spiro-oxathiazole synthesis

oxathiazole-2,2-dioxide. Synthesis of these compounds consisted of the reaction between cyanohydrin **91** and chlorosulfonyl isocyanate (CSI) leading, via intermediate **A**, exclusively to the carbamoyl derivative **123** (60%) (Scheme 16) [35, 36]. The latter seems to be the intermediate which, on subsequent treatment with $NaHCO_3$, may lead to the oxazolone derivative. Thus, treatment of **91** with CSI followed by addition of $NaHCO_3$ in situ led to the spiro derivative **124** with 76% yield.

The spiro moiety at the 3-position of compound **124** existed in different tautomeric forms **I–III** shown in Fig. 10. NMR analysis demonstrated that tautomer **II** was the most preponderant one.

On the other hand, the action of sulfamoyl chloride on cyanohydrin **91** using dimethylaminopyridine in dry dioxane yielded, via intermediate **B**, the corresponding 3-spirooxathiazole-*S*,*S*-dioxide **125** in 63% yield (Scheme 17).

Similar to what was observed for compound **124**, the spiro derivative **125** was supposed to exist as tautomers **I–III** (Fig. 11); however, NMR analysis identified tautomer **II** as the only form.

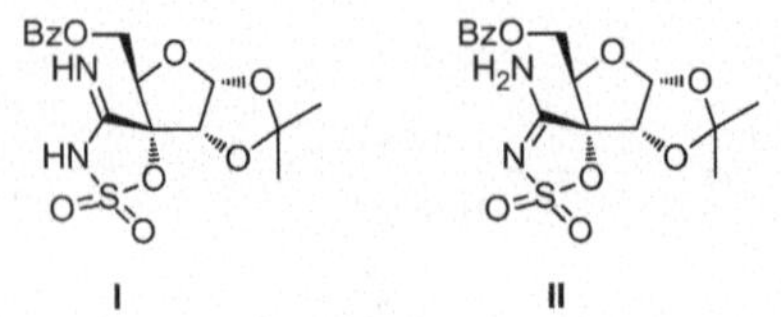

Fig. 11 Tautomeric forms of oxathiazole

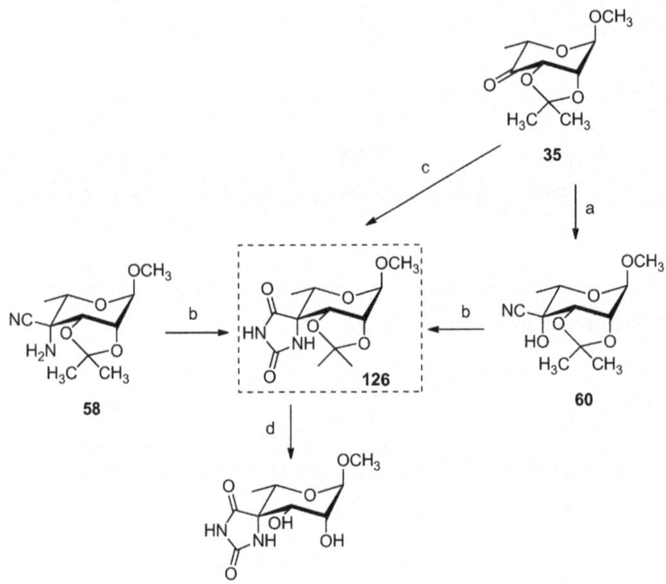

a. KCN, $(NH_4)_2CO_3$, 50%aq. EtOH, 60°C, 2h; b. $(NH_4)_2CO_3$, 50%aq. EtOH, reflux, 12h; c. KCN, $(NH_4)_2CO_3$, aq EtOH; d. AcOH, 90°C, 45min

Scheme 18 Spiro-hydantoin synthesis

3.4 *Hydantoin*

In some cases, cyanohydrin-like intermediates can be used in an efficient reaction pathway for the synthesis of spiranic heterocycles through a subsequent conversion to another intermediate. Kooś et al. demonstrated the effective achievement of the 4′-spiro hydantoin (**126**) by a two-step procedure consisting in reaction of **35** with KCN (80% yield) followed by heating the cyanohydrin **60** with $(NH_4)_2CO_3$ to give **126** in 80% yield (Scheme 18). The authors highlighted that this two-step procedure (conditions a + b) was more advantageous than direct transformation of the ketose **35** with KCN and $(NH_4)_2CO_3$ in aqueous ethanol (conditions c) where **126** was obtained in 35% yield or than starting from the aminonitrile **58** (conditions b) where **126** was obtained only in 25% yield [17].

3.5 Oxaphospholene

Phosphorus-containing molecules are popular targets for the development of new biologically active compounds. We investigated the synthesis of new families of nucleosidic reverse transcriptase (RT) inhibitors, analogs of TSAO derivatives, with an O-P bond in the spiro ring at position 3″. Such compounds, named P-TSAO-T, would also be unusual analogs of nucleoside cyclic phosphates [37, 38].

The synthesis strategy for accessing phosphorus compounds can be considered as a simple transposition of the methodology used for TSAO. In this case, the cyclization reaction would form a carbanion next to phosphorus. But the presence of phosphorus instead of sulfur resulted in a difference in the reactivity.

The phosphonylation step was performed by the reaction of cyanohydrins with methyl methylphosphonochloridate, which was prepared from dimethyl methylphosphonate ($CH_3PO(OCH_3)_2$) and PCl_5. The reaction of cyanohydrins **91** and **15** with methyl methylphosphonochloridate and DMAP in pyridine allowed the exclusive formation of the kinetic products **128** and **129** in 90% and 56% yields, respectively, as mixtures of two diastereomers which could not be separated by chromatography (Scheme 19).

During attempts for ring closure in the presence of *t*-BuOK, DBU, NaH, or HMDS, only the starting material was recovered from reactions of **128** or **129**. In the presence of Cs_2CO_3 or BuLi, a mixture of several derivatives was obtained with the major products characterized as mixtures of D-*xylo*- and D-*ribo*-cyanohydrins resulting from the phosphonate degradation under basic conditions. With these results in hand, we envisaged that introduction of an electron-withdrawing group α to the phosphorus atom should facilitate the cyclization. A series of phosphonochloridate derivatives with electron-withdrawing groups (COOEt, CN, Ph, and COOMe) was prepared and reacted with **91** to give, e.g., **130**. Upon treatment of derivative **130** with LDA or NaH, the desired oxaphospholene **131** was obtained in 56% or 70% yield, respectively (Scheme 20).

Using NaH, it was possible to carry out the synthesis in a one-pot procedure. Starting from cyanohydrins **91** and **15**, oxaphospholenes were obtained with yields ranging from 41 to 70% as mixtures of separable (**131** and **132**) or inseparable diastereomers (**133**) or as a pure compound (**134**) (Fig. 12).

When using the benzylphosphonochloridate **135,** it was not possible to carry out a one-pot reaction, as the cyclization step with NaH did not occur. The cyclized

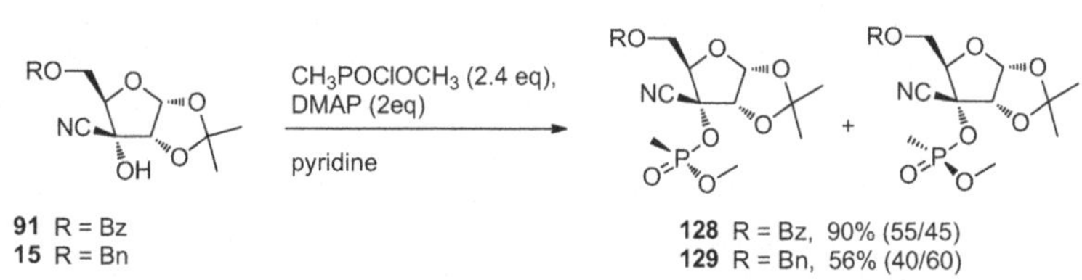

Scheme 19 Phosphonylation of cyanohydrins

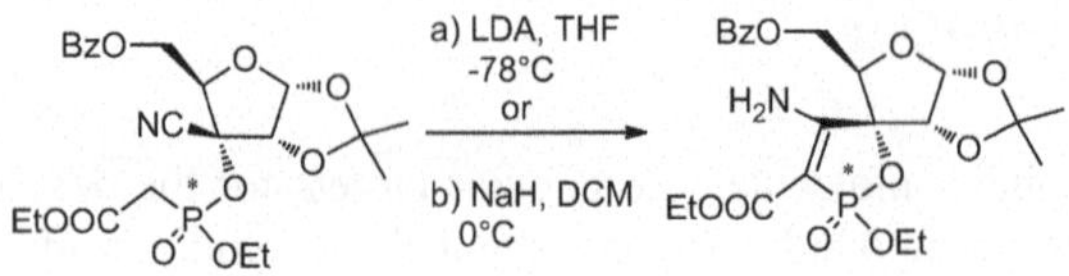

130

131

a) 56% (55/45)
b) 70% (55/45)

Scheme 20 Carbanion intramolecular cyclization

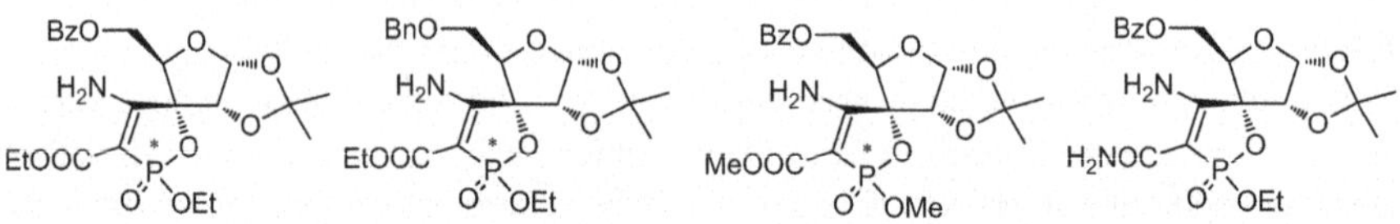

131 (70% - 55/45) **132** (42% - 70/30) **133** (41% - 60/40) **134** (53% - 100/0)

Fig. 12 Spiro-oxaphospholene derivatives

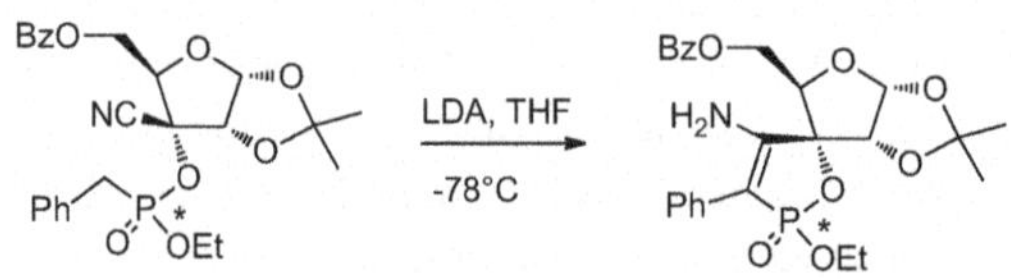

135 **136** (78% - 60/40)

Scheme 21 Spiro-oxaphospholene from benzylphosphonochloridate

compound **136** was obtained only with LDA (78% yield) as a mixture of two diastereomers (Scheme 21).

4 Spiro-heterocycles from α-Aminonitriles

The use of α-aminonitriles to access spiro-heterocycles on carbohydrate scaffolds has been widely described. Syntheses of imidazoline [5], imidazolidin-2-one or 2-thione [39, 40], oxopiperazine [25, 41], cyclic peptide analogs [25], hydantoin, thiohydantoin [42, 43], isothiazole [44], and azaphospholene [37] are found in the literature (Fig. 13).

It is possible to identify two major routes for the synthesis of these compounds (Scheme 22). The first is a reduction of the nitrile group and then functionalization of amines; in the second, a functionalization of the amine function takes place in a first step, and then the expected heterocycles are obtained after cyclization.

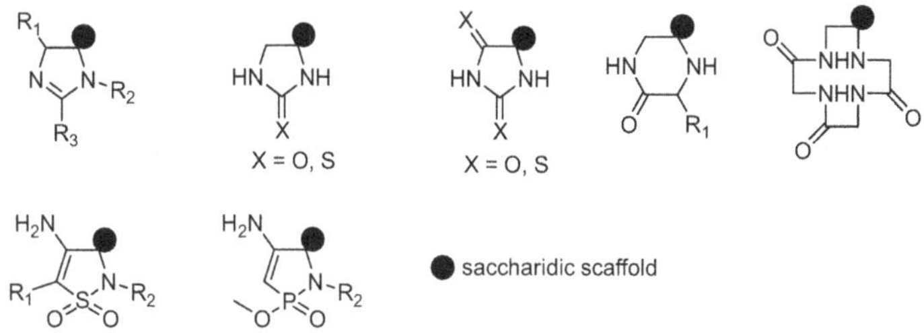

Fig. 13 Spiro-heterocycles described on saccharidic scaffold from aminonitriles

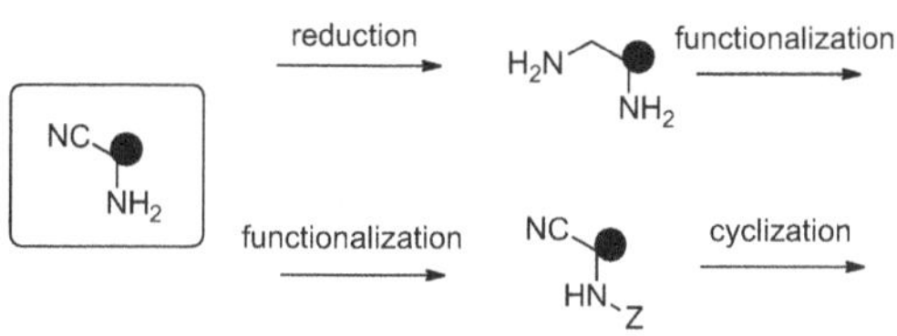

Scheme 22 Synthesis strategies

We will first focus on the strategy of reducing the nitrile and then achieving functionalization of the diamine obtained. Next we will discuss the synthetic routes starting with a functionalization of the amino group followed by a cyclization reaction.

4.1 *Strategy Reduction/Functionalization*

This synthetic strategy allowed to achieve the formation of 5- or 6-membered rings of imidazoline, imidazolidin-2-one or 2-thione, and oxopiperazine type and also an access to cyclic peptides.

4.1.1 Imidazoline

Merino-Montiel described in 2012 the syntheses of spiro imidazoline at C3 position starting from aminonitrile **2** [5]. After reduction of the aminonitrile by $LiAlH_4$, the obtained diamine **137** was condensed with benzamidine or acetamidine hydrochloride, which led to the formation of the protected cyclized compounds **138a,b** and the partially deprotected ones **139a,b**, respectively, in a 4:1 ratio (Scheme 23). By liquid/liquid extraction, it was possible to separate compounds **138a** and **139a**.

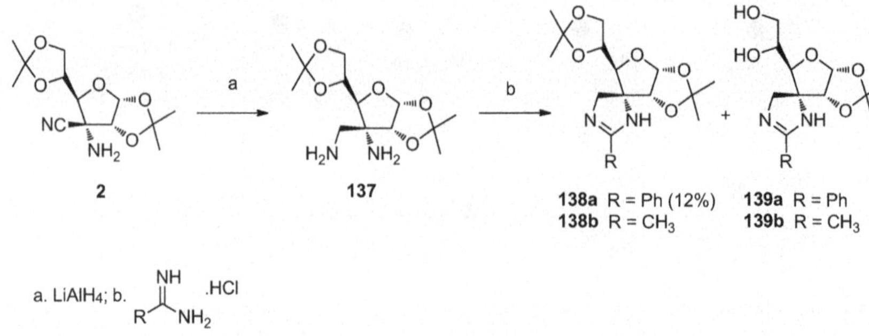

Scheme 23 Spiro-imidazoline synthesis

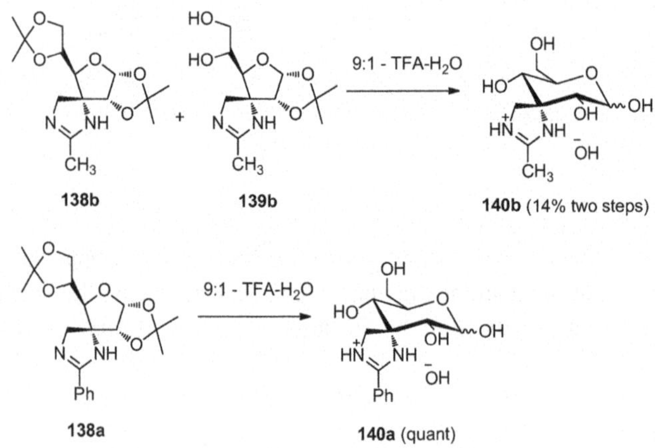

Scheme 24 Access to deprotected spiro-imidazoline derivatives

Imidazoline **138a** could be purified by chromatography and obtained pure with 12% yield.

The mixture of the two derivatives **138b** and **139b** could not be separated and purified, and so it was used directly for the deprotection step consisting of an acid treatment to remove the isopropylidene groups to give compound **140b** in a pyranose form with 14% yield starting from **137** (Scheme 24). Starting from **138a** compound **140a** was obtained with a quantitative yield.

Inhibitory properties of compounds **140a** and **140b** were evaluated against different glycosidases and against glycogen phosphorylase; both of them had poor inhibition properties.

From diamine **137**, the authors also described the coupling reaction with isothiocyanates (Scheme 25) [5].

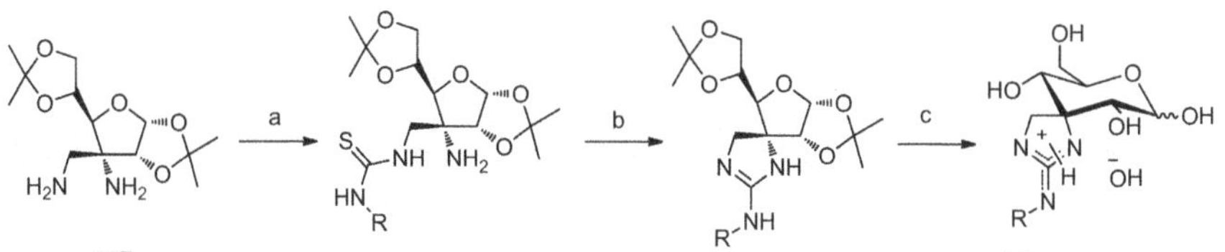

137

141a R = Ph (quant)
141b R = Bn (71%)
141c R = C_6H_{11} (quant)

142a R = Ph (50%)
142b R = Bn (quant)
142c R = C_6H_{11} (67%)

143a R = Ph (quant)
143b R = Bn (96%)
143c R = C_6H_{11} (91%)

a. R-NCS; b. Yellow HgO, anhydrous THF; c. TFA-H_2O 9:1

Scheme 25 Spiro-imidazoline synthesis using isothiocyanates

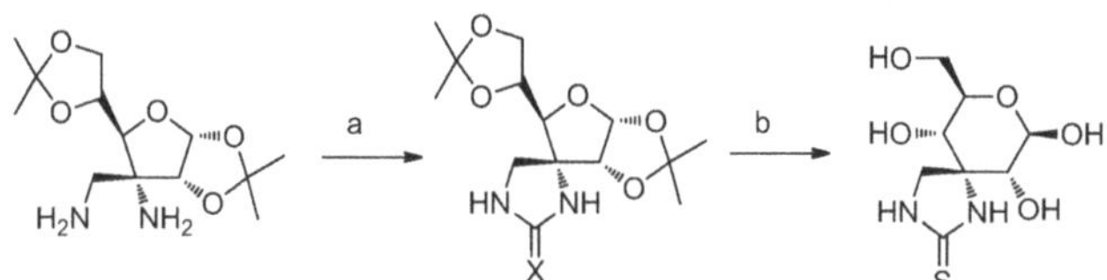

137 (66%)

144 X = O (50%)
145 X = S (75%)

146

a. IM_2CX; b. HCl, MeOH

Scheme 26 Spiro imidazolidine-2-one or 2-thione synthesis

The reaction on the primary amine allowed the formation of thioureas **141** with very good yields. These compounds then underwent a reaction of cyclodesulfurization in the presence of yellow HgO which resulted in the formation of spiro amino-imidazoline **142** with yields ranging from 50% to quantitative. The reaction might have proceeded via a carbodiimide to react with the free NH_2 on C3 as a nucleophile. The deprotected analogs **143** were obtained with very good yields after acid hydrolysis.

4.1.2 Imidazolidine-2-one or 2-Thione

Our group also described the synthesis of imidazolidine-2-one or 2-thione derivatives in 2002 [39]. Thus, condensation of the diamine **137** with carbonyldiimidazole or thiocarbonyldiimidazole directly gave spiro imidazolidine-2-one or 2-thione **144** and **145** with 50% and 75% yield, respectively. Further deprotection of the hydroxyl groups was performed by hydrochloric acid treatment (Scheme 26).

Access to spiro imidazolidine-2-thione **145** was also described in 2009 according to an identical synthetic pathway [40]. Further deprotection of the hydroxyl groups of compound **145** was performed by hydrochloric acid treatment (Scheme 26).

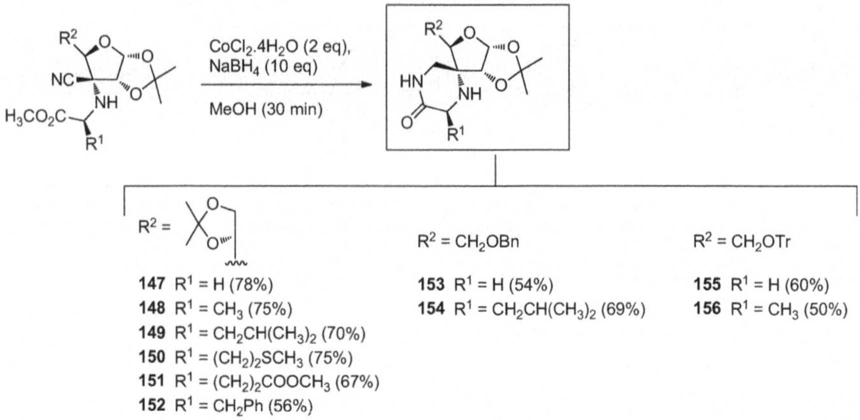

Scheme 27 Spiro-oxopiperazine synthesis

4.1.3 Oxopiperazine

Starting from aminonitrile derivatives obtained by condensation of ester protected amino acids on ulose previously described in part I [25], it was possible to form spirooxopiperazines (Scheme 27) [25, 41]. The nitrile group was reduced in the presence of $NaBH_4$-$CoCl_2$, and the primary amine obtained spontaneously reacted with the ester to form the heterocycle. Compounds **147–156** could thus be obtained with good yields ranging from 50 to 78%.

4.1.4 Cyclic Peptides

Starting from oxopiperazine derivative **147**, the spirocyclic tetrapeptide **161** was synthesized (Scheme 28) [25]. In the presence of NaOH and *N*-protected glycine *p*-nitrophenyl ester, compound **157** was obtained in 56% yield. Peptide coupling between the COOH function and methyl ester protected alanine with DCC gave compound **158** with 67% yield. Compound **159**, obtained quantitatively after deprotection of COOH and NH_2, was then cyclized in the presence of DPPA (diphenyl phosphoryl azide) and TEA to form the spiro tetrapeptide derivative **160** with 56% yield. Acidic treatment in the presence of TFA and H_2O removed the isopropylidenes to isolate **161** with 94% yield.

4.2 Strategy Functionalization/Cyclization

In this strategy, the amine function was functionalized as a carbamoyl ester, an amide, a lactam, a succinimide, a sulfonate, or a phosphonate, and then an

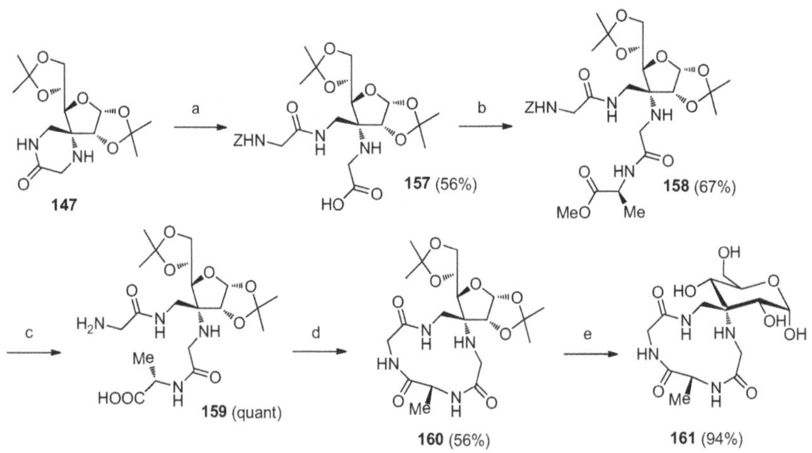

a. Z-Gly-O-PNP (1.2 eq), NaOH (3 eq), H_2O-dioxane; b. L-Ala-OMe (2eq), DCC (2 eq), N-hydroxysuccinimide (2 eq) DMF-MeCN -10°C to RT (24h); c. i) K_2CO_3, MeOH, H_2O, ii) H_2, Pd/C MeOH or EtOAc (12h); d.DPPA (1.2 eq), TEA (2 eq), DMF; e. TFA/H_2O 9:1

Scheme 28 Cyclic peptides synthesis

intramolecular cyclization reaction allowed the formation of hydantoin, thiohydantoin, imidazoline, isothiazole, or azaphospholene ring.

4.2.1 Hydantoin and Thiohydantoin

In 2001, our group described the synthesis of hydantoin and thiohydantoin derivatives [42, 43]. The reaction of aminonitrile **2** with either benzyl or phenylchloroformate allowed the formation of carbamoyl esters **162** and **169** with comparable yields of 45% and 44%, respectively (Scheme 29). Otherwise, the basic treatment to form the carboxamidoisocyanate **163**, which then spontaneously cyclized to give spirohydantoin, gave much better results from compound **162** (80%) than from compound **169** (20%).

By treatment with HCl, the partially deprotected compound **165** was obtained with a 52% yield. The totally deprotected spirohydantoin **166** was obtained with a 97% yield by reaction with a TFA-H_2O mixture.

From spirohydantoin **167**, prepared in an analogous way as described for **164** derivative, **168** was obtained with a quantitative yield (Scheme 29.)

The synthesis of spirohydantoin **164** has also been described by direct condensation of carbon dioxide or ammonium carbonate on aminonitrile with 80% yield.

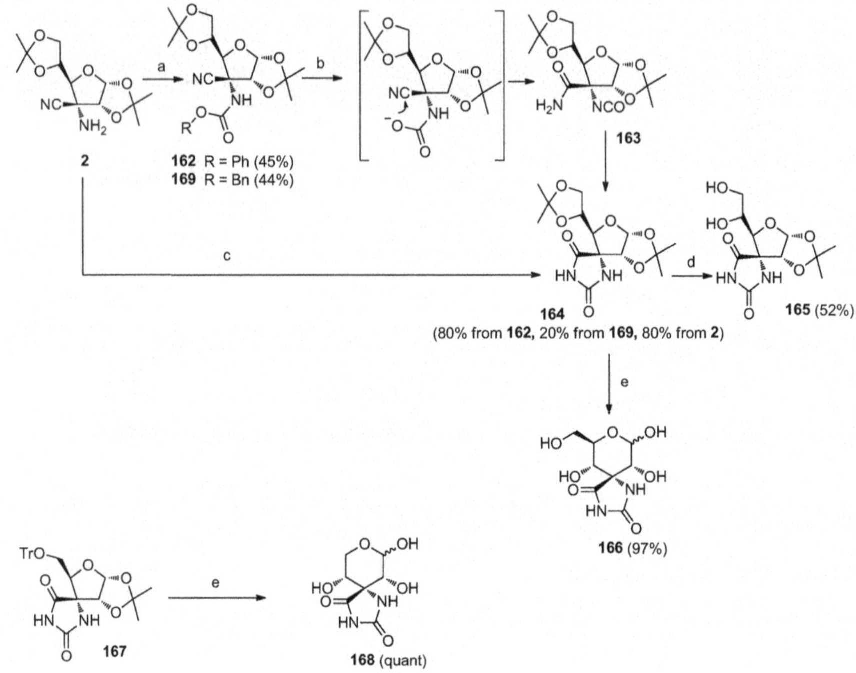

a. RO(CO)Cl, K_2CO_3, acetone/H_2O; b. NaOH, H_2O, dioxane; c. CO_2, 75 atm, MeOH or $(NH_4)_2CO_3$, MeOH-H_2O; d. HCl (1M); e. TFA-H_2O 9:1

Scheme 29 Spiro-hydanthoine synthesis

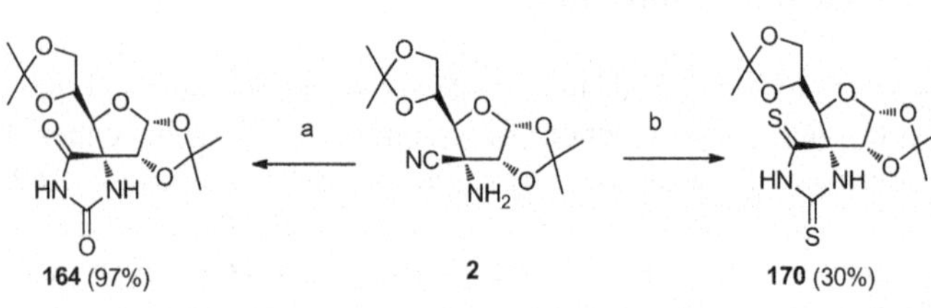

a. $(NH_4)_2CO_3$ (10 eq), MeOH, H_2O; b. CS_2, K_2CO_3

Scheme 30 Spiro-thiohydanthoine synthesis

This spirohydantoin **164** was also synthesized with a better yield of 97% using a very large excess of ammonium carbonate [39]. Starting from aminonitrile **2**, the authors also described the synthesis of spirothiohydantoin **170** by reaction of CS_2 in the presence of potassium carbonate with a low yield of 30% (Scheme 30).

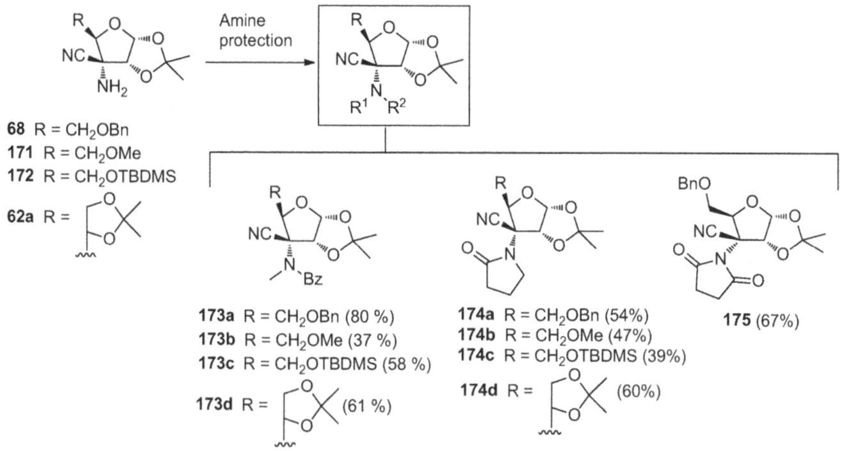

Scheme 31 Functionalization of aminonitriles as amides, lactams, or succinimide derivatives

4.2.2 Imidazoline

Spiroimidazoline derivatives were obtained after cyclopropanation reaction on the nitrile [45]. Prior to the cyclopropanation key step, the primary amine was protected as amide (**173a–d**), lactam (**174a–d**), or succinimide (**175**). A variety of substrates was obtained (Scheme 31).

A two-step procedure was used to obtain amide derivatives (**173a–d**): reaction with benzoyl chloride and then methylation with methyl iodide. For the lactam derivatives (**174a–d**), condensation of amine with 4-bromobutyryl chloride gave amide, and an intramolecular nucleophilic substitution in the presence of NaH allowed the formation of the desired compounds with moderate yields. Succinimide derivative **175** was obtained by reaction with succinic anhydride first and then thionyl chloride.

The cyclopropanation key step was performed using the conditions described by Szymoniak et al. [46]. A titanium cyclopropane was formed first by the reaction of EtMgBr with $Ti(OiPr)_3Me$, and after addition on the nitrile, a cyclic titanium iminate (**A**) was obtained. Treatment with $BF_3{\cdot}OEt_2$ allowed the formation of cyclopropyl derivatives **176** and **177** with poor yields starting from amides and lactams (Scheme 32).

Starting from succinimide **175**, after activation of the nitrile as cyclic titanium iminate (**A**), no cyclopropyl derivative was recovered. The imine derivative **178** was formed first, and then isomerization into enamine **B** took place before reaction of the amine with one carbonyl group of succinimide to give the tricyclic compound **179** which was isolated with 23% yield (Scheme 33).

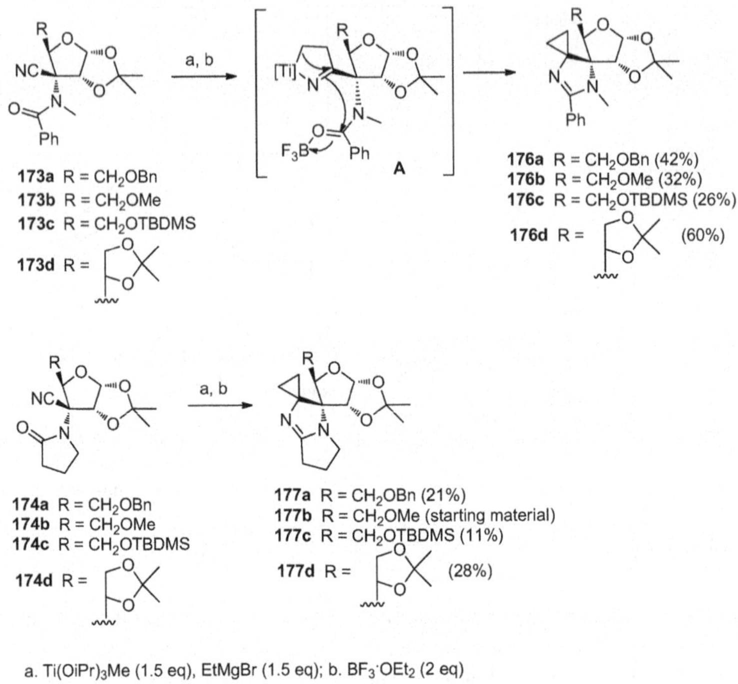

Scheme 32 Cyclopropanation key step on amides or lactams derivatives

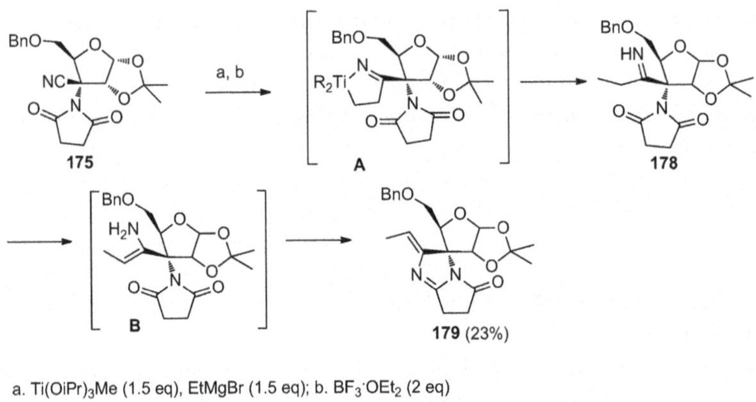

Scheme 33 Cyclopropanation key step on succinimide derivative

4.2.3 Isothiazole

From the aminonitrile it is possible to introduce an isothiazole unit on the monosaccharide (Scheme 34) [44]. The amine was first functionalized as a methanesulfonate.

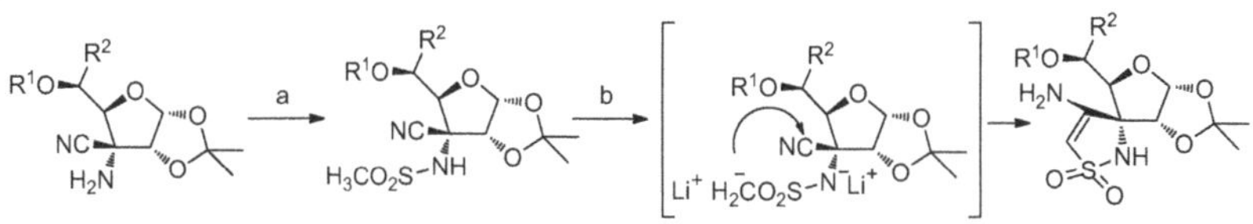

2 $R^1 = R_2 = C(CH_3)_2OCH_2$
67a R^1 = Tr, R^2 = H
68 R^1 = Bn, R^2 = H
69 R^1 = Bz, R^2 = H

180 $R^1 = R^2 = C(CH_3)_2OCH_2$ (80%)
181 R^1 = Tr, R^2 = H (92%)
182 R^1 = Bn, R^2 = H (98%)
183 R^1 = Bz, R^2 = H (98%)

184 $R^1 = R^2 = C(CH_3)_2OCH_2$ (60%)
185 R^1 = Tr, R^2 = H (98%)
186 R^1 = Bn, R^2 = H (98%)
187 R^1 = Bz, R^2 = H (0%)

a. MsCl, DMAP, pyridine; b. BuLi, THF

Scheme 34 Spiro-isothiazole synthesis

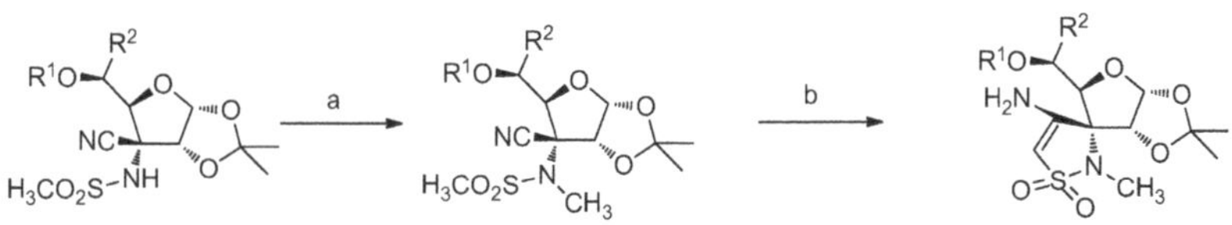

180 $R^1 = R^2 = C(CH_3)_2OCH_2$
181 R^1 = Tr, R^2 = H
182 R^1 = Bn, R^2 = H
183 R^1 = Bz, R^2 = H

188 $R^1 = R^2 = C(CH_3)_2OCH_2$ (95%)
189 R^1 = Tr, R^2 = H (98%)
190 R^1 = Bn, R^2 = H (96%)
191 R^1 = Bz, R^2 = H (98%)

192 $R^1 = R^2 = C(CH_3)_2OCH_2$ (40%^{ii})
193 R^1 = Tr, R^2 = H (61%^{i}, 90%^{ii})
194 R^1 = Bn, R^2 = H (35%^{i}, 33%^{ii})
195 R^1 = Bz, R^2 = H (25%^{ii})

a. MeI, K_2CO_3, acetone; b. (i) NaH, CH_3CN or (ii) Cs_2CO_3, CH_3CN

Scheme 35 *N*-methylated spiro-isothiazole synthesis

On this *N*-H form (compounds **180–182**), a carbanion-mediated sulfonamide intramolecular cyclization (CSIC) occurred using BuLi. Compounds **184**, **185**, and **186** were obtained with good to very good yields (60–98%), respectively. One equivalent of BuLi was necessary to form the amine salt, and a second equivalent was used to form the carbanion able to react on CN to give the isothiazole. Starting from **183**, no cyclized derivative was formed, and even the starting material was not recovered.

The reaction was also studied starting from *N*-CH_3 compounds **188–191** obtained by reacting methyl iodide with *N*-H compounds **180–183** (Scheme 35). From these *N*-CH_3 derivatives, the cyclization reaction was studied in the presence of NaH and Cs_2CO_3. Apart from the trityl derivative **189** which, in the presence of Cs_2CO_3, gave the cyclized **193** with a very good yield of 90%, the other derivatives cyclized to give the corresponding **192**, **194**, and **195** in modest yields ranging from 25 to 61%. However, it is important to point out that compound **183** which did not afford the isothiazole derivative in the presence of BuLi, led, after methylation and reaction with Cs_2CO_3, to the spiro isothiazole analog **194**.

Following the same strategy, starting from methansulfonamido-nitrile or benzylsulfonamido-nitrile, a series of isothiazole derivatives were prepared (**196–199**) (Fig. 14) [47, 48]. The CSIC key step reaction was performed using different bases (K_2CO_3, $CsCO_3$, LDA, and n-BuLi).

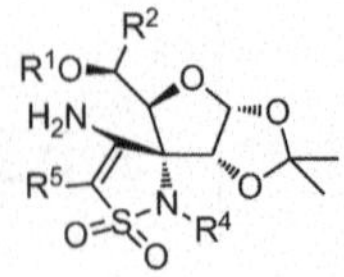

196 R^1 = Tr, R^2 = H	**197** $R^1 = R^2 = C(CH_3)_2OCH_2$	**198** R^1 = Bn, R^2 = H	**199** R^1 = Bz, R^2 = H
a R^5 = Ph, R^4 = H	**a** R^5 = Ph, R^4 = H	**a** R^5 = Ph, R^4 = Me	**a** R^5 = Ph, R^4 = Me
b R^5 = H, R^4 = Bn	**b** R^5 = H, R^4 = Bn	**b** R^5 = Ph, R^4 = Bn	**b** R^5 = Ph, R^4 = Bn
c R^5 = H, R^4 = Allyl	**c** R^5 = H, R^4 = Allyl	**c** R^5 = H, R^4 = Bn	**c** R^5 = H, R^4 = Bn
d R^5 = Ph, R^4 = Me	**d** R^5 = Ph, R^4 = Me	**d** $R^5 = CH_3$, $R^4 = CH_3$	
e R^5 = Ph, R^4 = Bn	**e** R^5 = Ph, R^4 = Bn	**e** $R^5 = CH_3$, R^4 = Bn	
f R^5 = Ph, R^4 = Allyl	**f** R^5 = Ph, R^4 = Allyl		

Fig. 14 Spiro-isothiazole derivatives

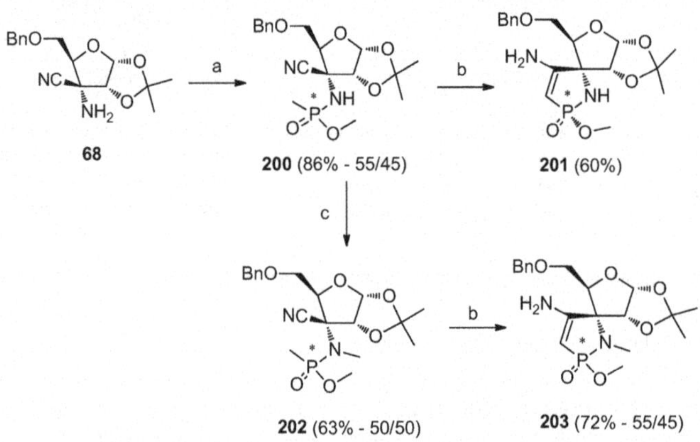

a. $CH_3POClOCH_3$ (2.4 eq), DMAP (2eq), pyridine; b. LDA (4 eq), THF, -78°C; c. MeI, NaH, DMF

Scheme 36 Spiro-azaphospholene synthesis

4.2.4 Azaphospholene

A phosphonate version of the CSIC reaction was studied a few years ago by our group to introduce an azaphospholene heterocycle on the saccharidic scaffold (Scheme 36) [37, 38]. The phosphonylation step was performed using methyl methylphosphonochloridate with DMAP on aminonitrile **68**. The methylphosphoramidyl-α-D-ribofuranose **200** was obtained as a mixture of two diastereomers in 86% yield. After cyclization with LDA, the azaphospholene derivative **201** was isolated with 60% yield. Starting from the *N*-methylated derivative **202**, the CPIC (carbanion-mediated phosphonate intramolecular cyclization) allowed the formation of the cyclized derivative **203** with 70% yield always as a mixture of two diastereomers not separable by chromatography.

5 Conclusion

We presented the syntheses of aminonitrile- or cyanohydrin-type compounds on saccharidic scaffolds that served as precursors to introduce spiro-heterocycles. The vast majority of derivatives found in the literature concerned the functionalization of position 3 of pentose units allowing access to heterocycles with oxygen, sulfur, nitrogen, and/or phosphorus.

References

1. Bourgeois JM (1973) Synthesis of branched sugars and branched amino sugars. Helv Chim Acta 56:2879–2882
2. Tronchet JMJ, Bourgeois JM (1973) Isomerizations in sugar chemistry. IV. Action of nucleophilic agents on 3-C-cyanomethylene-3-deoxy-1,2:5,6-di-O-isopropylidenefuranoses. Comparison with corresponding furanos-3-uloses. Carbohydr Res 29:373–385
3. Bourgeois JM (1975) Synthesis of branched amino sugars. III. Synthesis and reactions of α-amino-nitrile derivatives of di-O-isopropylidene-1,2:5,6-α-D-ribo-hexofuranos-3-ulose. Helv Chim Acta 58:363–372
4. Rosenthal A, Cliff BL (1976) Branched-chain sugars. Reaction of 1,2:5,6-di-O-isopropylidene-α-D-ribo-hexofuranos-3-ulose with sodium cyanide and methyl nitroacetate. Can J Chem 54:543–547
5. Merino-Montiel P, López Ó, Fernández-Bolaños JG (2012) Unprecedented spiro-annelated sugar iso-ureas, guanidines and amidines as new families of glycosidase inhibitors. RSC Adv 2:11326–11335
6. Hricoviniova Z, Hricovini M (2013) A new type of rearrangement in branched-chain carbohydrates: isomerization of 3-C-branched aldoses. Carbohydr Res 370:1–8
7. Calvo-Mateo A, Camarasa MJ, Diaz-Ortiz A, de las Heras FG, Alemany A (1988) Synthesis of 3′-C-cyano-3′-deoxypentofuranosylthymine nucleosides. Tetrahedron 44:4895–4903
8. Perez-Perez MJ, San-Felix A, Camarasa MJ, Balzarini J, de Clercq E (1992) Synthesis of {1-[2′,5′-bis-O-(tert-butyldimethylsilyl)-β-D-xylo- and β-D-ribofuranosyl]thymine}-3′-spiro-5-″-{4″-amino-1″,2″-oxathiole-2″,2″-dioxyde} (TSAO). A novel type of specific anti-HIV agents. Tetrahedron Lett 33:3029–3032
9. Perez-Perez MJ, Camarasa MJ, Diaz-Ortiz A, San Felix A, de las Heras FG (1991) Stereospecific synthesis of branched-chain sugars by a novel aldol-type cyclocondensation. Carbohydr Res 216:399–411
10. Velazquez S, Chamorro C, Perez-Perez M-J, Alvarez R, Jimeno M-L, Martin-Domenech A, Perez C, Gago F, de Clercq E, Balzarini J, San-Felix A, Camarasa M-J (1998) Abasic analogs of TSAO-T as the first sugar derivatives that specifically inhibit HIV-1 reverse transcriptase. J Med Chem 41:4636–4647
11. Ingate S, Perez-Perez M-J, de Clercq E, Balzarini J, Camarasa M-J (1995) Synthesis and anti-HIV-1 activity of novel TSAO-T derivatives modified at the 2′- and 5′-positions of the sugar moiety. Antivir Res 27:281–299
12. Ingate S, de Clercq E, Balzarini J, Camarasa M-J (1995) Novel L-lyxo and 5′-deoxy-5′-modified TSAO-T analogs: synthesis and anti-HIV-1 activity. Antivir Res 32:149–164

13. Zhang HY, Yu HW, Ma LT, Min JM, Zhang LH (1998) Synthesis of 2-C-(4-aminocarbonyl-2-thiazoyl)-1,4-anhydro-L-xylitols and their fluoro derivatives. Tetrahedron Asymmetry 9:141–149
14. Zhang M, Zhang H, Yang Z, Ma L, Min J, Zhang L (1999) Synthesis of 3-C-(methyl β-D-xylofuranosid-3-yl)-5-phenyl-1,2,4-oxadiazole. Carbohydr Res 318:157–161
15. Baer HH, Radatus B (1986) Two synthesis of 3-amino-3-deoxy-α-D-altropyranosyl 3-amino-3-deoxy-α-D-altropyranoside, a new analog of α,α-trehalose, involving reduction of a diazide and reductive amination of a diketone. Carbohydr Res 157:65–81
16. Steiner B, Langer V, Bella M, Koos M (2013) Cyanohydrins from methyl 6-deoxy-2,3-O-isopropylidene-α-L-lyxo-hexofuranosid-4-ulose via Bucherer-Bergs and Strecker reactions. Carbohydr Res 369:31–37
17. Koos M, Steiner B, Langer V, Gyepesova D, Durik M (2000) Preparation and structure determination of two sugar amino acids via corresponding hydantoin derivatives. Carbohydr Res 328:115–126
18. Nguyen Van Nhien A, Cordonnier R, Le Bas M-D, Delacroix S, Soriano E, Marco-Contelles J, Postel D (2009) Highly functionalized, enantiomerically pure furo[x,y-c]pyrans via alkylidenecarbenes derived from sugar templates: synthesis and mechanism study via computational chemistry. Tetrahedron 65:9378–9394
19. Cordonnier R, Nguyen Van Nhien A, Soriano E, Marco-Contelles J, Postel D (2010) Experimental and computational investigation of the unexpected formation of β-substituted polyoxygenated furans from conveniently functionalized carbohydrates. Tetrahedron 66:736–742
20. Yu J, Zhang S, Li Z, Lu W, Cai M (2005) Stereoselective synthesis, structural characterization, and properties of 1,2-O-isopropylidene-3-C-(5-phenyl-1,2,4-oxadiazol-3-yl)-β-D-psicopyranose. Bioorg Med Chem 13:353–361
21. Czernecki S, Dieulesaint A, Valery JM (1986) Chain extension of carbohydrates. I. Cyanoamination of 1,2:3,4-di-O-isopropylidene-α-D-galacto-hexodialdo-1,5-pyranose. J Carbohydr Chem 5:469–474
22. Czernecki S, Valery JM (1988) Chain-extension of carbohydrates. Part III. Stereochemical studies of the synthesis of α-amino-nitriles related to lincosamine. Carbohydr Res 184:121–130
23. Steiner B, Koos M, Langer V, Gyepesova D, Smrcok L (1998) 4-Amino-4-cyano-4,6-dideoxy sugar derivatives from methyl 6-deoxy-2,3-O-isopropylidene-α-L-lyxo-hexopyranosid-4-ulose via Strecker-type reaction. Carbohydr Res 311:1–9
24. Postel D, Nguyen Van Nhien A, Pillon M, Villa P, Ronco G (2000) Stereoselective synthesis of α-aminonitriles at non-anomeric positions of monosaccharides. Tetrahedron Lett 41:6403
25. Ducatel H, Nguyen Van Nhien A, Postel D (2008) Novel glycine like amino acids from glyco-α-aminonitriles as building blocks for peptide synthesis. Tetrahedron Asymmetry 19:67–81
26. Bera S, Chatterjee B, Mondal D (2016) Construction of quaternary stereo-centers on carbohydrate scaffolds. RSC Adv 6:77212–77242
27. Postel D, Nguyen Van Nhien A, Marco JL (2003) Chemistry of sulfonate- and sulfonamide-stabilized carbanions – the CSIC reactions. Eur J Org Chem 19:3713–3726
28. Calvo-Mateo A, Camarasa M-J, Diaz-Ortiz A, de las Heras FG (1988) Novel aldol-type cyclocondensation of O-mesyl (methylsulfonyl) cyanohydrins. Application to the stereospecific synthesis of branched-chain sugars. J Chem Soc Chem Commun 16:1114–1115
29. Camarasa MJ, Perez-Perez MJ, San-Felix A, Balzarini J, De Clercq E (1992) 3′-Spiro nucleosides, a new class of specific human immunodeficiency virus type 1 inhibitors: synthesis and antiviral activity of [2′,5′-bis-O-(tert-butyldimethylsilyl)-β-D-xylo- and -ribofuranose]-3′-spiro-5″-[4″-amino-1″,2″-oxathiole-2″,2″-dioxide] (TSAO) pyrimidine nucleosides. J Med Chem 35:2721–2727
30. Perez-Perez MJ, San-Felix A, Balzarini J, de Clercq E, Camarasa MJ (1992) TSAO analogs. Stereospecific synthesis and anti-HIV-1 activity of 1-[2′,5′-bis-O-(tert-butyldimethylsilyl)-β-D-

ribofuranosyl]-3′-spiro-5″-(4″-amino-1″,2″-oxathiole-2″,2″-dioxide)pyrimidine and pyrimidine-modified nucleosides. J Med Chem 35:2988–2995
31. Velazquez S, Jimeno M-L, Camarasa M-J, Balzarini J (1994) Synthesis of [1-[3′,5′-bis-O-(tert-butyldimethylsilyl)-β-D-arabino- and β-D-ribofuranosyl]cytosine]-2′-spiro-5″-(4″-amino-1″,2″-oxathiole-2″,2″-dioxide). Analogs of the highly specific anti-HIV-1 agent TSAO-T. Tetrahedron 50:11013–11022
32. Kiritsis C, Manta S, Dimopoulou A, Parmenopoulou V, Gkizis P, Balzarini J, Komiotis D (2014) Stereoselective facile synthesis of 2′-spiro pyrimidine pyranonucleosides via their key intermediate 2′-C-cyano analogs. Evaluation of their bioactivity. Carbohydr Res 383:50–57
33. Chamorro C, Luengo SM, Bonache M-C, Velazquez S, Perez-Perez M-J, Camarasa M-J, Gago F, Jimeno M-L, San-Felix A (2003) Synthesis and structural characterization of pyrimidine bi- and tricyclic nucleosides with sugar puckers conformationally locked into the eastern region of the pseudo-rotational cycle. J Org Chem 68:6695–6704
34. Cordeiro A, Quesada E, Bonache M-C, Velazquez S, Camarasa M-J, San-Felix A (2006) A cyclic enamine derived from 1,2-O-isopropylidene-α-D-xylofuranose as a novel carbohydrate intermediate to achieve skeletal diversity. J Org Chem 71:7224–7235
35. Alvarez R, Jimeno ML, Perez-Perez MJ, de Clercq E, Balzarini J, Camarasa MJ (1997) Synthesis and anti-human immunodeficiency virus type 1 activity of novel 3′-spiro nucleoside analogs of TSAO-T. Antivir Chem Chemother 8:507–517
36. Camarasa M-J, Jimeno M-L, Perez-Perez M-J, Alvarez R, Velazquez S, Lozano A (1999) Synthesis, NMR studies and theoretical calculations of novel 3-spiro-branched ribofuranoses. Tetrahedron 55:12187–12200
37. Moura M, Josse S, Postel D (2013) Preparation of Spiro[4.4]oxaphospholene and -azaphospholene systems from carbohydrate templates. J Org Chem 78:8994–9003
38. Moura M, Josse S, Postel D (2018) Synthesis of phosphorus analogs of TSAO-T. Tetrahedron 74:4721–4727
39. Nguyen Van Nhien A, Tomassi C, Ducatel H, Len C, Postel D (2002) Glyco-α-aminonitriles as intermediates in organic synthesis. J Maroc Chim Heterocycl 1:1–6
40. Gasch C, Illangua JM, Merino-Montiel P, Fuentes J (2009) Stereocontrolled synthesis of (5+5), (5+6) and (6+6) 3-spiropseudonucleosides. Tetrahedron 65:4149–4155
41. Nguyen Van Nhien A, Ducatel H, Len C, Postel D (2002) Novel conformationally restricted glycoamino acids from glyco-α-aminonitriles as potent turn mimics in peptide synthesis. Tetrahedron Lett 43:3805–3808
42. Nguyen Van Nhien A, Villa P, Ronco G, Postel D (2001) Spiro hydantoins from D-ribose as new potent enzymatic inhibitors. J Pharm Pharmacol 53:939–943
43. Postel D, Nguyen Van Nhien A, Villa P, Ronco G (2001) Novel spirohydantoins of D-allose and D-ribose derived from glyco-α-aminonitriles. Tetrahedron Lett 42:1499–1502
44. Postel D, Nguyen Van Nhien A, Villa P, Ronco G (2001) New 4-amino-5H-2,3-dihydroisothiazole 1,1-dioxides from sugar α-amino nitriles using the CSIC reaction. Tetrahedron Lett 42:593–595
45. Declerck D, Josse S, Nguyen Van Nhien A, Szymoniak J, Bertus P, Postel D (2012) Reactivity of sugar α-aminonitrile derivatives under titanium-mediated cyclopropanation conditions. Tetrahedron 68:1145–1152
46. Bertus P, Szymoniak J (2001) New and easy route to primary cyclopropylamines from nitriles. Chem Commun 18:1792–1793
47. Nguyen Van Nhien A, Dominguez L, Tomassi C, Torres MR, Len C, Postel D, Marco-Contelles J (2004) Synthesis and transformations of [1,2-O-isopropylidene-α-D-erythro (and α-d-ribo)furanose]-3-spiro-3′-(4′-amino-5′H-2′,3′-dihydroisothiazole-1′,1′-dioxide) derivatives. Tetrahedron 60:4709–4727
48. Dominguez L, Nguyen Van Nhien A, Tomassi C, Len C, Postel D, Marco-Contelles J (2004) Synthesis of 4-amino-5-H-2,3-dihydroisothiazole-1,1-dioxide ring systems on sugar templates via carbanion-mediated sulfonamide intramolecular cyclization reactions (CSIC protocols) of glyco-α-sulfonamidonitriles. J Org Chem 69:843–856

Top Heterocycl Chem (2019) 57: 171–214
DOI: 10.1007/7081_2019_31

Published online: 24 February 2019

Recent Advances in the Chemistry and Biology of Spirocyclic Nucleosides

Martín Soto, Humberto Rodríguez-Solla, and Raquel Soengas

Contents

Abstract Nucleosides are structural subunits of nucleic acids, the macromolecules that convey genetic information in living cells. Human creativity has been put to light in the ability of drug researchers to draw on an understanding of the biochemistry of naturally occurring nucleosides and to build up synthetic nucleoside analogues, which belong to the most important class of antiviral drugs and are extensively used as anticancer agents and in the treatment of other diseases. In this regard, the potential benefits associated with the spirocyclic restriction of nucleosides sparkled considerable interest in the synthesis and application of these molecules as therapeutic agents. The field of spirocyclic nucleosides started to grow since the isolation of hydantocidin, a natural spironucleoside isolated from fermentation

M. Soto and H. Rodríguez-Solla
Department of Organic and Inorganic Chemistry, University of Oviedo, Oviedo, Spain

R. Soengas (✉)
Área de Química Orgánica, Research Centre CIAIMBITAL, Universidad de Almería, Almería, Spain
e-mail: rsoengas@ual.es

broths of *Streptomyces hygroscopicus*, which exhibits potent herbicidal activity. The biological activity of hydantocidin prompted considerable synthetic interest in this nucleoside and also in a variety of analogues. The present overview describes the convenient approaches that have been developed in the past two decades for accessing varied members of the family of spiro-functionalized nucleosides.

Keywords Nucleosides · Spirocyclic compounds · Spironucleosides

1 Introduction

Nucleosides are structural subunits of nucleic acids, the macromolecules that convey genetic information in living cells. In fact, they embrace a large family of great structural diversity and wide biological activity spectrum [1–6]. For many years, drug researchers were committed to a better understanding of the biochemistry of naturally occurring nucleosides. For this reason, a large number of modified nucleosides have been synthesized. It has been found that some of them have potent biological activities; in fact, nucleosides belong to the most important class of antiviral drugs and are extensively used as anticancer agents [7–12].

In order to discover new biologically active derivatives, nucleoside analogues can be divided into three categories, (1) phosphate-modified, (2) base-modified, and (3) sugar-modified compounds, among which the two latter are the most active. Thus, base- and sugar-modified nucleosides are valuable targets in medicinal chemistry [13].

In the past few years, the study of the conformational behavior of natural and modified nucleosides – in particular the sugar puckering – has grown as an important research field, since it is closely related to the nucleosides' metabolic pathways and their final interactions with the target polymerases [14–20]. The sugar puckerings of natural ribo- and deoxyribonucleosides are known to exist in dynamic equilibria between two major conformers: the North (N) and the South (S) types. In solution, both species coexist due to a fast interconversion of the N and S conformers [21]. However, after binding to an enzyme, only one conformer is expected to be present within the active site. Conformationally restricted nucleosides have generated considerable attention since they can adopt only determined conformations, thus being useful tools in the study of the interactions of nucleosides with their enzymatic targets. However, the use of conformationally restricted nucleoside analogues (LNAs, locked nucleic acids) is not only limited to structure-activity studies. Recently, it has been found that LNAs, in which the sugar moieties are locked in either the N or S conformation, enhance base stacking and backbone pre-organization [22–26]. In fact, such nucleosides showed interesting binding properties with target nucleic acids as well as nuclease resistance. In addition, some of these LNAs incorporated into oligodeoxyribonucleotides (ODN-s) have shown strong RNA-binding affinities, which led to their use in antisense [27] and RNA interference [28, 29] strategies.

Stimulated by the potential application of the resulting analogues as therapeutic agents, conformational restriction of the furanose ring of nucleosides, nucleotides, and oligonucleotides has been intensively pursued in recent years [30–32]. Interest in the spiro-functionalization of the sugar ring has risen [33], particularly since the discovery of the first natural spirocyclic nucleoside (+)-hydantocidin. The observation that hydantocidin shows potent plant-growth regulatory activities and low toxicity to microorganisms and mammals sparkled the interest in the synthesis of conformationally restricted spiro-functionalized nucleosides [34–36].

(+)-Hydantocidin is an example of what it is called a "spironucleoside." This term is used to designate a type of nucleoside in which the anomeric carbon of the sugar moiety belongs simultaneously to a pyranoid or furanoid sugar ring and to an aza-heterocyclic moiety (C-1′-spiro) [37]. When the spiranic carbon atom is not the anomeric carbon, the term "spiropseudonucleoside" is employed instead (C-2′-spiro, C-3′-spiro, and C-4′-spiro).

Most of the synthetic strategies for spirocyclic nucleosides are based on the use of carbohydrate derivatives to generate the desired stereochemistry in the sugar ring. However, a very diverse range of strategies are available for the synthesis of the characteristic spirocyclic base. Our main aim in this chapter is to draw together all of the synthetic information on spiro-functionalized nucleosides reported in the past two decades in a form which is easily consulted. The coverage is primarily from the point of view of organic chemists, so our intention is to describe in detail those strategies that have been employed to prepare spirocyclic nucleosides. Moreover, from a biological point of view, we have also mentioned, along this chapter, some properties of the most remarkable examples, including their activity such as herbicidal action, enzymatic inhibition, antiviral or anticancer activity, and pharmacological relevance.

2 Spironucleosides (C-1′-Spiro-Functionalized Nucleosides)

Spironucleosides are a family of conformationally restricted nucleosides in which the spirocyclic restriction is in the anomeric position [38]. This fixes the nucleotide base in a specific orientation around the *N*-glycosidic bond, thus altering the flexibility of the sugar moiety. Anomeric spirocyclic nucleosides gained considerable interest with the discovery of (+)-hydantocidin, a natural spironucleoside isolated from fermentation broths of *Streptomyces hygroscopicus* SANK 63584 [39, 40], Tu-2474 [41], and A1491 [42]. Hydantocidin has potent herbicidal and plant-growth regulatory activity with highly selective toxicity between plants and animals [43]. Biochemical studies have shown that hydantocidin is a proherbicide which is phosphorylated at the 5′-position in vivo and inhibits adenylosuccinate synthetase (AdSS) [44], an enzyme that plays a key role in the de novo purine synthesis in plants [45]. These observations have generated considerable interest, not only in the synthesis of (+)-hydantocidin itself but also in a variety of its analogues, with the notion that important pharmaceutical leads can be found among modified nucleoside derivatives. A number of anomeric spirocyclic nucleosides have subsequently appeared in the literature, being hydantoin or diketopiperazine analogues, but also barbiturates and more diverse spiroheterocyclic subunits.

This massive research on the synthesis and biology of hydantocidin analogues was awarded with the discovery of a glucopyranose spirohydantoin, a very active inhibitor of glycogen phosphorylase with a K_i value of 3.1 μM [46]. Glycogen phosphorylase (GP) is a key enzyme in the regulation of muscle, brain, and hepatic glycogen metabolism and catalyzes the first step in the intracellular degradation of glycogen [47–49]. Inhibition of glycogen phosphorylase is believed to assist in shifting the equilibrium between glycogen degradation and glycogen synthesis in favor of the latter in both muscle and liver [50–53]. Therefore, GP inhibitors may be clinically useful for the treatment of diabetes mellitus, in particular the non-insulin-dependent diabetes mellitus (NIDDM or type II diabetes) [54, 55].

2.1 Hydantoins

The most representative member of the hydantoins' family is (+)-hydantocidin **1**, a natural spironucleoside displaying potent herbicidal activity (Fig. 1).

The interesting biological profile of hydantocidin **1** prompted many research groups to investigate the synthesis of this spironucleoside. The first total synthesis of **1** was developed by Mio and co-workers in 1991 [56]. Since Mio's pioneering work, several examples have been reported in the literature for the synthesis of hydantocidin [57].

In spite of the considerable synthetic work focused on the synthesis of hydantocidin and its analogues, the problem of accessing multigram quantities of hydantocidin for a detailed biological evaluation remained unresolved. Aiming at overcoming this limitation, Shiozaki et al. reported in 2002 a synthesis of hydantocidin from dichloroolefin **2** [58]. After oxidation of **2** with *m*-chloroperbenzoic acid (*m*CPBA) and chromatographic separation, treatment with sodium azide afforded azide **3** (Scheme 1). Finally, formation of the hydantoin ring was accomplished from azide **3** via the corresponding urea, to give after hydrogenolysis and acidic hydrolysis, hyantocidin **1**.

The basic structure of hyantocidin is comprised of a spirohydantoin ring at the anomeric carbon of a D-ribofuranose. That is a total of four stereocenters, which could play a fundamental role in herbicidal activity. After achieving the synthesis of natural hydantocidin itself, Mio and co-workers also accomplished the synthesis of all the stereoisomers of the natural spironucleoside [59, 60] and also of several deoxy

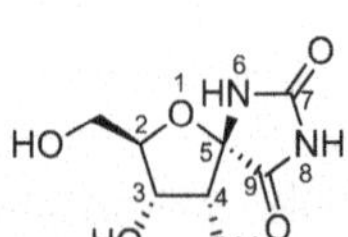

Fig. 1 Natural spironucleoside (+)-hydantocidin **1**

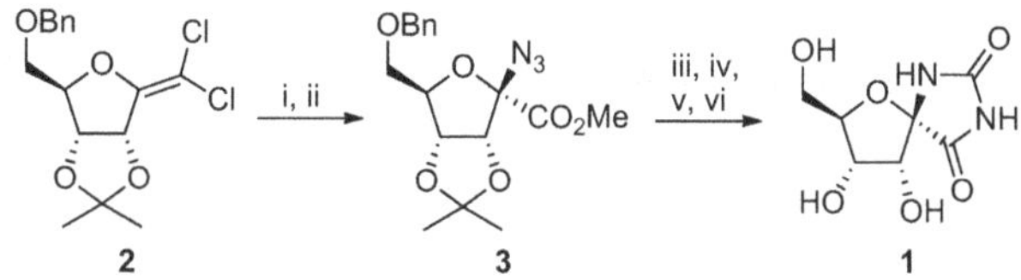

Scheme 1 *Reagents and conditions*: (i) *m*CPBA, MeOH, CH_2Cl_2, r.t., 16 h, 68%. (ii) NaN_3, DMF, 24°C, 16 h, 95%. (iii) 1. PPh_3, THF, 24°C, 2 h; 2. PMBNCO, THF, 24°C, 16 h, two steps 90%. (iv) 1. 1:20 aq 1 M HCl–THF, r.t., 15 min, quant.; 2. CAN, 2:1 MeCN/water, r.t., 20 min, 97%. (v) 0.2 M NH_3 in MeOH, 27°C, 4 h, 99%. (vi) 1. H_2, Pd/C, EtOAc, 24°C, 30 min; 2. 1:3 TFA/H_2O, 0°C, 2 h, quant

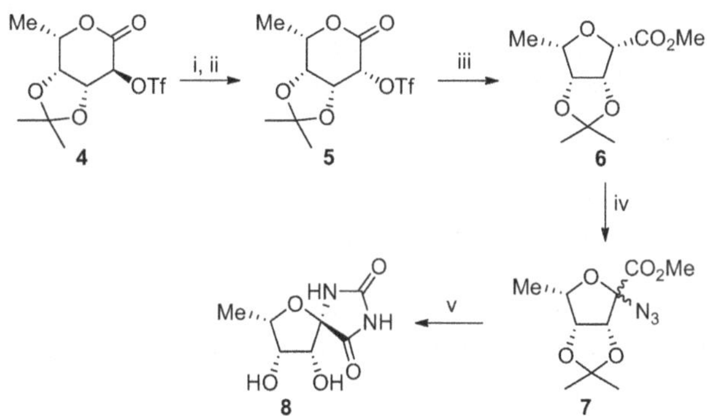

Scheme 2 *Reagents and conditions*: (i) CF_3COONa, DMF. (ii) Tf_2O, pyridine, −30°C, 84%. (iii) 1. 3% HCl in MeOH; 2. 2,2-dimethoxypropane, CSA, acetone, 97%. (iv) 1. NBS, CCl_4, $(PhCOO)_2$; 2. NaN_3, DMF, r.t., 74%. (v) 1. H_2, Pd/C, MeOH, 70%; 2. KOCN, AcOH, 74%; 3. 50% aq TFA, 94%

derivatives [61]. Even though the mode of herbicidal action of hydantocidin remained elusive at that time, Fleet and co-workers surmised that spirohydantoins of other sugars could also possess interesting biological properties, so they initiated extensive investigations targeting diverse hydantoins of pentoses and hexoses [62–65]. Thus, Fleet's group is responsible for the first report of the spirohydantoin of glucopyranose, which is a potent and specific inhibitor of the glucosyl transferase glycogen phosphorylase [66, 67]. Among the multiple contributions of Fleet's group to the chemistry of sugar hydantoins, we will take as an example the synthesis of L-*lyxo*-configured analogues of hydantocidin, reported in 2006 [68]. Thus, reaction of L-fucose-derived triflate **4** with sodium trifluoroacetate, followed by triflation of the intermediate inverted lactone, afforded triflate **5** (Scheme 2). Ring contraction in acidic conditions gave, after acetonation, carboxylate **6**. The azido group was introduced by means of a radical bromination of **6** and azide displacement of the intermediate bromide. Treatment of the resulting inseparable mixture of epimeric azidoesters **7** with potassium cyanate in acetic acid afforded the corresponding urea, which on reaction with potassium *tert*-butoxide gave, after acidic removal of the isopropylidene group, lyxofuranose hydantoin **8**.

Since (+)-hydantocidin possess a *N,O*-hemiacetal functionality at the anomeric position, it could be easily isomerized to the thermodynamically more stable 5-epimer. Aiming at avoiding epimerization, the first synthesis of a carbocyclic hydantoin was reported by Fleet et al. in 1993 [69]. Since then, several examples of this type of hydantoins were reported, as, for example, carbocyclic hydantocidin **13** and its 3′,4′-diepimer **14** reported by Pham and co-workers. Thus, starting from ethyl 2-butynoate **9** and *N,N*′-diprotected-5-methylenehydantoin **10**, a phosphine-catalyzed [3 + 2]-cycloaddition afforded the spiroheterocyclic ester **11**, which was then isomerized to ester **12** on treatment with base (Scheme 3) [70]. Acid hydrolysis of **12**, followed by reduction of the resulting acid and *cis*-dihydroxylation afforded, after removal of the protecting groups, the desired carbocyclic hydantoins **13** and **14**.

In an attempt to determine the basic structural requirements for maintaining the herbicidal activity, investigations were then directed to the synthesis of hydantocidin analogues resulting from modifications in the structure of the hydantoin ring of the parent molecule [71]. However, it was very soon reached the conclusion that important structural changes invariably led to a drastic reduction of the herbicidal action, so the investigations were then focused on derivatives with minimal modifications in which the basic structure remained unchanged. In this regard, Sano and co-workers described the first thiohydantoin derivative, resulting from the replacement of the C7 carbonyl group with a thiocarbonyl group [72].

The observation that the introduction of the sulfur atom did not diminish the herbicidal action sparkled the interest in thiohydantoins [73]. In this regard, Somsák and co-workers reported the synthesis and biological evaluation of several thio-analogues of hydantocidin [74–79]. As a representative example, the synthesis of glucopyranose thiohydantoin **17** is depicted in Scheme 4 [77]. Thus, partial hydrolysis of the nitrile in glucopyranosyl cyanide **15** with HBr in acetic acid, followed by photobromination [80, 81], furnished derivative **16**. Desired spiro-thiohydantoin **17**

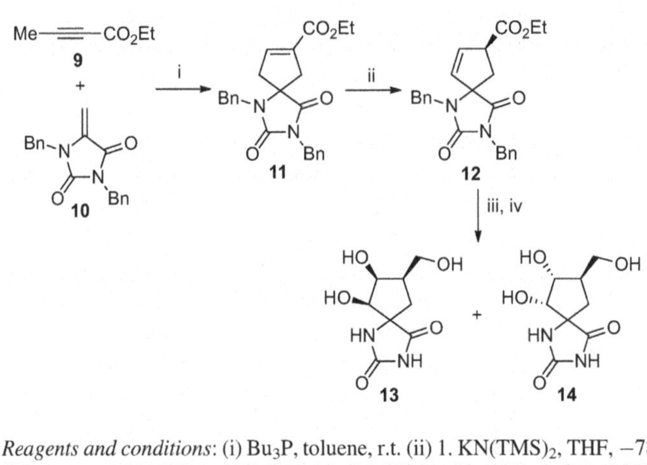

Scheme 3 *Reagents and conditions*: (i) Bu_3P, toluene, r.t. (ii) 1. $KN(TMS)_2$, THF, −78°C, 10 min; 2. AcOH, −78°C to r.t., 99%. (iii) 1. 10% HCl, MeCN, 90°C, 15 h, 99%; 2. BH_3·DMS, THF, 0°C, 6 h, 95%; 3. $K_2OsO_4·2H_2O$, NMO, acetone/H_2O (4:1), r.t., 5 days, 37%; 4. Ac_2O, pyridine, MeCN, r.t., 10 h, 91%. (iv) 1. NBS, C_6H_5Cl, 125°C, 14 h; 2. 10% HCl, THF, reflux, 4 h, 95%

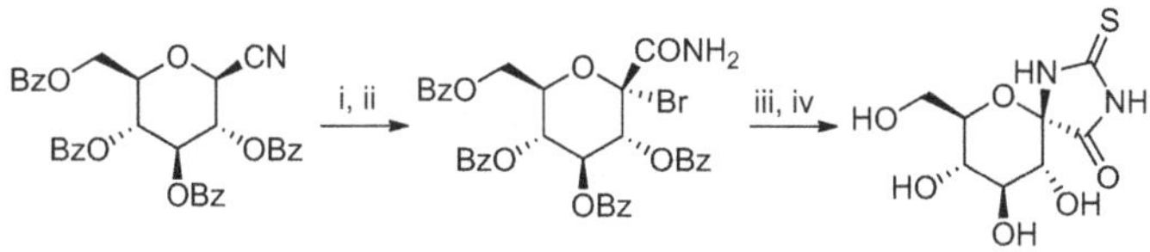

Scheme 4 *Reagents and conditions*: (i) HBr, AcOH, r.t., 94%. (ii) Br_2, $CHCl_3$, $h\nu$, reflux, 89%. (iii) NH_4SCN, CH_3NO_2, 80°C, 79%. (iv) NaOMe, MeOH, reflux, 92%

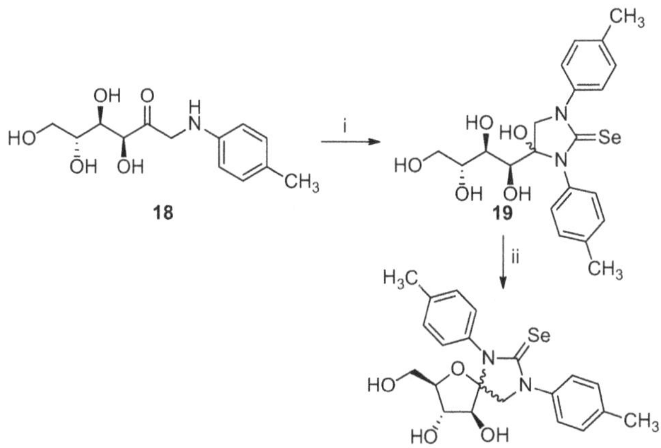

Scheme 5 *Reagents and conditions*: (i) PhNCSe, DMF, r.t., 72 h, 98%. (ii) AcOH, EtOH/H_2O (2:1), reflux, 1 h, 60%

was easily obtained from amide **16** on reaction with ammonium thiocyanate in nitromethane.

On the other hand, the first selenium-containing hydantocidin analogue was reported by Maza et al. in 2009 [82]. Reaction of *N*-arylfructosamine **18** with *p*-methylphenyl isoselenocyanate afforded selone **19**, which under acidic conditions gave spiranic derivative **20** as the major compound (Scheme 5).

2.2 *Diketopiperazines*

The interesting biological profile of hydantocidin led to extensive synthetic studies toward not only to hydantocidin itself but also to a large variety of analogues, including those derived from the substitution of the hydantoin ring by other spirocyclic units. Among them, diketopiperazines attracted much interest on account of their wide range of chemotherapeutic applications. Due to the biological relevance of diketopiperazines, Fleet and co-workers investigated the incorporation of a spirodiketopiperazine ring into the anomeric position of various sugars [83–87].

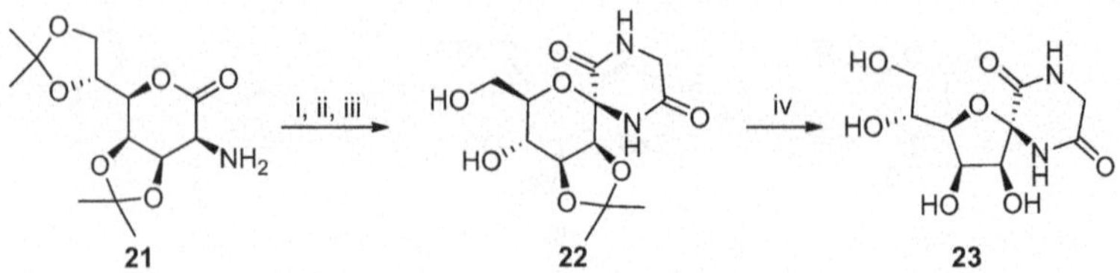

Scheme 6 *Reagents and conditions*: (i) Cbz-Gly-OH, ClCOOEt, Py, MeCN/THF (1:l), 86%. (ii) 1. AcOH/H_2O (1:l), quant.; 2. *N*-bromophthalimide, MeCOONa, 54%. (iii) H_2, Pd. MeOH. (iv) 1. CF_3COOH, H_2O; 2. *t*-BuOK, DMF

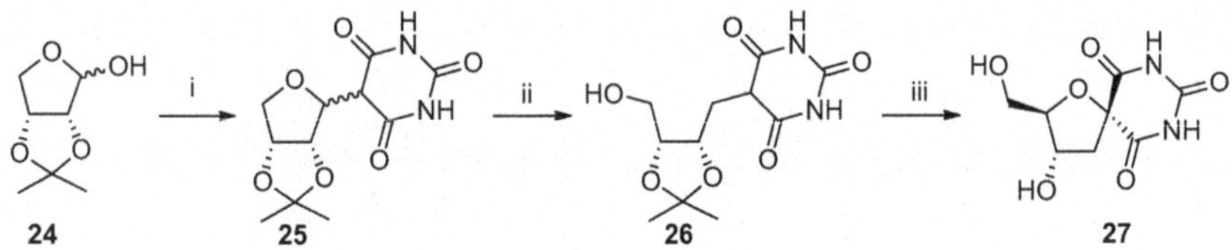

Scheme 7 *Reagents and conditions*: (i) barbituric acid, Na_2CO_3. (ii) H_2, Pd-C. (iii) 1. Br_2, TBDMSCl, imidazole; 2. TMSCl, MeOH

As an example, Scheme 6 displays the synthesis of mannopyranose diketopiperazine **23** [88]. Starting from aminolactone **21**, coupling with Cbz-glycine, followed by acidic hydrolysis of the 6,7-*O*-isopropylidene group, oxidation, and catalytic hydrogenation, afforded spirodiketopiperazine **22**. Finally, acidic hydrolysis followed by reaction with base furnished mannofuranose diketopiperazine **23**.

2.3 Barbiturates

The main problem that hydantoins and diketopiperazines show is their lack of stability, which is attributed to a spiro-epimerization in basic media. To avoid this, hydantoin or diketopiperazine rings can be replaced by a barbiturate moiety. Similar to the hydantoin heterocycle, the barbiturate ring system possesses thymine-like hydrogen-bonding capacity against adenine derivatives [89] and is present in a wide range of bioactive molecules [90].

Renard and co-workers reported in 2002 the synthesis of spiro barbituric derivatives **27** and **30** (Schemes 7 and 8) [91]. Thus, erythrolactol **24** was condensed with barbituric acid in the presence of sodium carbonate to give the erythrosyl barbituric acid derivative **25** (Scheme 7). After hydrogenolysis, bromination of the resulting alcohol **26** in the presence of *t*-butyldimethylsilyl chloride (TBDMSCl) followed by acidic hydrolysis afforded the desired spiro barbituric derivative **27**.

On the other hand, carbocyclic analogue **30** was prepared from diol **28** by a sequence of silylation, condensation of the resulting cyclopentane **29** with urea, and acidic hydrolysis (Scheme 8). Despite both barbiturates **27** and **30** displayed

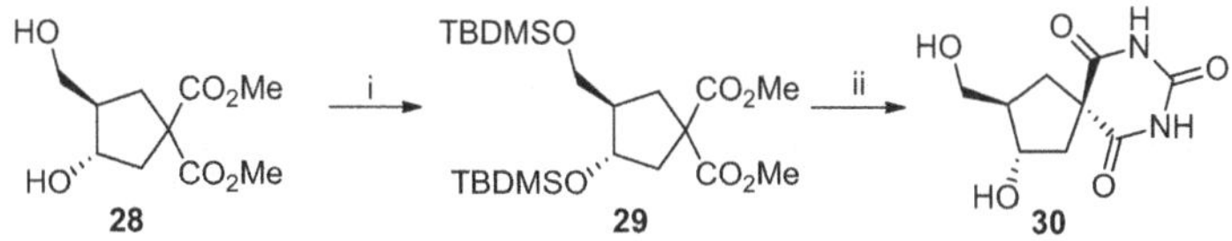

Scheme 8 *Reagents and conditions*: (i) TBDMSCl, imidazole. (ii) 1. Urea; 2. TMSCl, MeOH

enhanced hydrogen-bonding capacity with diacetyladenosine, spironucleoside **30** is a more suitable building block for the synthesis of modified oligonucleotides on account of its enhanced stability to nucleophilic ring opening.

2.4 *Miscellaneous Spiroheterocyclic Units*

In addition to diketopiperazines and barbiturates, several other analogues with very diverse spiroheterocyclic rings were synthetized [92–98].

Thus, Gasch et al. reported the stereoselective synthesis of several pyranoid and furanoid spiroheterocyclic nucleosides [99, 100]. For example, reaction of psicofuranose derivative **33** with trimethylsilyl isothiocyanate in the presence of trimethylsilyl triflate provided the corresponding thioxo-oxazolidine **32** (Scheme 9) [99]. On the other hand, *N*-glycosylation of **33** with trimethylsilyl isothiocyanate in the presence of a Lewis acid afforded oxazolidinone **31** as the major product. For the synthesis of spiro-*C*-glycosides **37** and **38**, spiroketal **33** was transformed into psicofuranosyl cyanides **34** and **35** on reaction with trimethylsilyl cyanide and a Lewis acid. Reduction with hydride, followed by treatment with thiophosgene and triethylamine, furnished spiro compounds **37** and **38**. Finally, the reaction of **33** with trimethylsilyl acetonitrile afforded derivative **36** as the major product.

The same researchers also described the synthesis of spironucleosides from spiroketal derivative **40** (Scheme 10) [99]. Thus, *N*-glycosylation of **40** with trimethylsilyl azide afforded anomeric azide **39**, which on catalytic hydrogenation, desilylation, and treatment with thiocarbonyldiimidazole furnished spironucleoside **42**. *C*-Glycosylation of **40** with trimethylsilyl cyanide in the presence of a Lewis acid, followed by treatment with tetrabutylammonium fluoride (TBAF), gave the cyanide **43**. From **43**, spiro compound **44** was easily available on hydride reduction, reaction with thiocarbonyldiimidazole, and triethylamine-promoted intramolecular cyclization. Finally, treatment of **40** with trimethylsilyl acetonitrile and a Lewis acid furnished spiro-derivative **41**.

In their route to α-homonojirimycin analogues, Silva et al. planned the ring opening of the iminosugar **45** by a stabilized Wittig reagent, followed by base-catalyzed recyclization to afford derivative **46** [101]. However, treatment of compound **45** with (methoxycarbonylmethylene)triphenylphosphorane in the presence of a catalytic amount of benzoic acid yielded spironucleoside analogue **47** (Scheme 11).

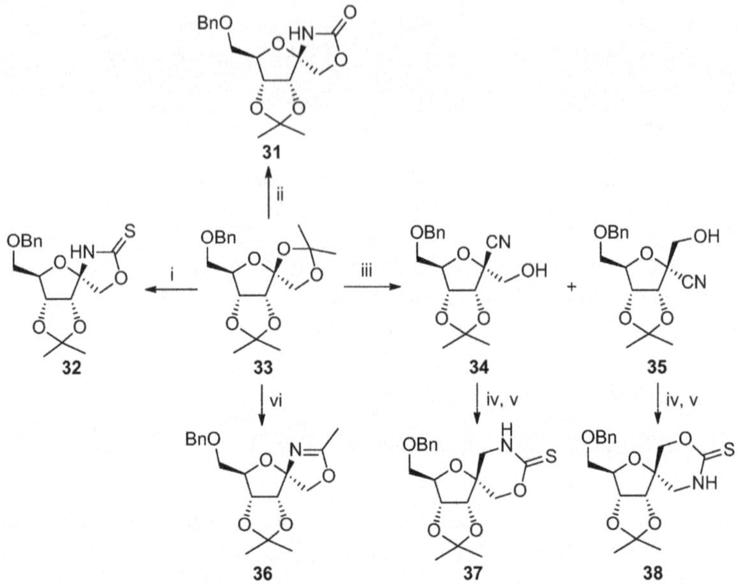

Scheme 9 *Reagents and conditions*: (i) TMSNCS, TMSOTf, −20°C, 1 h, 10%. (ii) TMSNCO, TMSOTf, −20°C, 2 h, 22%. (iii) TMSCN, TMSOTf, −20°C, 2 h, 81%. (iv) 1. $LiAlH_4$, Et_2O, 0°C to r.t., 2 h; 2. Cl_2CS, r.t., 6 h, 75%. (v) Et_3N, 80°C, 40 min, 85% (**37**) and 93% (**38**). (vi) $TMSCH_2CN$, TMSOTf, −20°C, 23%

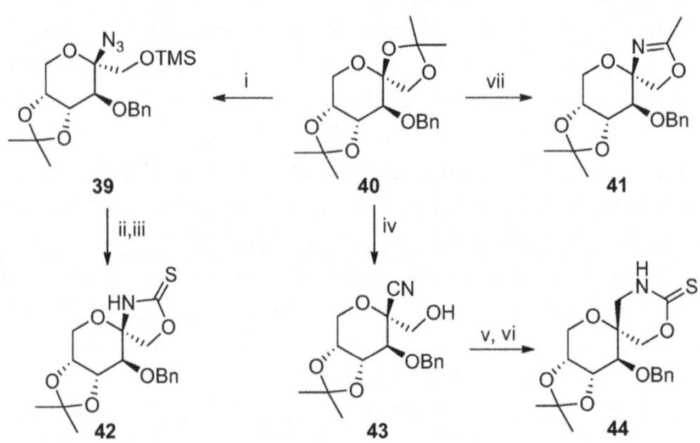

Scheme 10 *Reagents and conditions*: (i) 1. $TMSN_3$, 0°C, 5 min; 2. TMSOTf, 0°C, 5 min, 85%. (ii) H_2/Pd-C, r.t., 2 h, 90%. (iii) 1. $Bu_4NF{\cdot}3H_2O$, r.t., 1 h; 2. Im_2CS, r.t., 3 h, 83%. (iv) 1. TMSCN, −20°C, 5 min; 2. TMSOTf, 0°C, 5 min; 3. $Bu_4NF{\cdot}3H_2O$, r.t., 8 h, 55%. (v) 1. $LiAlH_4$, 0°C, 30 min; 2. Im_2CS. (vi) Et_3N, r.t., 10 h, 80%. (vii) $TMSCH_2CN$, TMSOTf, −20°C, 1 h, 75%

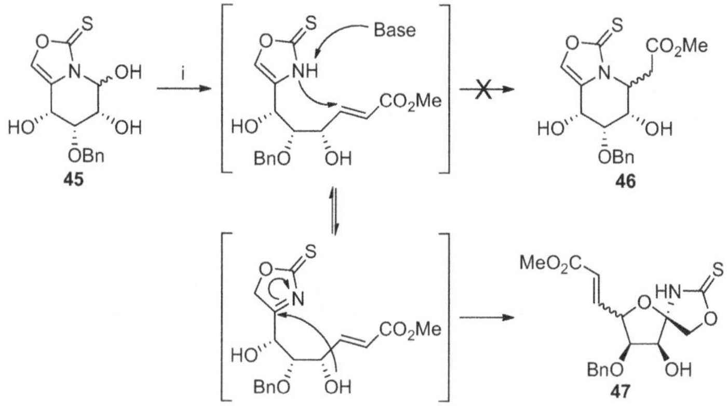

Scheme 11 *Reagents and conditions*: (i) $Ph_3P{=}CHCOOMe$, benzoic acid, THF, reflux, 8 h, 87%

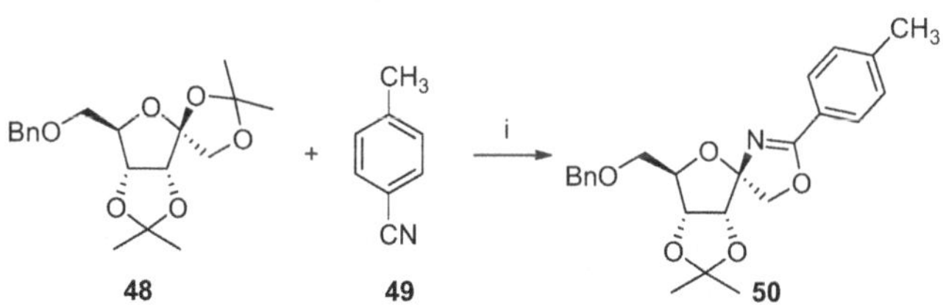

Scheme 12 *Reagents and conditions*: (i) TMSOTf, toluene, 0°C to r.t., 1 h, 22%

As yet another example of spiro-oxazoline furanosides, a very straightforward procedure for the synthesis of spirocyclic oxazoline **50** has been recently described [102, 103]. The synthetic route involves the trimethylsilyl trifluoromethanesulfonate (TMSOTf)-promoted nucleophilic addition of nitrile **49** to protected psicofuranose derivative **48** (Scheme 12).

Related to the spirodiketopiperazine skeleton, the spiro-derivative **54** was synthesized from mannono-lactone **51** in a route involving an indium-mediated Reformatsky reaction with ethyl α-bromoisobutyrate as the key step (Scheme 13) [104]. The resulting ulosonate **52** was acetylated followed by reaction with trimethylsilyl azide in the presence of a Lewis acid to give azide **53**. Catalytic hydrogenation and reaction of the resulting anomeric mixture of amino esters with phenyl isocyanate afforded, after treatment with base, spirocycle **54**.

In 2004, Taillefumier et al. reported the preparation of a novel class of spironucleosides based on a 1,4-diazepine-2,5-dione moiety [105]. Starting from *exo*-glycal **55**, aza-Michael addition of benzylamine, followed by catalytic hydrogenation of the benzyl group of the resulting amino ester **56** and coupling of the free amine with Cbz-Ala-OH, furnished dipeptide **57** (Scheme 14). After saponification of the methyl ester and hydrogenolysis of the Cbz group of compound **57**, base-promoted cyclization finally gave the desired spiro-derivative **58**.

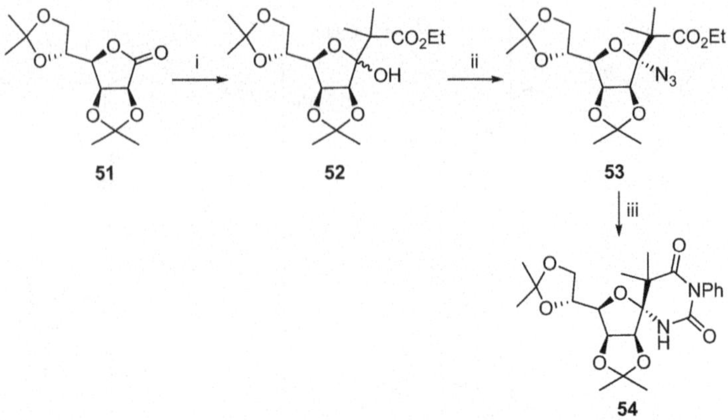

Scheme 13 *Reagents and conditions*: (i) In, $BrC(CH_3)_2CO_2Et$, THF, US, r.t. (ii) 1. Ac_2O, Et_3N, CH_2Cl_2, r.t., 14 h; 2. $TMSN_3$, TMSOTf, powdered MS, CH_2Cl_2, r.t., 14 h, 71%. (iii) 1. Pd/C, MeOH, r.t., 14 h; 2. PhCNO, toluene, r.t., 3 h, 85%; 3. NaOMe, r.t., 14 h, quant

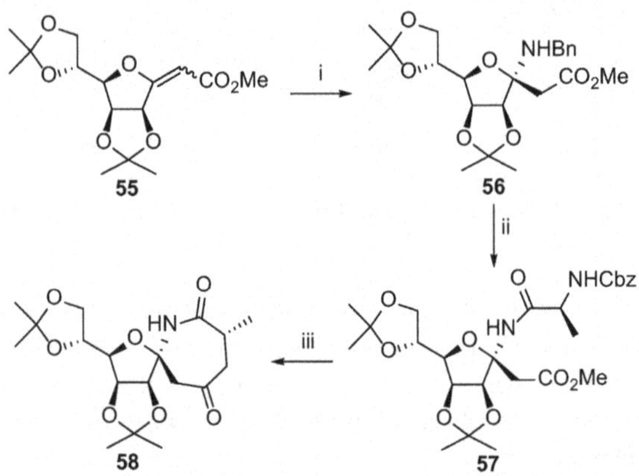

Scheme 14 *Reagents and conditions*: (i) Neat $BnNH_2$, 48 h. (ii) 1. H_2/1 atm, 10% Pd–C, EtOH/EtOAc: 1.5/1; 2. Cbz-Ala-OH, PyBOP, Et_3N, DMF, r.t., 14 h. (iii) 1. K_2CO_3, MeOH/H_2O: 10/1, r.t., 48 h; 2. H_2/1 atm, 10% Pd–C, EtOH/EtOAc: 1.5/1; 3. Diphenylphosphorylazide (DPPA); Et_3N, DMF, 0°C to r.t.; 14 h, 47%

On the other hand, Zhang et al. described the synthesis of spironucleosides with a six-membered 1,3-oxazinan-2-one ring [106]. The synthetic route, depicted in Scheme 15, involves a microwave-assisted cycloaddition of exo-glycal **59** and nitrone **60** to afford the spiroisoxazolidines **61** and **62**. Zinc-mediated reductive cleavage of the O–N bond, followed by reaction with triphosgene furnished, after catalytic hydrogenation, the spiro-oxazinanone glycosides **63** and **64**.

Shi and co-workers described the synthesis of morpholinone spironucleosides, which were shown to be effective for the asymmetric epoxidation of olefins [107, 108]. Starting from aminoketose **65**, reaction with 2-bromo-2-methylpropanoyl

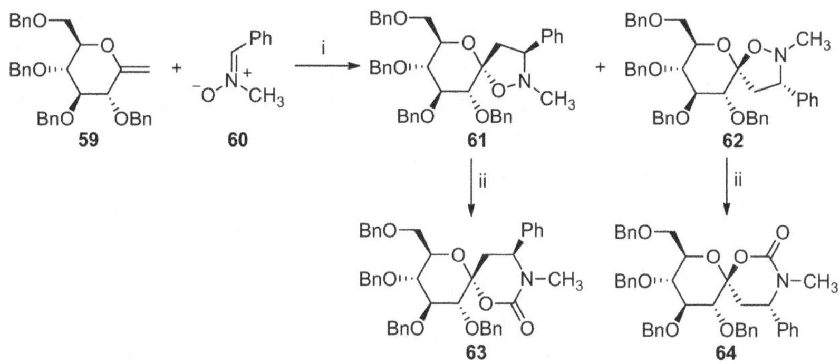

Scheme 15 *Reagents and conditions*: (i) diglyme, MW, 200°C, 2 min, 73%. (ii) 1. Zn, 85% AcOH, r.t.; 2. triphosgene, Et_3N, 0°C to r.t., CH_2Cl_2; 3. MeOH, $Pd(OH)_2/C$, H_2, 34% of **63** and 28% of **64**

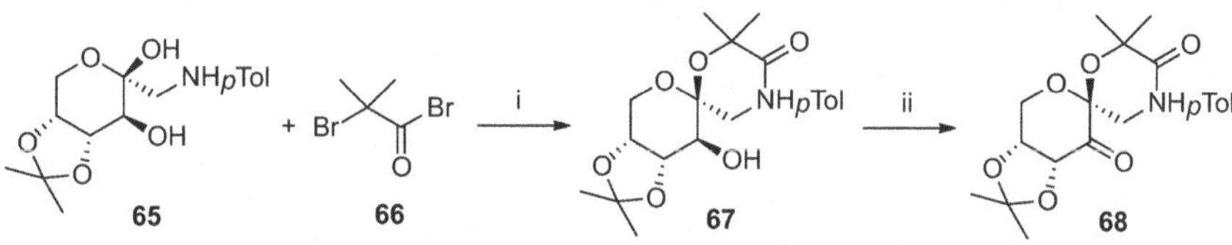

Scheme 16 *Reagents and conditions*: (i) 1. 2-bromo-2-methylpropanoyl bromide, $HNaCO_3$, CH_2Cl_2; 2. NaH, THF, 45%. (ii) PDC, CH_2Cl_2, 95%

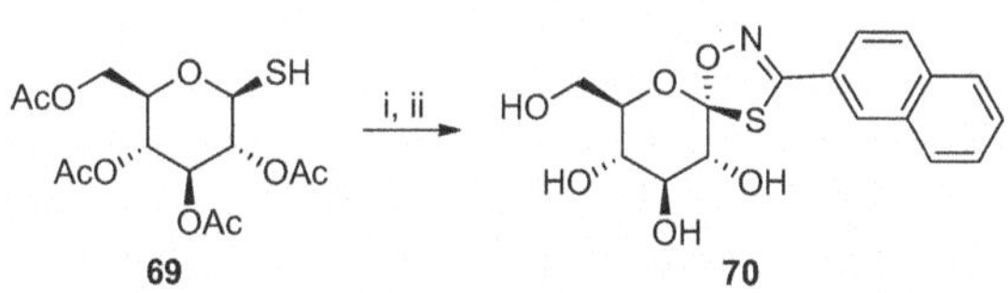

Scheme 17 *Reagents and conditions*: (i) NaphC(Cl) = NOH, Et_3N, CH_2Cl_2, r.t., 78%. (ii) 1. NBS, hν, CCl_4, reflux, 45 min, 36%; 2. NaOMe, MeOH, r.t., 94%

bromide **66**, followed by base-promoted cyclization, afforded spiro compound **67** (Scheme 16). Oxidation of **67** with PDC finally gave the desired morpholinone spironucleoside **68**.

Somsák, Praly et al. developed a series of oxathiazol spironucleosides of great pharmacological relevance [109, 110]. For example, they reported the synthesis of a 2-naphthyl-substituted glucopyranosylidene-spiro-oxathiazole, a very potent inhibitor of rabbit muscle glycogen phosphorylase b (K_i 160 nM) [111]. Thus, reaction of the 1-thioglucopyranose **69** with *N*-hydroxy-2-naphthimidoyl chloride, followed by photochemical spirocyclization of the resulting nitrile oxide, afforded spironucleoside **70** (Scheme 17).

The interesting biological profile of glucopyranosylidene-spiro-oxathiazoles sparkled the interest in the synthesis of several derivatives [112–114]. For example, Praly and co-workers reported the preparation of glucose-based spiro-isoxazolines

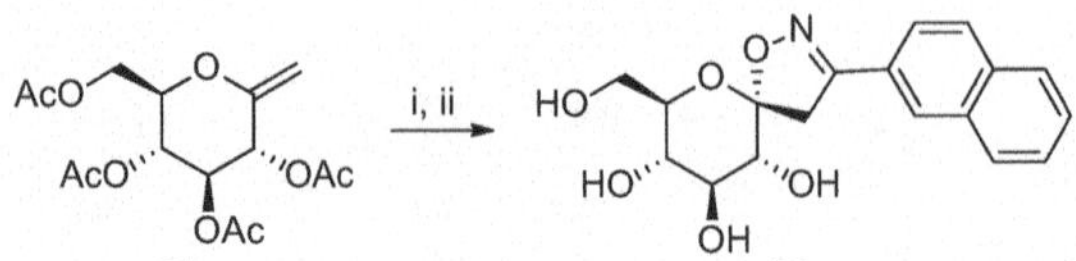

71 72

Scheme 18 *Reagents and conditions*: (i) (NaphC(Cl) = NOH, Et_3N, CH_2Cl_2, r.t., 94%. (ii) NaOMe, MeOH, rt, 3 h, 78%

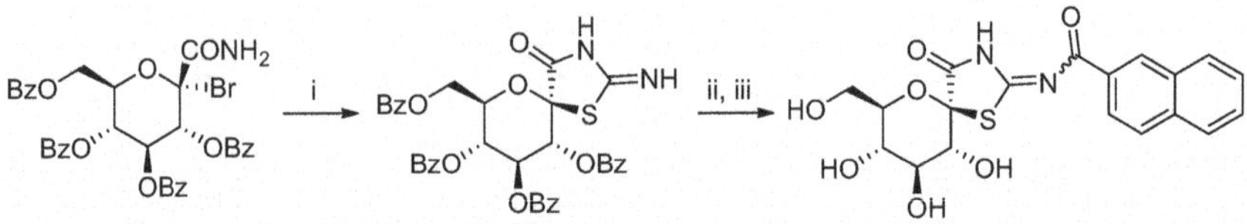

73 74 75

Scheme 19 *Reagents and conditions*: (i) thiourea, EtOH, MW, 1 h, 97%. (ii) 2-Naphthoyl chloride, Py, r.t., 89%. (iii) NaOMe, MeOH, r.t., 50%

by means of a regio- and stereoselective [3 + 2]-cycloaddition between *exo*-glucal **71** and a nitrile oxide generated in situ from *N*-hydroxy-2-naphthimidoyl chloride and triethylamine (Scheme 18) [115, 116]. The new family of spironucleosides **72** were potent inhibitors of muscle glycogen phosphorylase b, with K_i values ranging from 0.63 to 92.5 μM.

On the other hand, Somsák's group also described the preparation and biological evaluation of glucopyranosylidene-spiro-iminothiazolidinones [117]. Treatment of 1-bromo-amide **73** with thiourea under microwave heating conditions afforded the spiro-thiazolidinone **74** (Scheme 19). Reaction of **74** with 2-naphthoyl chloride, followed by debenzoylation under Zemplén conditions, yielded 2-naphthoylimino-thiazolidinone **75**, inhibitor of glycogen phosphorylase in the micromolar range ($K_i = 10$ μM).

2.5 Polycyclic Spironucleosides

Another subtype of spironucleosides are those in which the base linked to the anomeric position of the sugar is a polycyclic system, giving rise to conformationally fixed models. This family of derivatives is very useful to elucidate the glycosidic torsion angle of nucleosides.

Gimisis et al. reported the synthesis of several polycyclic spironucleosides [37, 118]. A relevant example is the synthesis of orthoamidic polycyclic spironucleoside **78**, depicted in Scheme 20 [119]. Starting from 1′-*C*-cyano-pyrimidine nucleoside **76**, treatment with methyllithium afforded spironucleoside **77**, which on reaction with ammonium fluoride in refluxing MeOH gave the corresponding desilylated nucleoside **78**.

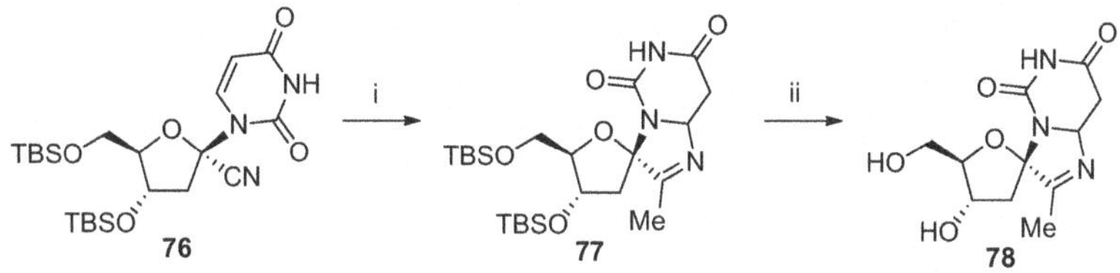

Scheme 20 *Reagents and conditions*: (i) MeLi, THF, −78°C, 20 min, 52%. (ii) NH_4F, MeOH, reflux, 14 h, 90%

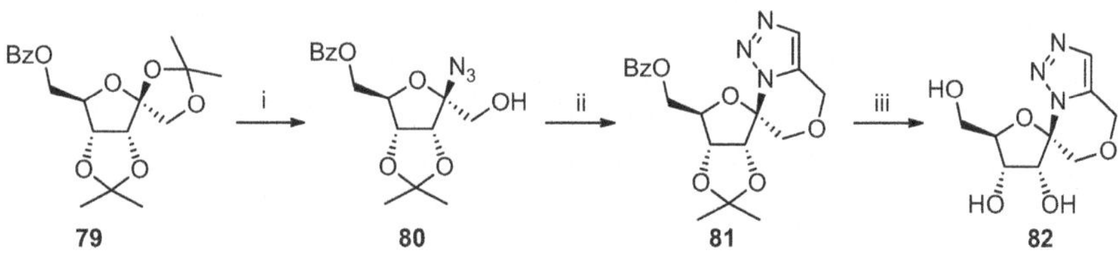

Scheme 21 *Reagents and conditions*: (i) 1. $TMSN_3$, TMSOTf, 4 Å MS, CH_3CN, 0°C, 5 min, 89%; 2. AcOH, MeOH, acetone, 98%. (ii) 1. Propargyl bromide, CH_3CN, 0°C, 2 h; 2. toluene, reflux, 24 h, 51%. (iv) 1. NH_3, MeOH; 2. Dowex-H^+, MeOH/H_2O 8:2, 50°C, 56% over two steps

Dell'Isola and co-workers reported the synthesis spirocyclic triazolooxazine nucleoside **82** from D-psicopyranose derivative **79** (Scheme 21) [120].

Reaction of **79** with trimethylsilyl azide in the presence of trimethylsilyl triflate afforded, after acidic hydrolysis, anomeric azide **80**. *O*-Alkylation of compound **80** with propargyl bromide, followed by intramolecular 1,3-dipolar cycloaddition of the resulting propargyl ether intermediate in refluxing toluene, gave the protected nucleoside **81**. Finally, deacylation and hydrolysis of the isopropylidene group furnished the desired polycyclic spironucleoside **82**.

Also related to the structure of polycyclic spironucleosides are pyrrole spiroketal alkaloids (PSAs), a family of natural products that have been reported in the last decade [121]. The synthesis of PSA capparisine B (**87**) was achieved from *N*-fructos-1-yl-pyrrole **83** [122] in a route involving bis-hydroxymethylation of the pyrrole and oxidation to afford dialdehyde **84** (Scheme 22) [123]. After selective protection of one formyl group with a neopentyl glycol (NPG) moiety and benzylation, acidic hydrolysis yielded the key intermediate **85**. Reduction of one formyl group in **85** with sodium borohydride, followed by acid-catalyzed cyclization of the intermediate alcohol, afforded compound **86** as the major product. Dehydroxylation of **86** through a modified Barton-McCombie reaction, followed by and debenzylation, finally furnished the natural product **87**.

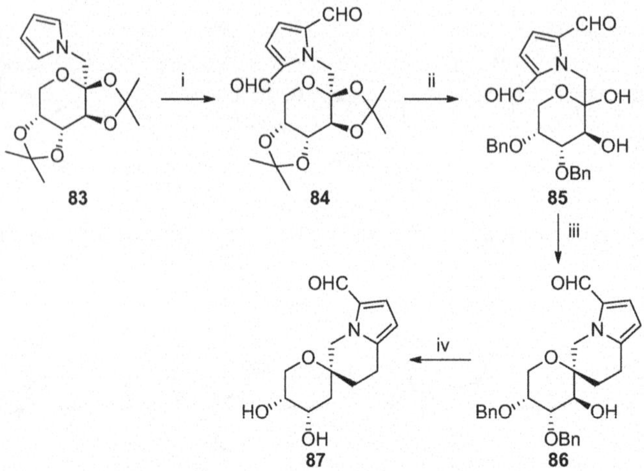

Scheme 22 *Reagents and conditions*: (i) 1. HCHO, K_2CO_3, H_2O, MW; 2. MnO_2, acetone, 87% over two steps; 3. DDQ, MeCN, H_2O, 85%. (ii) 1. NPG, PTSA, CH_2Cl_2, 97%; 2. BnCl, NaH, DMF, 88%; 3. AcOH, H_2O, 100°C, 93%. (iii) 1. $NaBH_4$, MeOH, 90%; 2. PTSA, CH_2Cl_2, 67%. (iv) 1. Imidazole, NaH, CS_2, MeI, THF, 88%; 2. Benzoyl peroxide, Et_3SiH, dioxane, 100°C, 78%; 3. $TiCl_4$, CH_2Cl_2, 81%

3 Spiropseudonucleosides

3.1 C-2′-Spiropseudonucleosides

In the search for effective treatment options for hepatitis C virus (HCV), nucleoside derivatives that are inhibitors of the HCV NS5B RNA-dependent RNA polymerase have been extensively investigated. In this regard, Jonckers hypothesized that 2′-deoxy-2′-spirocyclopropylcytidine **92** could provide a useful and particularly attractive analogue in the class of HCV-inhibiting nucleosides, because the 2′-carbon is not a stereogenic center [124]. With the aim of evaluating this nucleoside analogue as HCV NS5B inhibitor, these researchers also developed a new and more efficient procedure for the multigram-scale synthesis of **92**. Starting from uridine **88**, protection of the 3′- and 5′-OH groups as a cyclic disilylether, followed by Dess-Martin oxidation of the free 2′-OH group and Wittig olefination of the resulting ketone using methyltriphenylphosphonium bromide, afforded vinylic analogue **89** (Scheme 23). After benzoylation, the creation of the spirocyclopropyl fragment was achieved through a 1,3-dipolar cycloaddition reaction of diazomethane. The resulting mixture of spiropyrazolines **90** was subjected to a photochemically induced nitrogen extrusion reaction to provide 2′-spirocyclopropane derivative **91**. The target nucleoside **92** was easily obtained from intermediate **91** by debenzoylation, ammonolysis, and desilylation. Regarding the biological activity, 2′-deoxy-2′-spirocyclopropylcytidine **92** inhibits HCV NS5B polymerase with an EC_{50} of 7.3 μM, and no associated cytotoxicity was observed.

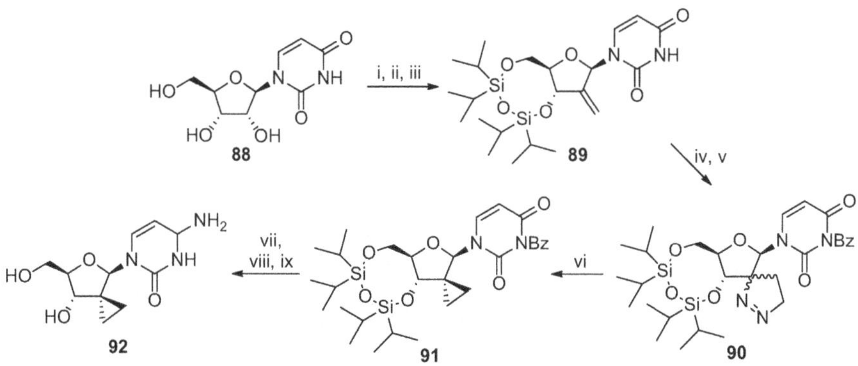

Scheme 23 *Reagents and conditions*: (i) TIPDSCl, Py, r.t., 12 h, 70%. (ii) Dess-Martin periodinane, CH_2Cl_2, r.t., 12 h, 87%. (iii) methyltriphenylphosphonium bromide, *sec*-BuLi, THF, r.t., 12 h, 60%; (iv) BzCl, Hünig's base, CH_2Cl_2, r.t., 12 h, 86%. (v) CH_2N_2, diethyl ether, r.t., 48 h, 84%. (vi) hν, toluene/CH_3CN, benzophenone, r.t., 3 h, 84%. (vii) 1. NH_3, MeOH, r.t., 2 h, 40%; 2. 2,4,6-triisopropylbenzenesulfonyl chloride, DMAP, Et_3N, CH_3CN, r.t., 12 h. (viii) NH_3/H_2O, CH_3CN, r.t., 3 h, 93%. (ix) TBAF, THF, 62%

Considering the interesting biological profile of the 2′-spirocyclopropanic nucleosides, Shen and Hong envisioned the synthesis of a novel class of nucleosides comprising 2′-spirocyclopropyl-5′-deoxyphosphonic acid furanosyl nucleoside analogues in order to search for more effective therapeutics against human immunodeficiency virus (HIV) and to provide analogues for probing the conformational preferences of enzymes associated with the nucleoside kinases [125]. 2′-Spirocyclopropyl furanosyl donor was prepared from monosilyl-cyclopropanoid **93** [126]. Swern oxidation of **93**, followed by Wittig reaction, hydroboration-oxidation, and Swern reaction, afforded aldehyde **94** (Scheme 24). Carbonyl addition reaction by vinylmagnesium bromide and PMB protection, followed by desilylation and oxidation, furnished the allylic alcohol **95**. Removal of the PMB protective group produced, after acetylation, lactol intermediate **96**. The synthesis of adenine nucleoside was carried out by condensation of **96** with silylated 6-chloropurine using trimethylsilyl trifluoromethanesulfonate (TMSOTf) as a catalyst in dichloroethene (DCE) to give protected 6-chloropurine derivatives **97a** and **97b**. Cross-metathesis of **97b** with diethylphosphonate using the second-generation Grubbs catalyst gave vinylidene phosphonate nucleoside analogue **98**. Ammonolysis and hydrolysis of **98** afforded adenosine 2′-spirocyclopropyl phosphonic acid **100**. On the other hand, saturation of the vinylidene phosphonate by catalytic transfer hydrogenation, followed by ammonolysis and hydrolysis, gave the 2′-spirocyclopropyl phosphonic acid analogue **99**. Adenine phosphonic acid analogue **99** shows significant anti-HIV activity ($EC_{50} = 7.9\ \mu M$).

On view of the potent antiviral activity of **99**, the same authors synthetized a series of related 2′-spirocyclopropyl-4′-deoxythreosyl phosphonic acid nucleosides [127]. Starting from glycosyl donor **101**, easily available from 1,4-dihydroxy-2-butene, nucleoside analogues **104** and **105** were readily obtained (Scheme 25). Thus,

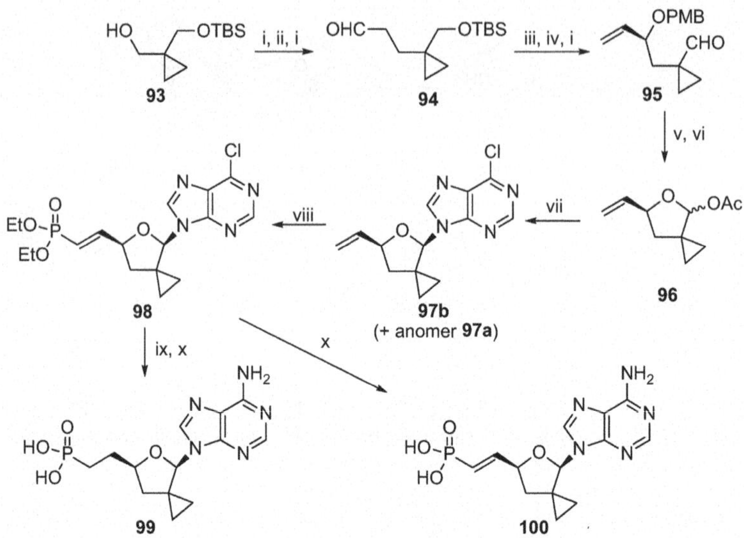

Scheme 24 *Reagents and conditions*: (i) $(COCl)_2$, DMSO, TEA, CH_2Cl_2, 93%. (ii) 1. *n*-BuLi, Ph_3PCH_3I, PPh_3, THF; 2. BH_3/THF; 3. NaOH, H_2O_2, H_2O, 60% (two steps). (iii) VinylMgBr, THF, 76%; (iv) 1. PMBCl, DMF, NaH, 67%; 2. TBAF, THF, 81%; (v) DDQ, CH_2Cl_2, H_2O, 63%. (vi) Ac_2O, pyridine, 86%. (vii) silylated 6-chloropurine, TMSOTf, DCE, **97a** 32% and **97b** 33%. (viii) Vinyldiethylphosphonate, Grubbs cat.(II) CH_2Cl_2, 61%. (ix) Pd/C, cyclohexene, MeOH, 76%. (x) 1. NH_3/MeOH; 2. TMSBr, 2,6-lutidine, CH_3CN, 45% (over two steps)

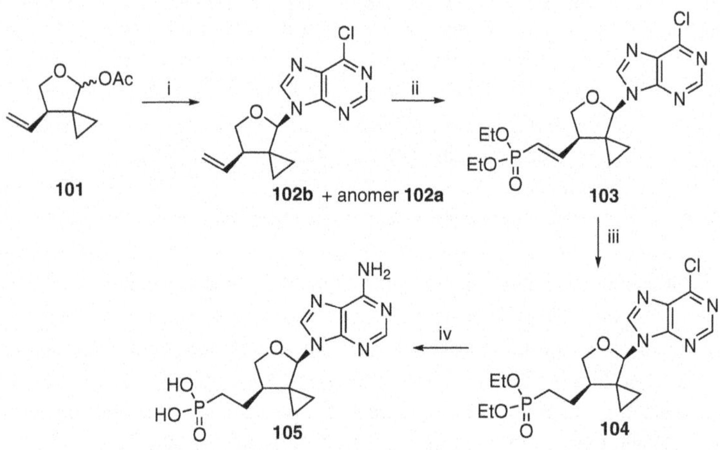

Scheme 25 *Reagents and conditions*: (i) silylated 6-chloropurine, TMSOTf, DCE 32% (**102a**) and 33% (**102b**). (ii) Diethyl vinylphosphonate, Grubbs cat.(II) CH_2Cl_2, 60%. (iii) Pd/C, cyclohexene, MeOH, 81%. (iv) 1. NH_3/MeOH; 2. TMSBr, 2,6-lutidine, CH_3CN, 44% two steps

condensation of compound **101** with silylated 6-chloropurine using TMSOTf as a catalyst afforded a separable mixture of anomers **102a** and **102b**. Cross-metathesis of **102b** with diethyl vinylphosphonate yielded **103**, which, on transfer catalytic hydrogenation followed by ammonolysis and hydrolysis of the resulting 6-chloropurine **104**, finally gave adenine phosphonic acid analogue **105**. However, none of the 4′-deoxythreosyl phosphonic acid analogues showed any anti-HIV activity or cytotoxicity up to 100 μM.

In 2003, Babu and co-workers developed a series of conformationally restricted 2′-spirooxetane and tetrahydrofuran nucleosides for their use as conformationally restricted mechanistic probes for ribonucleotide reductases [128]. Introduction of an allyl group into ribofuranose **106**, followed by coupling with persilylated thymine using $SnCl_4$ as promoter provided, after deprotection with saturated methanolic ammonia, 2′-*C*-allyl ribonucleoside **107** (Scheme 26). Silylation of **107** gave derivative **108**, which on hydroboration-oxidation, followed by selective mesylation and subsequent sodium hydride-induced cyclization afforded, after desilylation, the desired 2′-spirofurano ribonucleoside **112**. For the synthesis of the 2′-spirooxetane analogue, the terminal double bond of compound **108** was oxidatively cleaved using catalytic osmium tetroxide and sodium periodate as a co-oxidant and then reduced by treatment with sodium borohydride, resulting in the formation of ribonucleoside **109**. Selective mesylation followed by base-induced ring closure afforded, after desilylation, the 2′-spiro ribonucleoside **111**.

Using the chemistry developed by Babu and co-workers, Du et al. prepared 2′-spirocyclic ethers of nucleoside triphosphates **115** and **116** from nucleosides **113** and **114** (Scheme 27) [129]. The triphosphate analogue **116** demonstrated potent

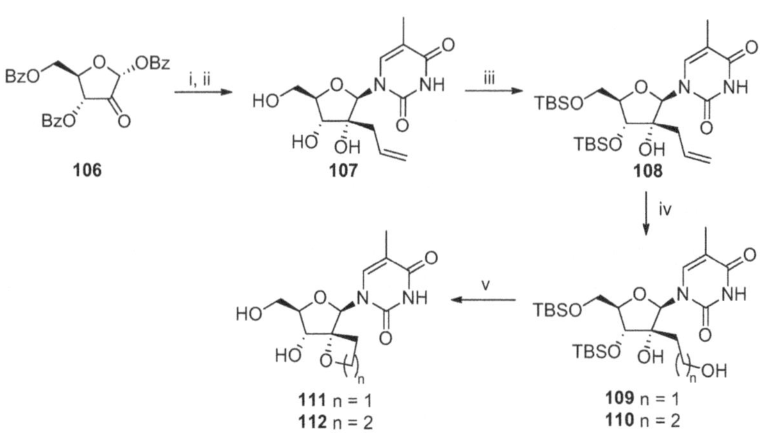

Scheme 26 *Reagents and conditions*: (i) allylMgBr, $CeCl_3$, THF, −78°C, 53%. (ii) 1. Thymine, $SnCl_4$, BSA, CH_3CN. 2. Saturated methanolic NH_3 (54% two steps). (iii) TBDMSCl, imidazole, DMF, 84%. (iv) For $n = 1$: BH_3 (7.8 M in 1,4-oxathiane), THF then aq. NaOH (2 M), H_2O_2, 55% of **109**. For $n = 2$: 1. OsO_4 (2.5 M in *n*-butanol), $NaIO_4$, THF, H_2O, 2. $NaBH_4$, THF, H_2O, 47% of **110**. (v) For $n = 1$: 1. MsCl, pyridine, CH_2Cl_2, 58%; 2. NaH, THF, 77% of **111**; For $n = 2$: MsCl, DMAP, Et_3N, CH_2Cl_2, 64% of **112**

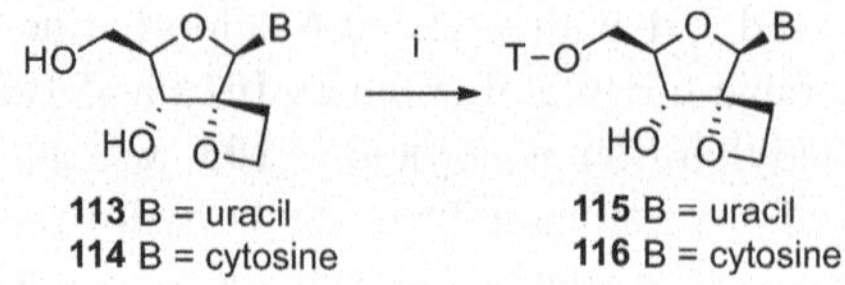

Scheme 27 *Reagents and conditions*: (i) $POCl_3/Bu_3N/Bu_3N{\cdot}$pyrophosphate/1,3-dimethyl-3,4,5,6-tetrahydro-2(1*H*)-pyrimidinone (DTP). *T* triphosphate

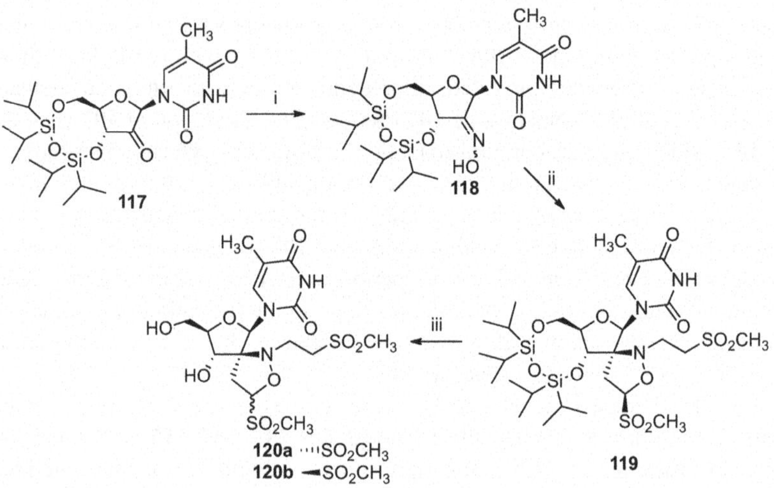

Scheme 28 *Reagents and conditions* (i) $H_2NOH{\cdot}HCl$, Py, r.t., 48 h, 67%. (ii) Methylvinylsulfone, toluene, 28 h, 66%. (iii) TBAF, THF, 0°C, 1 h, 58%

intrinsic biochemical activity against the NS5B polymerase, with $IC_{50} = 8.5\ \mu M$. On the other hand, phosphoramidate prodrug of 2′-oxetane uridine **113** demonstrated potent anti-HCV activity in vitro, and the corresponding triphosphate **115** retained similar potent activity against both wild-type and S282T HCV NS5B polymerase ($IC_{50} = 10.4\ \mu M$).

As stated in the introduction, several types of spironucleoside analogues can lock the sugar conformation in either the N or S conformation, thus being of interest in antisense and RNA interference strategies. In order to determine the effect it had on sugar dynamics, Versteeg et al. introduced a spiroisoxazolidine ring at the 2′-position of thymidine nucleoside analogues [130]. The synthesis (Scheme 28) began with the reaction of disiloxane-protected ketone **117** with hydroxylamine hydrochloride to give the 2′-oxime **118**, which on cycloaddition with methylvinylsulfone in refluxing toluene afforded compound **119** as a single isomer. Deprotection of **119** with fluoride ion occurred with concomitant epimerization at C7′ to give a 1:2 mixture of stereoisomers **120a** and **120b**. Proton relaxation measurements and coupling constant analysis indicated the sugar to be locked in the N conformation with substantial sugar rigidity.

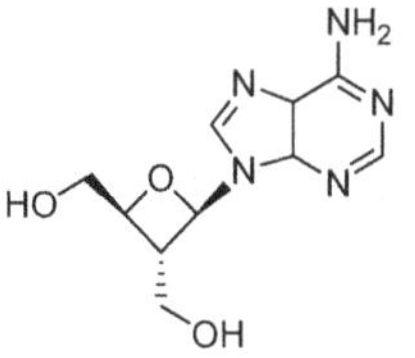

Fig. 2 Nucleoside oxetanocin-A2

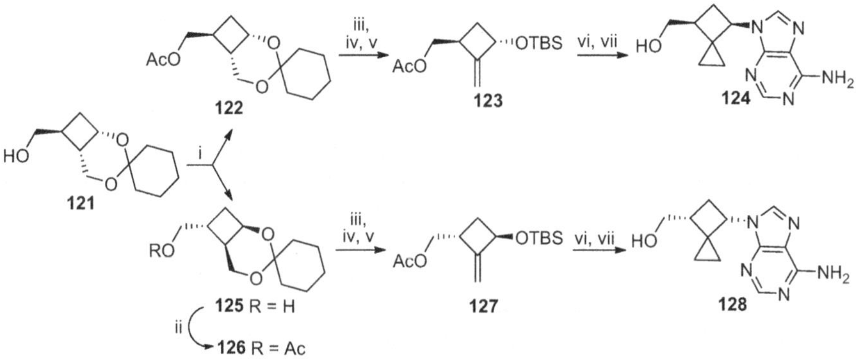

Scheme 29 *Reagents and conditions*: (i) *P. cepacia* lipase, vinyl acetate, 28°C, 2 h. (ii) Ac_2O, Py, CH_2Cl_2. (iii) 1. Amberlite IR 120 (H^+), MeOH, r.t., 1 h, 2. Ph_3CCl, Py, r.t., overnight; 3. TBDPS-Cl, imidazole, CH_2Cl_2, r.t., 1 h. (iv) 1. $BF_3{\cdot}OEt_2$, MeOH, CH_2Cl_2, r.t., 1 h; 2. $MeP(OPh)_3I$, DMF, 1 h, r.t.; (v) DBU, THF reflux, 24 h. (vi) 1. TBAF, THF, r.t., 1 h; 2. $(C_2H_5)_2Zn$, CH_2I_2, ether, 0–45°C. (vii) 1. 6-chloropurine, PPh_3, DIAD, THF, 0°C to r.t., 16 h. 2. NH_3, MeOH, 100°C, 24 h

Since the discovery of the potent antiviral activity of oxetanocin-A2 (Fig. 2), numerous studies have been devoted to explore the chemistry and biological activity of four-membered-ring containing nucleosides.

Bondada and co-workers reported in 2004 the synthesis of both enantiomers of the 2′-spirocyclopropane derivatives **124** and **128** (Scheme 29) [131]. Starting from compound **121**, enzymatic resolution with *P. cepacia* (PS) gave acetate **122** and alcohol **125**, which was then converted into acetate **126**. Acidic hydrolysis of compound **122**, followed by tritylation of the primary alcohol, silylation of the secondary alcohol, removal of the trityl group, and subsequent iodination, afforded an iodide intermediate, which on DBU-mediated elimination conducted to compound **123**. Desilylation, cyclopropanation under Simmons-Smith conditions, and condensation of the hydroxyl group with 6-chloropurine under Mitsunobu conditions afforded, after ammonolysis, nucleoside **124**. Following a similar procedure, acetate **126** was converted into the (*S*)-enantiomer **128**.

3.2 C-3′-Spiropseudonucleosides

Searching for new nucleoside derivatives as inhibitors of HIV replication, Camarasa et al. synthesized a series of C-3′-spirooxathiole nucleosides [132]. Among them, [1-[2′,5′-bis-*O*-(*tert*-butyldimethylsilyl)-D-ribofuranosylthymine]-3′-spiro-5″-[4″-amino-1″,2″-oxathiole 2″,2″-dioxide] (TSAO) **129** exhibited potent (EC_{50} = 0.034 μg/mL, CC_{50} = 7.7 μg/mL) and selective inhibition on HIV-1 replication [133] (Fig. 3).

As a consequence of TSAO's highly interesting biological activity, great efforts were devoted to the development of analogues in which the cytotoxicity would be decreased without compromising the antiviral activity. To achieve this, the most common strategy is the introduction of slight modifications in either the base or the sugar ring [134–137]. However, modifications in the sultone moiety were also reported. For example, Camarasa's group developed a series of TSAO analogues functionalized on the sultone moiety via Pd-catalyzed cross-coupling reaction (Scheme 30) [138, 139]. The required iodo precursor **130** was prepared by treatment of TSAO **129** with elemental iodine and ceric ammonium nitrate (CAN). Stille cross-coupling of **130** with allyl stannane gave nucleoside **131**. Nevertheless, no antiviral activity at subtoxic concentration was achieved.

In 2005, a series of (TSAO) derivatives substituted at the 4″-amino group of the sultone moiety with different carbonyl functionalities were designed and synthesized [140]. For example, reaction of **132** with an excess of ethoxycarbonyl isocyanate afforded the corresponding 4″-*N*-alkyl- and -acyl-substituted ureido derivative **133** (Scheme 31). On contrary, when **132** reacted with ethoxycarbonylmethyl isocyanate

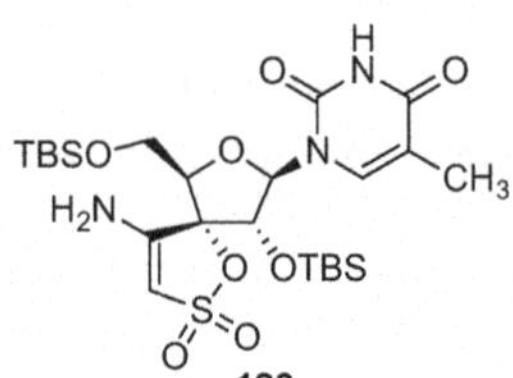

129

Fig. 3 Structure of TSAO

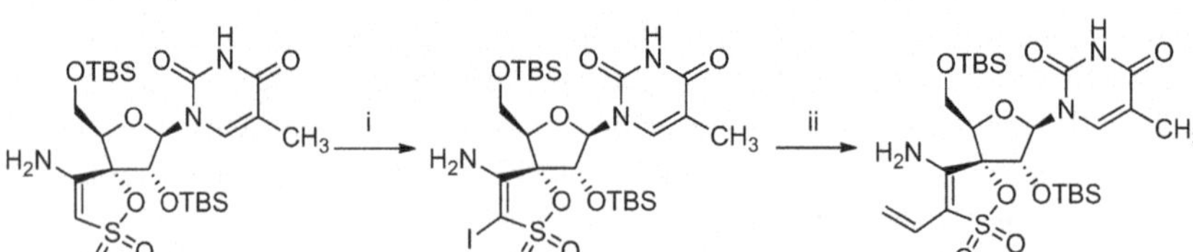

129 **130** **131**

Scheme 30 *Reagents and conditions* (i) I_2, CAN, Et_3N, 80°C. (ii) *n*-$Bu_3Sn(C{=}CH_2)$, $Pd_2(dba)_3$, $AsPh_3$, CuI, NMP

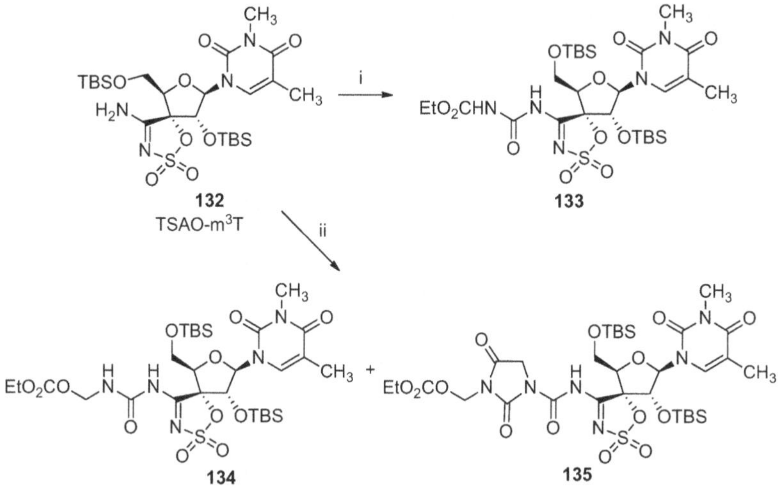

Scheme 31 *Reagents and conditions*: (i) $OCNCH_2CO_2Et$, CH_3CN, 80°C, 80%. (ii) $CH_3CH_2OCOCH_2NCO$, NEt_3, CH_3CN, 100°C, **134** (52%), **135** (30%)

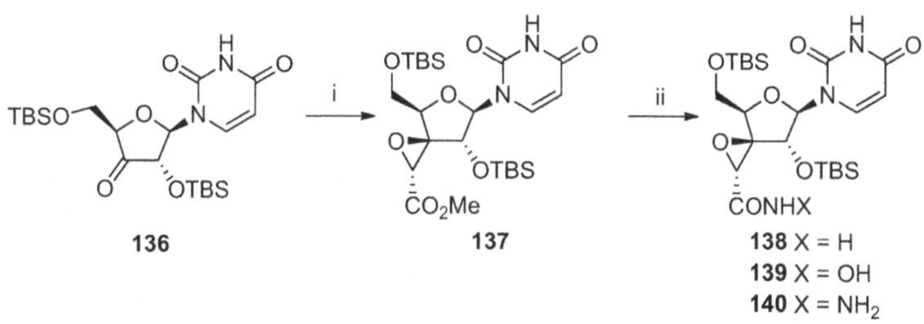

Scheme 32 *Reagents and conditions*: (i) $BrCH_2COEt$, LiHMDS, $CeCl_3$, THF, −78°C, 78%. (ii) For X = H: NH_3, MeOH, r.t., 16 h, 90%; for X = OH: NH_2OH, Et_3N, MeOH, r.t., 6 h, 79%; for X = NH_2: NH_2NH_2, MeOH, r.t., 2 h, 87%

in a sealed tube in the presence of a catalytic amount of triethylamine at 100°C, the corresponding *N*-substituted ureido derivative **134** was isolated together with the unexpected cyclic compound **135**. The compounds were evaluated for their inhibitory effect in cell cultures of both wild-type and TSAO-resistant HIV-1 strains. The most relevant biological feature is that some of the studied compounds also showed inhibition of human cytomegalovirus (HCMV) replication at subtoxic concentrations. This had never been observed previously for TSAO derivatives. In particular, compound **134** represents the first TSAO derivative with dual anti-HIV-1 and -HCMV activity.

As yet another example of 3′-spiropseudonucleosides, Tronchet et al. reported the synthesis and biological evaluation of a series of 3′-spirooxirane derivatives [141]. For example, 3′-ketonucleoside derivative **136** was submitted to a Darzens reaction with methyl bromoacetate and using lithium bis(trimethylsilyl)amide as a base to give compound **137**, which upon ammonolysis afforded **138** (Scheme 32).

In the same way, treatment of **137** with hydroxylamine afforded **139**, whereas hydrazinolysis led to **140**. Following the same synthetic route, the 3′-spiro-epimeric oxiranes were also prepared [140]. The new 3′-spirooxirane nucleoside analogues were moderately cytotoxic with slightly selective antiproliferative activities in some cases. None of these compounds were active against HIV, but some other antiviral activities against herpes simplex virus type 2 (HSV-2), cytomegalovirus (CMV), Epstein-Barr virus (EBV), or varicella-zoster virus (VZV), in the micromolar range, were noted.

In all these nucleoside analogues, the base is located in the anomeric position, and the spirocyclic ring is considered as a part of the sugar moiety, a feature introduced in the sugar structure in order to add a conformational restriction. However, there are a couple of examples in the literature of 3′-spiropseudonucleosides, which contain a spirocyclic base in the 3′-position of the sugar ring instead of position 1′. For example, Gash and co-workers reported a general procedure for the preparation of 3′-spiropseudonucleosides with diverse heterocyclic bases starting from a D-ribo-ketose **141** (Scheme 33) [142]. Compound **141** was transformed into **144** with sodium cyanide, followed by reduction with lithium aluminum hydride. This reduction was also applied to obtain compound **145**, when **141** was previously transformed into the imine by reaction with ammonia. Compounds **148** and **149**

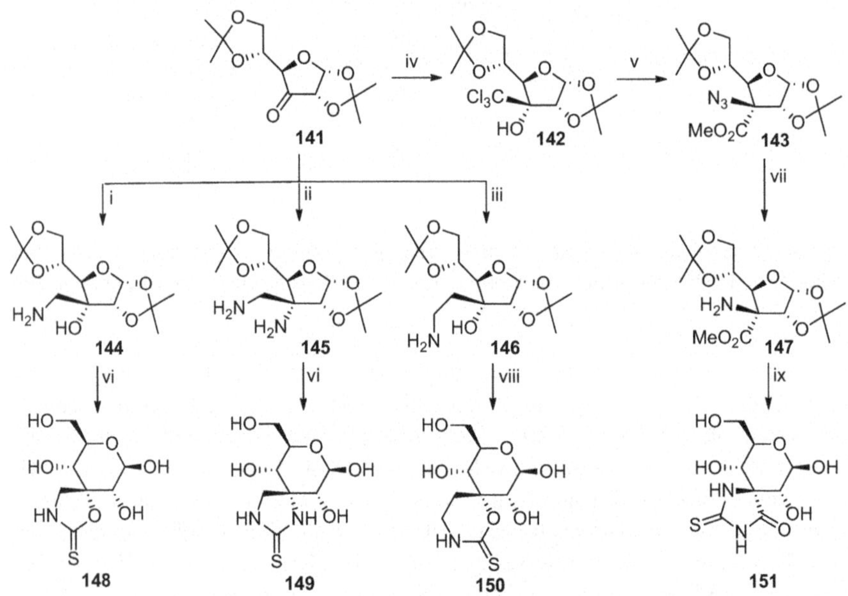

Scheme 33 *Reagents and conditions*: (i) 1. NaCN; 2. $LiAlH_4$, Et_2O, 0°C to r.t. (ii) 1. NH_3, NaCN; 2. $LiAlH_4$, Et_2O, 0°C to r.t. (iii) 1. $TMSCH_2CN$, TBAF, THF, r.t.; 2. $LiAlH_4$, Et_2O, 0°C to r.t. (iv) $CHCl_3$/LiHMDS, THF, −78°C. (v) NaN_3/DBU,18-crown-6, MeOH, 50°C; (vi) 1. Im_2CS, CH_2Cl_2, r.t.; 2. HCl/MeOH, 60°C. (vii) H_2/Pd, MeOH, r.t. (viii) 1. $CSCl_2$, $CaCO_3$, CH_2Cl_2/H_2O, r.t.; 2. Et_3N, CH_2Cl_2, r.t.; 3. HCl/MeOH, 60°C. (ix) 1. $CSCl_2$, $CaCO_3$, CH_2Cl_2/H_2O, r.t.; 2. Et_3N, CH_2Cl_2, r.t.; 3. HCl/MeOH, 60°C

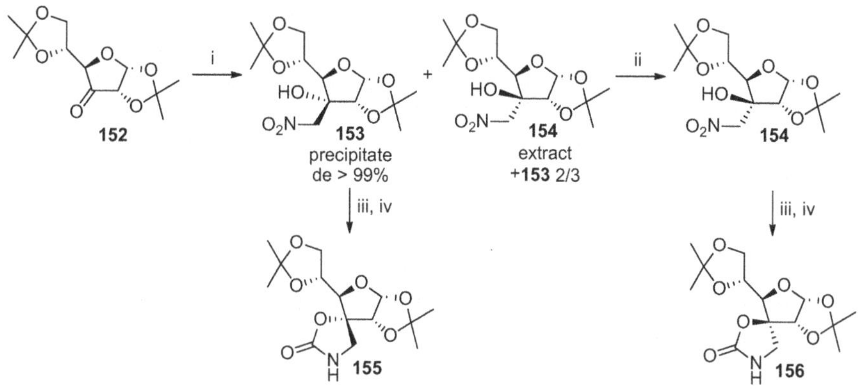

Scheme 34 *Reagents and conditions*: (i) $MeNO_2$, NaOH, MeOH. (ii) 1. Ac_2O, DMSO; 2. NaOH. (iii) H_2, 35 atm, Pd/C. (iv) 1. CbzOSuc, K_2CO_3; 2. NaH, DMF

were isolated from **144** and **145** on treatment with Im_2CS and $CaCO_3$. A similar protocol was employed to generate compound **150**. Derivative **141** was reacted with $TMSCH_2CN$, followed by reduction with hydride to afford derivative **146**. Reaction with thiophosgene conducted to the formation of **150**. On the other hand, compound **141** reacted with chloroform and base to generate alcohol **142**, which, on treatment with sodium azide, led to the formation of azide **143**. Reduction with H_2/Pd conducted to the formation of **147**. Once again, reaction with thiophosgene afforded the desired spiro compound **151**.

More recently, Turks et al. reported the synthesis of 3′-spirooxazolidinone pseudonucleosides from glucose-derived ketone **152** using a Henry reaction as the key step (Scheme 34) [143].

Thus, nitromethane addition to **152** in the presence of sodium hydroxide in aqueous methanol afforded the mixture of nitroalkanols **153** and **154**. Major isomer **153**, isolated from the mixture by precipitation, was reduced employing H_2 over Pd/C, and the resulting amine was transformed into oxazolidinone **155** by formation of the intermediate benzyl carbamate followed by cyclization in the presence of base. On the other hand, when the mixture of nitroalkanols **153** and **154** was submitted to an Albrecht-Moffatt dehydration-rehydration sequence produced diastereomerically pure **154** in a good yield. Nitrosugar **154** was transformed into oxazolidinone **156** using the same protocol as for epimer **153**.

3.3 C-4′-Spiropseudonucleosides

Since the first example of a C-4′-homologated nucleoside reported in 1992 [144], this class of derivatives have attracted much interest [145], motivated by several factors. On one hand, the torsion angle about the C-4′–C-5′ bond is fixed, resulting in the adoption of rather different spatial orientations. Moreover, DNA and RNA fragments have a considerably large void space in the region below C-4′, which can accommodate several methylene groups [146]. Finally, DNA strand cleavage has been shown to occur via C-4′-radicals, which are not formed from 4′-C-branched nucleosides lacking the 4′-hydrogen. Of all the possibilities of substitution at the C-4′-position, the spirocyclization is particularly interesting, mainly due to conformational restrictions inherent to the spiro system. Paquette et al. carried out extensive research in this field, developing a new family C-4′-spirocyclic nucleosides which were shown to display significant antiviral activity. For example, starting from easily accessible unsaturated lactone **157** [147], dihydroxylation with catalytic osmium tetroxide in the presence of *N*-methyl morpholine-*N*-oxide (NMO) furnished lactone **158**, which was then reduced with DIBAL-H and directly acetylated to generate triacetate **159**. Reaction of **159** with bis-*O*-silylated thymine in the presence of trimethylsilyl triflate, followed by treatment of the resulting deprotected alcohol **160** with methanolic potassium carbonate, finally afforded 4′-spironucleoside **161** (Scheme 35) [148].

Unlike the *cis* series (e.g., **161**), when considering the synthesis of spirocyclic nucleosides featuring a hydroxyl at C-5′ *trans* with respect to the ring oxygen, the methoxymethyl (MOM) protecting group proved to be problematic when dealing with the somewhat more elevated steric congestion. To circumvent the problem, MOM was replaced by a *p*-methoxybenzyl (PMB) group; the desired *trans*-1-oxaspiro[4.4]nonanyl nucleoside analogue was easily obtained following a similar sequence than for the *cis* series. Thus, reaction of triacetate **162** with bis-*O*-silylated uracil in the presence of trimethylsilyl triflate, followed by oxidation and deprotection of the resulting nucleoside **163**, afforded 4′-spironucleoside **164** (Scheme 36) [149].

Paquette's group also described the use of 2-phenylthio-substituted lactones in the synthesis of 4′-spironucleosides [150]. For example, reduction of lactone **165** with diisobutylaluminum hydride in the presence of chlorotrimethylsilane, followed by acetylation, gave anomeric mixture of **166**, which on coupling with persilylated

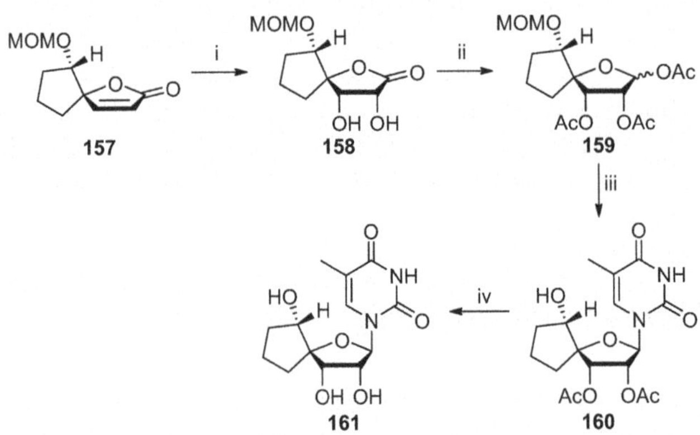

Scheme 35 *Reagents and conditions*: (i) OsO_4, NMO, THF, Py, 93%. (ii) 1. DIBAL-H; 2. Ac_2O, Py, 61%. (iii) Silylated thymine, TMSOTf, THF, 76%. (iv) K_2CO_3, MeOH

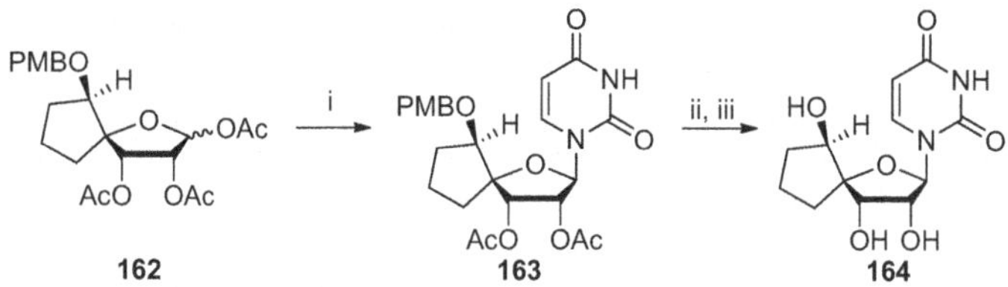

Scheme 36 *Reagents and conditions*: (i) silylated uracil, TMSOTf, CH_2Cl_2, 54%. (ii) DDQ, H_2O, CH_2Cl_2, 91%. (iii) K_2CO_3, MeOH, 99%

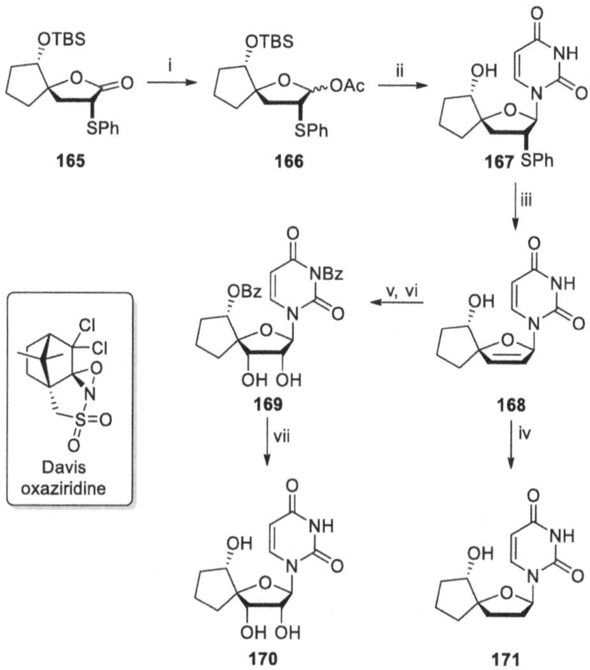

Scheme 37 *Reagents and conditions*: (i) 1. DIBAL, TMSCl, −78°C; 2. Ac_2O, Py, 66% over two steps. (ii) Silylated uracil, $SnCl_4$, CH_3CN, −50°C to r.t., 56%. (iii) 1. Davis oxaziridine; 2. xylene, Py, Δ, 85%. (iv) H_2, 50 psi, 10% Pd/C, EtOH, 89%. (v) BzCl, Py, 78%. (vi) 1. OsO_4, THF, Py; 2. H_2S, THF, 21%. (vii) NH_3, MeOH, 96%

uracil or thymine promoted by $SnCl_4$ afforded derivative **167** (Scheme 37). Elimination of the 2-phenylthio substituent in **167** was next accomplished by oxidation with the Davis oxaziridine reagent followed by thermal elimination of phenylsulfenic acid, to generate didehydrodideoxy nucleoside **168**. Saturation of the double bond was achieved by catalytic hydrogenation at 50 psi to furnish nucleosides **171**. Additionally, benzoylation and osmium tetroxide-promoted dihydroxylation afforded **169**, which on debenzoylation gave nucleoside **170**.

In a recent implementation of their synthetic strategies toward 4′-spiropseudonucleosides, Paquette and co-workers described a direct enantioselective pathway that furnishes exclusively the β-anomer of a 4′-spironucleoside [151]. Thus, starting from known dihydroxy lactone **172**, refluxing in 2,2-dimethoxypropane in the presence of a catalytic quantity of *p*-toluenesulfonic acid, followed by reduction with hydride, afforded lactol **173** (Scheme 38). Treatment of **173** with tris(dimethylamino) phosphine and carbon tetrachloride and reduction of the resulting mixture in a solution of lithium in liquid ammonia gave product **174**. Desilylation of **174** was followed by the formation of 3,5-*O*-(tetraisopropyldisiloxane-1,3-diyl) glycal **175** which was submitted to *N*-iodosuccinimide (NIS)-initiated glycosidation with persilylated thymine, resulting in the formation of a single isomer **176**. The targeted nucleoside **177** was obtained quantitatively following sequential deiodination and desilylation.

This class of spiropseudonucleosides developed in Paquette's group, resulting from substitution of the C-5′-hydroxymethyl group in natural nucleosides by a spirocyclopentanol ring, have attractive structural advantages and a very interesting biological profile. Closely related to those derivatives reported by Paquette's group, nucleoside **180** was prepared using a completely different synthetic strategy [152]. Starting from D-glucose-derived substrate **178** (Scheme 39), ring closing metathesis (RCM) furnished **179**. Removal of the 1,2-acetonide protection from **179** by acid treatment and subsequent peracetylation generated an anomeric mixture of triacetates, which was subjected to nucleosidation by 2,4-bis(trimethylsiloxy) pyrimidine in the presence of a Lewis acid to provide, after removal of the protecting groups, nucleoside **180** as the major product.

As a direct consequence of the potent antitumor and antiviral properties of several naturally occurring carbocyclic nucleosides [153], many efforts have been dedicated to the preparation of different types of these nucleoside analogues [154–158]. In order to achieve rigidification of the molecular architecture, the use of a spirocyclic restriction has been considered. Starting from lactone **181** [159], Luche reduction yielded separable alcohols **182** and **183** in a 1:1 diastereomeric ratio (Scheme 40). The Mitsunobu protocol allowed the transformation of **182** into its epimer **183**. After catalytic hydrogenation, reaction of alcohol **183** with *N*-benzoylthymine and diisopropyl azodicarboxylate-PPh_3, followed by silyl group deprotection and ammonolysis, gave the targeted spirocyclic carbanucleoside **184**.

Scheme 38 *Reagents and conditions*: (i) 2,2-dimethoxypropane, *p*-TsOH, 83°C, quant. (ii) DIBAL, Et_2O, −78°C, quant. (iii) 1. CCl_4, P$(NMe_2)_3$, THF, −78 to 0°C; 2. excess Li-NH_3 then NH_4OH, 73%. (iv) TBAF, THF, 96%. (v) $TIPDSCl_2$, collidine, AgOTf, DMF, 0°C, 62%. (vi) Silylated uracil, NIS, CH_2Cl_2, r.t., 70%; (vii) 1. Bu_3SnH, AIBN, toluene, 70°C; 2. TBAF, THF, quant

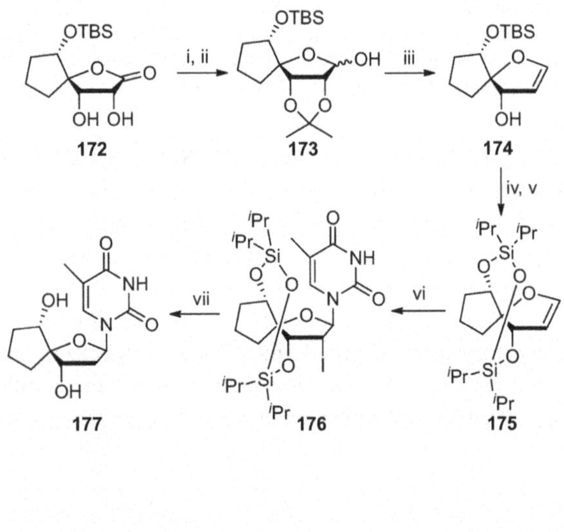

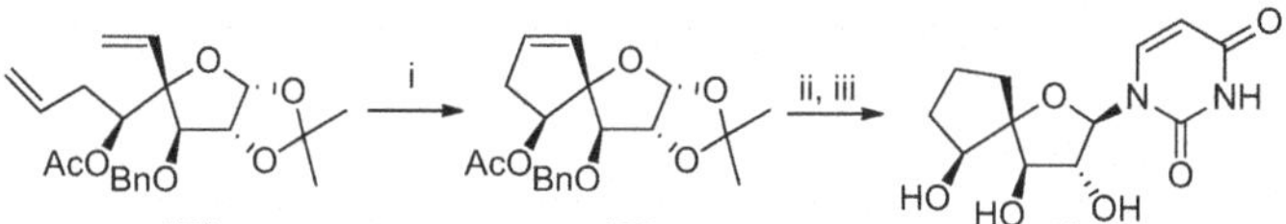

Scheme 39 *Reagents and conditions*: (i) Grubbs I cat., CH_2Cl_2, r.t., 5 h, 79%. (ii) 1. H_2SO_4 (4%), MeCN:H_2O (3:1), r.t., 12 h; 2. Ac_2O, Py, r.t., 12 h; 3. 2,4-Bis(trimethylsiloxy) pyrimidine, MeCN, TMSOTf, reflux, 6 h, 58% over three steps. (iii) 1. K_2CO_3, MeOH, r.t., 30 min, 95%; 2. Pd/C (10%), cyclohexene, EtOH, reflux, 6 h, 54%

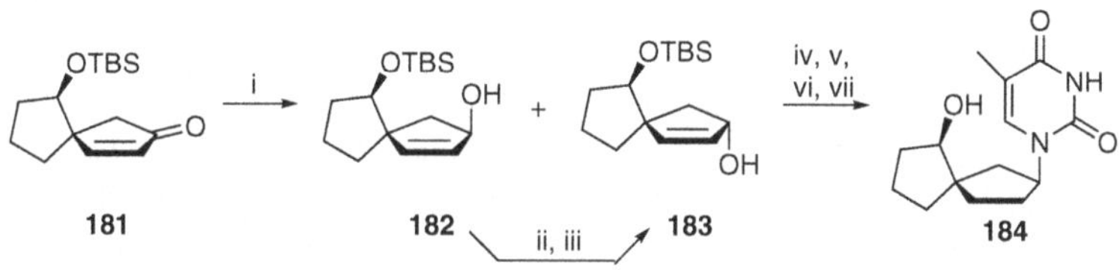

Scheme 40 *Reagents and conditions*: (i) $NaBH_4$, $CeCl_3$, r.t., 99%. (ii) Ph_3P, DIAD, $PhCO_2H$, r.t., 90%. (iii) NaOH, aq MeOH, r.t., 90%. (iv) H_2, Pd/C, r.t., 72%. (v) *N*-benzoyl-thymine, DIAD, Ph_3P, THF, r.t., 30%. (vi) NH_3, MeOH, r.t., 74%. (vii) TBAF, THF, r.t., 75%

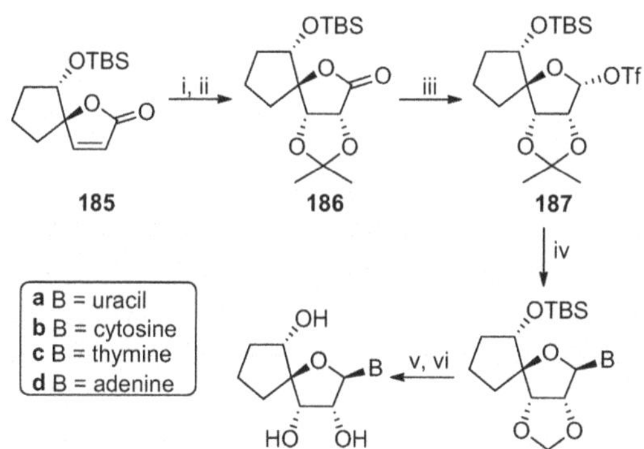

Scheme 41 *Reagents and conditions*: (i) OsO_4, H_2O, acetone, Py, H_2S, r.t., 68%. (ii) 2,2-Dimethoxypropane, *p*-TsOH, r.t., quant. (iii) 1. L-Selectride, THF, −78°C; 2. Tf_2O, Py, CH_2Cl_2, r.t., 80% over two steps. (iv) KH, base, DMF, r.t., **188a** (10%), **188b** (61%), **188c** (14%), **188d** (41%). (v) TBAF, THF, r.t. (vi) MeOH, *p*-TsOH, r.t., **189a** (88%), **189b** (quant.), **189c** (90%), **189d** (85%)

On the other hand, the fully hydroxylated 4′-spirocarbocyclic nucleoside analogues were synthetized by a modification of the previous strategy [160]. Thus, **185** undergoes osmium tetroxide oxidation in an entirely stereoselective manner to furnish the corresponding diol derivative (Scheme 41). After formation of the corresponding acetonide **186**, L-selectride reduction and triflation of the resulting alcohol afforded triflate **187**. S_N2 displacement of **187** by nucleobases yielded nucleosides **188a–d**, which, after desilylation and acidic hydrolysis, afforded nucleoside analogues **189a–d**.

It is known that the substitution of the oxygen in the furanoid ring of nucleosides by other heteroatoms can provoke a profound effect on biological activity. For example, several sulfur mimics have been recognized as potent antiviral and anticancer agents [161–165]. These findings have ignited the interest for the synthesis of C-4′-spiropseudothionucleosides to be used as biochemical probes [166].

For example, reduction of spironolactone **190** with lithium aluminum hydride afforded a separable mixture of epimeric alcohols **191** and **192** (Scheme 42). After silylation, selective oxidation of the resulting sulfides to sulfoxide **193** was performed by treatment with sodium periodate supported on silica gel. Glycosylation of **193** by silylated thymine (generated in situ) in the presence of trimethylsilyl triflate and catalytic zinc(II) iodide furnished thionucleoside analogue **194**.

The interesting features of C-4′ spirocyclopentanic nucleosides prompted the interest in the synthesis of C-4′ spirocyclic nucleosides containing other spirofused rings at C-4′. Thus, Roy and co-workers described a series of C-4′ spiropseudonucleosides containing oxetane, thietane, and azetidine rings [167]. Starting from dimesylate **195**, easily available from 4-*C*-hydroxymethyl-branched xylose, treatment with sodium sulfide afforded the spirothietane **196** (Scheme 43). On the other hand, heating compound **195** in benzylamine at 110°C furnished the spiroazetidine **197**, while heating **195** with diethyl malonate in the presence of an excess of NaH gave oxetane **198**. The nucleobase was successfully installed on **196** and **197** by cleavage of the acetonide group, followed by acetylation and treatment of the resulting acetates **199** and **200** with 2,4-bis-(trimethylsilyloxy)pyrimidine in the presence of TMSOTf to afford the nucleoside derivatives **202** and **203**. Attempts to install a base on **201** were unsuccessful.

Preparation of **206**, the methyl homologue of **202**, follows essentially the same procedure but using 4-*C*-hydroxymethyl-5-*C*-methyl-branched mesylate **204** as the starting material [168]. Thus, treatment of **204** with sodium sulfide, followed by acetonide removal and acetylation, afforded the spirothietane **205**. Finally, treatment with 2,4-bis-(trimethylsilyloxy)pyrimidine in the presence of TMSOTf gave the nucleoside derivative **206** (Scheme 44).

Scheme 42 *Reagents and conditions*: (i) $LiAlH_4$, Et_2O, r.t., 85%. (ii) TBSOTf, 2,6-lutidine, r.t., 95%. (iii) 10% $NaIO_4$/SiO_2, CH_2Cl_2, hexane, r.t., 95%. (iv) 1. Thymine, TMSOTf, ZnI_2, Et_3N, toluene, r.t.; 2. TBAF, THF, r.t., 40% over two steps

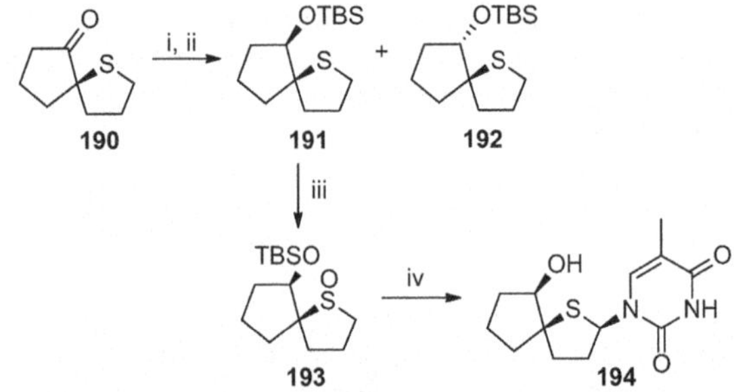

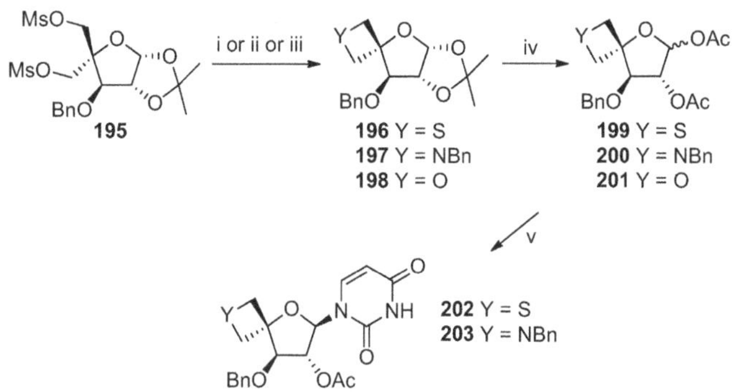

Scheme 43 *Reagents and conditions*: (i) Na_2S, DMF, 110°C, 76% (**196**). (ii) $BnNH_2$, 110°C, 71% (**197**). (iii) NaH, DMF, 100°C, 55% (**198**). (iv) 1. TFA/H_2O (3:2), r.t., 6 h; 2. Ac_2O, Py, DMAP, r.t., 12 h, 70–80% over two steps. (v) Silylated uracil, TMSOTf, CH_3CN, reflux, 6 h, 55–60%

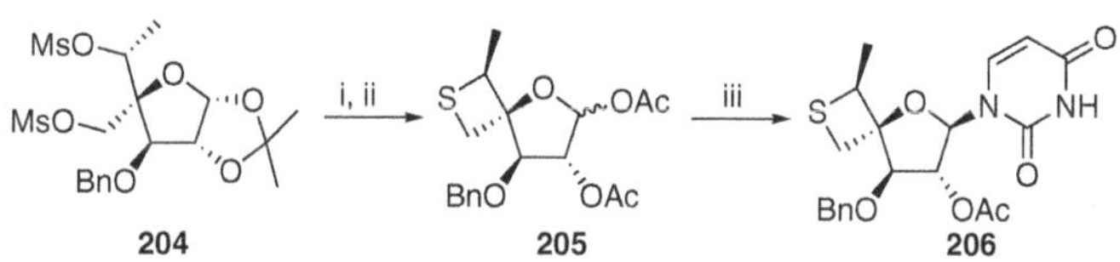

Scheme 44 *Reagents and conditions*: (i) Na_2S, DMF, 120°C, 4 h, 77%. (ii) 1. H_2SO_4 (4%), CH_3CN/H_2O (3:2), r.t., 12 h; 2. Ac_2O, Py, r.t., 12 h. (iii) Silylated uracil, TMSOTf, CH_3CN, reflux, 8 h, 60%

More recently, a similar procedure was used to synthesize highly constrained nucleoside analogue **211** (Scheme 45) [169]. Thus, conversion of tritosylate **207** into the di-*O*-acetate **208**, followed by reaction with silylated *N*6-benzoyl adenine in the presence of TMSOTf, yielded the protected nucleosides **209** and **210** as a 7:3 mixture. Reaction of isomer **209** with ammonium hydroxide in methanol finally afforded constrained nucleoside analogue **211**.

As another example of 4′-*C*-spirofused systems, the spirocyclic oxepane nucleoside analogue **212** was synthesized from D-glucose (Scheme 46) [170]. Swern oxidation of **212**, easily available from D-glucose, followed by reaction of the corresponding aldehyde with *N*-benzylhydroxylamine, afforded the spiroannulated product **213**. Acetonide deprotection and acetylation, followed by Lewis acid-catalyzed nucleobase coupling using 2,4-bis(trimethylsiloxy)pyrimidine, furnished the spironucleoside **214**. After cleavage of the N–O bond by hydrogenolysis and subsequent peracetylation gave the spironucleoside **215**.

Recently, a series of 4′-spirocyclic phosphono-nucleosides were designed to mimic the monophosphate of R-1479, a known nucleoside inhibitor of HCV NS5B [171]. The synthetic route used a triphenylphosphine-iodine cyclization reaction as the key step to form the tetrahydrofuran 4′-spirocycle and efficiently afforded the desired 4′-spirocyclic phosphono-cytidine analogues in 11 steps (Scheme 47) [172]. Thus, reaction of 4′-*C*-hydroxymethyl-ribose derivative **216**

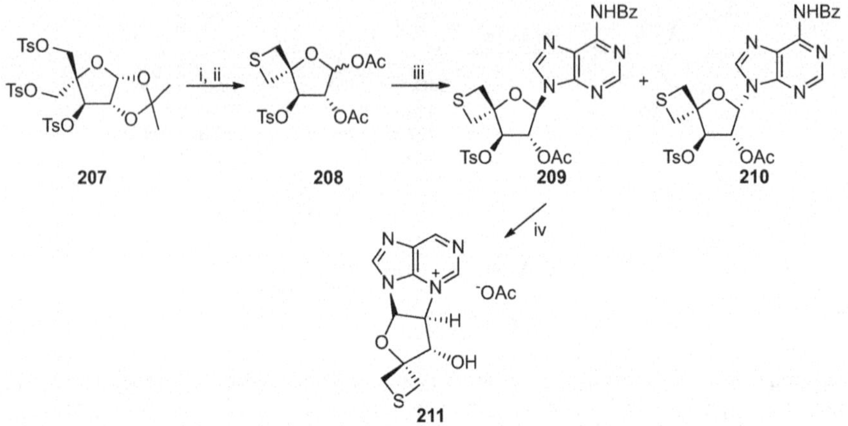

Scheme 45 *Reagents and conditions*: (i) KSAc, DMF, 120°C. (ii) H_2SO_4, Ac_2O, AcOH, r.t. (iii) 1. *N*6-benzoyl adenine, HMDS; 2. TMSOTf, CH_3CN, reflux. (iv) NH_4OH, MeOH

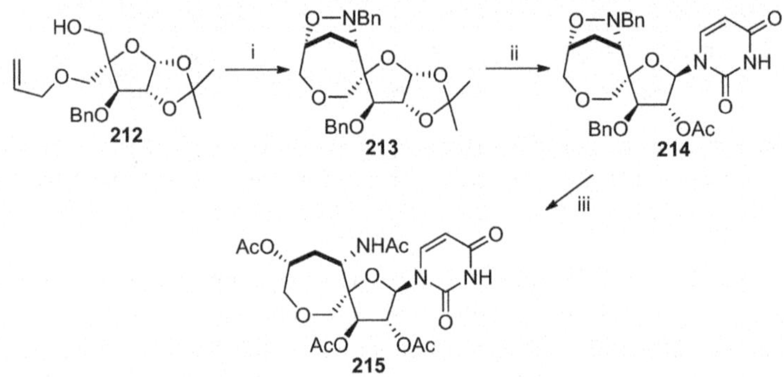

Scheme 46 *Reagents and conditions*: (i) 1. $(COCl)_2$, DMSO, CH_2Cl_2, Et_3N, −65°C. 2. BnNHOH, EtOH, r.t., 24 h, reflux, 2 h, 54%. (ii) 1. H_2SO_4 (4%), CH_3CN/H_2O (3:2), r.t., 12 h; 2. Ac_2O, Py, r.t., 15 h. 3. Silylated uracil, TMSOTf, CH_3CN, r.t., 5 h, 63% (iii) 1. Pd/C (10%), cyclohexene, EtOH, reflux, 5 h. 2. Ac_2O, Py, r.t., 15 h, 44%

with dimethoxytrityl chloride (DMTrCl) in pyridine, followed by removal of the silyl protecting group and subsequent oxidation of the hydroxymethyl group, yielded aldehyde **217**, which on Wittig reaction furnished phosphonate ester **218**. Hydroboration of olefin **218** under Negishi conditions occurred with concomitant removal of the DMTr protecting group to give the corresponding diol, which on cyclization by treatment with triphenylphosphine and iodine in the presence of imidazole gave the 4′-spiro-tetrahydrofuran **219**. Acetylation of compound **219**, followed by Vorbrüggen coupling with silylated *N*-Ac-cytidine in the presence of tin(IV) chloride, gave nucleoside **220**. Finally, removal of the protecting groups by oxidation, basic hydrolysis, and trimethylsilylbromide-mediated removal of the ethyl groups from the phosphonate afforded the desired compounds **221** and **222**.

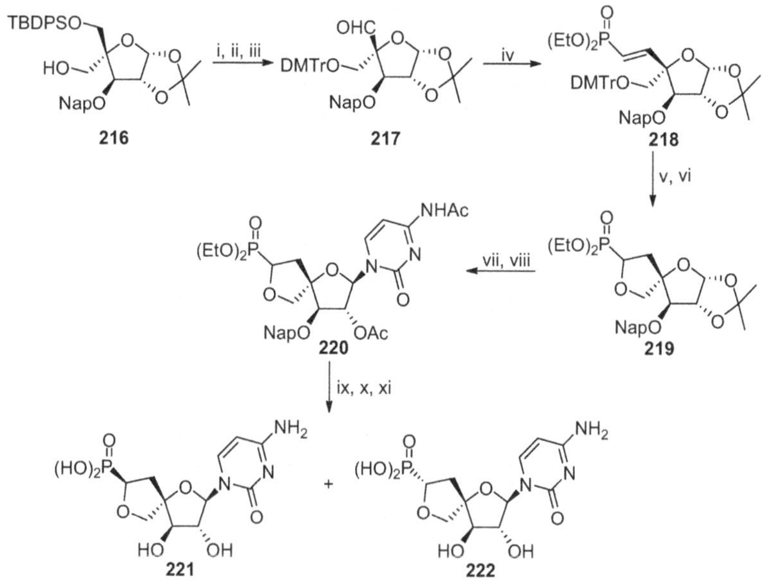

Scheme 47 *Reagents and conditions*: (i) DMTrCl, pyridine, 25°C, 16 h, 90%. (ii) TBAF, THF, 25°C, 48 h, 66%. (iii) $(n\text{-Bu})_4NBr$, TEMPO, $KHCO_3$, NBS, CH_2Cl_2/H_2O (1:1), 0°C, 1 h, 95%. (iv) $Ph_3P{=}CH\text{-}PO(OEt)_2$, DMF, 90°C, 16 h, 65%. (v) 1. $BH_3{\cdot}THF$, THF, 25°C, 12 h; 2. H_2O_2, NaOAc, H_2O, 55°C, 0.5 h, 36%. (vi) 1. PPh_3, imidazole, toluene, reflux, 5 min; 2. I_2, reflux, 0.5 h, 26%. (vii) $AcOH/Ac_2O/H_2SO_4$ (5.5:2.0:0.2, v/v, 0.08 M solution), 25°C, 20 h, 52%. (viii) 1. *N,O*-bis(trimethylsilyl)acetamide, *N*-Ac cytidine, MeCN, reflux, 2 h; 2. Silylated *N*-acetyl-cytidine, $SnCl_4$, 25°C, 15 h, 40%. (ix) DDQ, CH_2Cl_2/H_2O (20:1, v/v), 25°C, 48 h, 22%. (x) NH_4OH/MeOH (1:1, v/v), 25°C, 1.5 h, 43%. (xi) TMSBr, DMF, 0–25°C, 4 h, 8% (**221**), 4.4% (**222**)

4 Conclusions

In view of their broad biological activity as antidiabetic, antiviral, and antitumoral agents, spirocyclic nucleosides are synthetic targets of great interest. Accordingly, the past several years have witnessed explosive developments in spirocyclic nucleoside chemistry. The increasing number of spirocyclic nucleoside analogues opens the door to new, more potent and effective drugs. In this regard, more nucleoside analogues are still required, especially those designed on the basis of structure-activity relationship studies.

References

1. de Clercq E (2010) Highlights in the discovery of antiviral drugs: a personal retrospective. J Med Chem 53:1438–1450. https://doi.org/10.1021/jm900932g
2. Mehellou Y, De Clercq E (2010) Twenty-six years of anti-HIV drug discovery: where do we stand and where do we go? J Med Chem 53:521–538. https://doi.org/10.1021/jm900492g

3. de Clercq E (2009) The history of antiretrovirals: key discoveries over the past 25 years. Rev Med Virol 19:287–299. https://doi.org/10.1002/rmv.624
4. de Clercq E (2009) Looking back in 2009 at the dawning of antiviral therapy now 50 years ago in historical perspective. Adv Virus Res 73:1–53. https://doi.org/10.1016/S0065-3527(09)73001-5
5. De Clercq E (2010) In search of a selective therapy of viral infections. Antivir Res 85:19–24. https://doi.org/10.1016/j.antiviral.2009.10.005
6. Gandhi V, Plunkett W (2006) Clofarabine and nelarabine: two new purine nucleoside analogs. Curr Opin Oncol 18:584–590. https://doi.org/10.1097/01.cco.0000245326.65152.af
7. Elgemeie GH (2003) Thioguanine, mercaptopurine: their analogs and nucleosides as antimetabolites. Curr Pharm Des 9:2627–2642. https://doi.org/10.2174/1381612033453677
8. Miura S, Izuta S (2004) DNA polymerases as targets of anticancer nucleosides. Curr Drug Targets 5:191–195. https://doi.org/10.2174/1381612033453677
9. Parker WB, Secrist JA, Waud WR (2004) Purine nucleoside antimetabolites in development for the treatment of cancer. Curr Opin Investig Drugs 5:592–596
10. Ying C, de Clercq E, Neyts J (2003) Selective inhibitors of hepatitis B virus replication. Curr Med Chem Anti-Infect Agents 2:227–240. https://doi.org/10.2174/1568012033483042
11. Sharma PL, Nurpeisov V, Hernandez-Santiago B, Beltran T, Schinazi RF (2004) Nucleoside inhibitors of human immunodeficiency virus type 1 reverse transcriptase. Curr Top Med Chem 4:895–921. https://doi.org/10.2174/1568026043388484
12. Otto M (2004) New nucleoside reverse transcriptase inhibitors for the treatment of HIV infections. Curr Opin Pharmacol 4:431–436. https://doi.org/10.1016/j.coph.2004.06.001
13. Ichikawa E, Kato K (2001) Sugar-modified nucleosides in past 10 years, a review. Curr Med Chem 8:385–423. https://doi.org/10.2174/0929867013373471
14. Ford H, Dai F, Mu L, Siddiqui MA, Nicklaus MC, Anderson L, Marquez VE, Barchi JJ (2000) Adenosine deaminase prefers a distinct sugar ring conformation for binding and catalysis: kinetic and structural studies. Biochemistry 39:2581–2592. https://doi.org/10.1021/bi992112c
15. Hernandez S, Ford H, Marquez VE (2002) Is the anomeric effect an important factor in the rate of adenosine deaminase catalyzed hydrolysis of purine nucleosides? A direct comparison of nucleoside analogues constructed on ribose and carbocyclic templates with equivalent heterocyclic bases selected to promote hydration. Bioorg Med Chem 10:2723–2730. https://doi.org/10.1016/S0968-0896(02)00099-8
16. Marquez VE, Choi Y, Comin MJ, Russ P, George C, Huleihel M, Ben-Kasus T, Agbaria R (2005) Understanding how the herpes thymidine kinase orchestrates optimal sugar and nucleobase conformations to accommodate its substrate at the active site: a chemical approach. J Am Chem Soc 127:15145–15150. https://doi.org/10.1021/ja053789s
17. Costanzi S, Joshi BV, Maddileti S, Mamedova L, Gonzalez-Moa MJ, Marquez VE, Harden TK, Jacobson KA (2005) Semirational design of (north)-methanocarba nucleosides as dual acting A_1 and A_3 adenosine receptor agonist: novel prototypes for cardioprotection. J Med Chem 48:8103–8107. https://doi.org/10.1021/jm050726b
18. Besada P, Shin DH, Costanzi S, Ko H, Mathé C, Gagneron J, Gosselin G, Maddileti S, Harden TK (2005) Human $P2Y_6$ receptor: molecular modeling leads to the rational design of a novel agonist base on a unique conformational preference. J Med Chem 48:8108–8111. https://doi.org/10.1021/jm050911p
19. Boyer PL, Julias JG, Marquez VE, Hugues SH (2005) Fixed conformation nucleoside analogues effectively inhibit excision-proficient HIV-1 reverse transcriptases. J Mol Biol 345:441–450. https://doi.org/10.1016/j.jmb.2004.10.021
20. Besada P, Shin DH, Costanzi S, Ko H, Mathé C, Gagneron J, Gosselin G, Maddileti S, Harden TK, Jacobson KA (2006) Structure-activity relationships of uridine 5′-diphosphate analogues at the human $P2Y_6$ receptor. J Med Chem 49:5532–5543. https://doi.org/10.1021/jm060485n
21. Altona C, Sundaralingam M (1972) Conformational analysis of the sugar ring in nucleosides and nucleotides. New description using the concept of pseudorotation. J Am Chem Soc 94:8205–8212. https://doi.org/10.1021/ja00778a043

22. Mathé C, Périgaud C (2008) Recent approaches in the synthesis of conformationally restricted nucleoside analogues. Eur J Org Chem 1489–1505. https://doi.org/10.1002/ejoc.200700946
23. Kvaerna L, Wightman R, Wengel J (2001) Synthesis of a novel bicyclic nucleoside restricted to an S-type conformation and initial evaluation of its hybridization properties when incorporated into oligodeoxynucleotides. J Org Chem 66:5106–5112. https://doi.org/10.1021/jo015602v
24. Freitag M, Thomasen H, Christensen N, Petersen M, Nielsen P (2004) A ring-closing metathesis based synthesis of bicyclic nucleosides locked in S-type conformations by hydroxyl functionalised 3′-4′-trans linkages. Tetrahedron 60:3775–3786. https://doi.org/10.1016/j.tet.2004.03.019
25. Wengel J (1999) Synthesis of 3′-C- and 4′-C- branched oligodeoxynucleotides and the development of locked nucleic acid (LNA). Acc Chem Res 32:301–310. https://doi.org/10.1021/ar980051p
26. Singh S, Kumar R, Wengel J (1998) Synthesis of 2′-amino-LNA: a novel conformationally restricted high-affinity oligonucleotide analogue with a handle. J Org Chem 63:10035–10039. https://doi.org/10.1021/jo9814445
27. Kurreck J (2003) Antisense technologies. Improvement through novel chemical modifications. Eur J Biochem 270:1628–1644. https://doi.org/10.1046/j.1432-1033.2003.03555.x
28. Paroo Z, Corey D (2004) Challenges in RNAi in vivo. Trends Biotechnol 22:390–394
29. Manoharan M (2004) RNA interference and chemically modified small interfering RNAs. Curr Opin Chem Biol 8:570–579. https://doi.org/10.1016/j.cbpa.2004.10.007
30. Marquez VE, Ezzitouni A, Russ P, Siddiqui MA, Ford H, Feldman RJ, Mitsuya H, George C, Barchi JJ (1998) HIV-1 reverse transcriptase can discriminate between two conformationally locked carbocyclic AZT triphosphate analogues. J Am Chem Soc 120:2780–2789. https://doi.org/10.1021/ja973535+
31. Herdewijn P (1996) Targeting RNA with conformationally restricted oligonucleotides. Liebigs Ann Chem 1337–1348. https://doi.org/10.1002/jlac.199619960902
32. Kool ET (1997) Preorganization of DNA: design principles for improving nucleic acid recognition by synthetic oligonucleotides. Chem Rev 97:1473–1488. https://doi.org/10.1021/cr9603791
33. Soengas RG, Silva S (2012) Spirocyclic nucleosides in medicinal chemistry: an overview. Mini Rev Med Chem 12:1485–1496. https://doi.org/10.2174/138955712803832681
34. Gimisis T, Chatgilialoglu C (1996) 1,5-radical translocation protocol for the generation of C-1′ radicals in nucleosides. Synthesis of spiro nucleosides through a rare 5-endo-trig cyclization. J Org Chem 61:1908–1909. https://doi.org/10.1021/jo952218n
35. Chatgilialoglu C, Gimisis T, Spada GP (1999) C-1′ radical-based approaches for the synthesis of anomeric spironucleosides. Chem Eur J 5:2866–2877. https://doi.org/10.1002/(SICI)1521-3765(19991001)5:10<2866::AID-CHEM2866>3.0.CO;2-6
36. Kittaka A, Asakura T, Kuze T, Tanaka H, Yamada N, Nakamura KT, Miyasaka T (1999) Cyclization reactions of nucleoside anomeric radical with olefin tethered on base: factors that induce anomeric stereochemistry. J Org Chem 64:7081–7093. https://doi.org/10.1021/jo990611d
37. Gimisis T, Castellari C, Chatglilialoglu C (1997) A new class of anomeric spironucleosides. Chem Commun 2089–2090. https://doi.org/10.1039/A705742C
38. El Kouni MH, Naguib FN, Panzica RP, Otter BA, Chu SH, Gosselin G, Chu CK, Schinazi RF, Shealy YF, Goudgaon N, Ozerov AA, Ueda T, Iltzsch MH (1996) Effects of modifications in the pentose moiety and conformational changes on the binding of nucleoside ligands to uridine phosphorylase from *Toxoplasma gondii*. Biochem Pharmacol 51:1687–1700. https://doi.org/10.1016/0006-2952(96)00213-4
39. Sankyo (1987) Europe Patent Appl 0,232,572,A2. see CA 107: 149218n
40. Sankyo (1988) US Patent 4,952,234,A28. see CA 115: 66804q
41. Ciba-Geigy AG (1990) DE Patent 4,129,616 A1
42. Mitsubishi Kasei Corp (1990) Jap Patent 04/222,589 A

43. Nakajima M, Itoi K, Takamatsu Y, Kinoshita T, Okazaki T, Kawakubo K, Shindo M, Honma T, Tohjigamori M, Haneishi T (1991) Hydantocidin: a new compound with herbicidal activity from *Streptomyces hygroscopicus*. J Antibiot 44:293–300. https://doi.org/10.7164/antibiotics.44.293
44. Heim DR, Cseke C, Gerwick BC, Murdoch MG, Green SB (1995) Hydantocidin: a possible proherbicide inhibiting purine biosynthesis at the site of adenylosuccinate synthetase. Pestic Biochem Physiol 53:138–145. https://doi.org/10.1006/pest.1995.1061
45. Smith JL (1995) Enzymes of nucleotide synthesis. Curr Opin Struct Biol 5:752–757
46. Gregoriou M, Noble ME, Watson KA, Garman EF, Krulle TM, de la Fuente C, Fleet GW, Oikonomakos NG, Johnson LN (1998) The structure of a glycogen phosphorylase glucopyranose spirohydantoin complex at 1.8 A resolution and 100 K: the role of the water structure and its contribution to binding. Protein Sci 7:915–927. https://doi.org/10.1002/pro.5560070409
47. Barford D, Johnson LN (1989) The allosteric transition of glycogen phosphorylase. Nature 340:609–616. https://doi.org/10.1038/340609a0
48. Johnson LN, O'Reilly M (1996) Control by phosphorylation. Curr Opin Struct Biol 6:762–769. https://doi.org/10.1016/S0959-440X(96)80005-4
49. Ortmeyer HK, Bodkin NL, Hansen BC (1997) Insulin regulates liver glycogen synthase and glycogen phosphorylase activity reciprocally in rhesus monkeys. Am J Phys 272:E133–E138. https://doi.org/10.1152/ajpendo.1997.272.1.E133
50. Martin JL, Veluraja K, Ross K, Johnson LN, Fleet GWJ, Ramsden NG, Bruce I, Orchard MG, Oikonomakos NG, Papageorgiou AC, Leonidas DD, Tsitoura HS (1991) Glucose analogue inhibitors of glycogen phosphorylase: the design of potential drugs for diabetes. Biochemistry 30:10101–10116. https://doi.org/10.1021/bi00106a006
51. Watson KA, Mitchell EP, Johnson LN, Son JC, Bichard CJF, Orchard MG, Fleet GWJ, Oikonomakos NG, Leonidas DD, Kontou M, Papageorgiou AC (1994) Design of inhibitors of glycogen phosphorylase: a study of α- and β-C-glucosides and 1-thio- β-D-glucose compounds. Biochemistry 33:5745–5758. https://doi.org/10.1021/bi00185a011
52. Board M, Bollen M, Stalmans W, Kim Y, Fleet GWJ, Johnson LN (1995) Effects of C-1 substituted glucose analogue on the activation states of glycogen synthase and glycogen phosphorylase in rat hepatocytes. Biochem J 311:845–852. https://doi.org/10.1042/bj3110845
53. Martin JL, Johnson LN, Withers SG (1990) Comparison of the binding of glucose and glucose 1-phosphate derivatives to T-state glycogen phosphorylase b. Biochemistry 29:10745–10757. https://doi.org/10.1021/bi00500a005
54. Osei K (1995) In: Leslie RDG, Robbins DC (eds) Diabetes, clinical science in practice. Cambridge University, Cambridge
55. DeFronzo RA (1988) Lilly lecture 1987. The triumvirate: beta-cell, muscle, liver. A collusion responsible for NIDDM. Diabetes 37:667–687
56. Mio S, Ichinose R, Goto K, Sugai S (1991) Synthetic studies on (+)-hydantocidin: a total synthesis of (+)-hydantocidin, a new herbicidal metabolite from microorganism. Tetrahedron 47:2111–2120. https://doi.org/10.1016/S0040-4020(01)96122-8
57. Soengas RG, da Silva G, Estévez JC (2017) Synthesis of spironucleosides: past and future perspectives. Molecules 22:2028–2063. https://doi.org/10.3390/molecules22112028
58. Shiozaki M (2002) Syntheses of hydantocidin and C-2-thioxohydantocidin. Carbohydr Res 337:2077–2088. https://doi.org/10.1016/S0008-6215(02)00217-3
59. Mio S, Shiraishi M, Sugai S (1991) Synthetic studies on (+)-hydantocidin (2): aldol addition approaches towards the stereoisomers of (+)-hydantocidin. Tetrahedron 47:2121–2132. https://doi.org/10.1016/S0040-4020(01)96123-X
60. Mio S, Ueda M, Hamura M, Kitakawa J, Sugai S (1991) Synthetic studies on (+)-hydantocidin (4): synthesis of stereoisomers of (+)-hydantocidin. Tetrahedron 47:2145–2154. https://doi.org/10.1016/S0040-4020(01)96125-3

61. Mio S, Sano H, Shindou M, Honma T, Sugai S (1991) Synthesis and herbicidal activity of deoxy derivatives of (+)-hydantocidin. Agric Biol Chem 55:1105–1109. https://doi.org/10.1080/00021369.1991.10870692
62. Burton JW, Son JC, Fairbanks AJ, Choi SS, Taylor H, Watkin DJ, Winchester BG, Fleet GWJ (1993) Anomeric spirohydantoins of mannofuranose: approaches to novel anomeric amino acids by oxidative ring contraction. Tetrahedron Lett 34:6119–6122. https://doi.org/10.1016/S0040-4039(00)61745-8
63. Brandstetter TW, Kim Y, Son JC, Taylor HM, Lilley PMDQ, Watkin DJ, Johnson LN, Oikonomakos NG, Fleet GWJ (1995) Spirohydantoins of glucofuranose: analogues of hydantocidin. Tetrahedron Lett 36:2149–2152. https://doi.org/10.1016/0040-4039(95)00198-L
64. Brandstetter TW, Wormald MR, Dwek RA, Butters TD, Platt FM, Tsitsanou KE, Zographos SE, Oikonomakos NG, Fleet GWJ (1996) A galactopyranose analogue of hydantocidin. Tetrahedron Asymmetry 7:157–170. https://doi.org/10.1016/0957-4166(95)00432-7
65. Estévez JC, Smith MD, Wormald MR, Besra GS, Brenan PJ, Nash RJ, Fleet GWJ (1996) Mimics of L-rhamnose: analogues of rhamnopyranose containing a constituent amino acid at the anomeric position. A rhamnopyranose analogue of hydantocidin. Tetrahedron Asymmetry 7:391–394. https://doi.org/10.1016/0957-4166(96)00014-6
66. Fuente C, Krülle TM, Watson KA, Gregoriou M, Johnson LN, Zographos SE, Oikonomakos NG, Fleet GWJ (1997) Glucopyranose analogues of spirohydantoins: specific inhibitors of glycogen phosphorylase. Synlett 485–487. https://doi.org/10.1055/s-1997-940277
67. Bichard CJF, Mitchell EP, Wormald MR, Watson KA, Johnson LN, Zographos S, Koutra DD, Oikonomakos NG, Fleet GWJ (1995) Potent inhibition of glycogen phosphorylase by a spirohydantoin of glucopyranose: first pyranose analogues of hydantocidin. Tetrahedron Lett 36:2145–2148. https://doi.org/10.1016/0040-4039(95)00197-K
68. Blériot Y, Simone MI, Wormald MR, Dwek RA, Watkin DJ, Fleet GWJ (2006) Sugar amino acids at the anomeric position of carbohydrates: synthesis of spirocyclic amino acids of 6-deoxy-L-lyxofuranose. Tetrahedron Asymmetry 17:2276–2286. https://doi.org/10.1016/j.tetasy.2006.07.027
69. Estévez JC, Smith MD, Lane AL, Crook S, Watkin DJ, Besra GS, Brennan PJ, Nash RJ, Fleet GWJ (1993) Synthesis of cyclopentane spirohydantoins by aldol cyclisations: an approach to highly substituted α-cyclopentane amino acids. Tetrahedron Lett 34:7953–7956. https://doi.org/10.1016/S0040-4039(00)61521-6
70. Pham TQ, Pyne SG, Skelton BW, White AH (2005) Synthesis of carbocyclic hydantocidins via regioselective and diastereoselective phosphine catalyzed [3 + 2]-cycloadditions to 5-methylenehydantoins. J Org Chem 70:6369–6377. https://doi.org/10.1021/jo050827h
71. Hanessian S, Sancéau JY, Chemla P (1995) Synthesis of surrogate structures related to the herbicidal agent hydantocidin. Tetrahedron 51:6669–6678. https://doi.org/10.1016/0040-4020(95)00324-2
72. Sano H, Mio S, Kitagawa J, Shindou M, Honma T, Sugai S (1995) Synthesis of spirothiohydantoin analogues of hydantocidin. Tetrahedron 51:12561–12572. https://doi.org/10.1016/0040-4020(95)00810-U
73. Ghoneim AA (2011) Synthesis of some nucleosides derivatives from L-rhamnose with expected biological activity. Chem Cent J 5:7. https://doi.org/10.1186/1752-153X-5-7
74. Ősz E, Somsák L, Szilágyi L, Dinya Z (1997) A straightforward route to hydantocidin analogues with pyranose ring structure. Tetrahedron 53:5813–5824. https://doi.org/10.1016/S0040-4020(97)00213-5
75. Ősz E, Somsák L, Szilágyi L, Kovács L, Docsa T, Tóth B, Gergely P (1999) Efficient inhibition of muscle and liver glycogen phosphorylases by a new glucopyranosylidene-spiro-thiohydantoin. Bioorg Med Chem Lett 9:1385–1390. https://doi.org/10.1016/S0960-894X(99)00192-4
76. Somsák L, Nagy V, Docsa T, Tóth B, Gergely P (2000) Gram-scale synthesis of a glucopyranosylidene-spirothiohydantoin and its effect on hepatic glycogen metabolism studied in vitro and in vivo. Tetrahedron Asymmetry 11:405–408. https://doi.org/10.1016/S0957-4166(99)00486-3

77. Somsák L, Nagy V (2000) A new, scalable preparation of a glucopyranosylidene-spiro-thiohydantoin: one of the best inhibitors of glycogen phosphorylases. Tetrahedron Asymmetry 11:1719–1727. https://doi.org/10.1016/S0957-4166(00)00107-5
78. Somsák L, Kovács L, Tóth M, Ősz E, Szilágyi L, Györgydeák Z, Dinya Z, Docsa T, Tóth B, Gergely P (2001) Synthesis of and a comparative study on the inhibition of muscle and liver glycogen phosphorylases by epimeric pairs of D-gluco- and D-xylopyranosylidene-spiro-(thio)hydantoins and *N*-(D-glucopyranosyl) amides. J Med Chem 44:2843–2848. https://doi.org/10.1021/jm010892t
79. Kandra L, Remenyik J, Batta G, Somsák L, Gyémánt G, Park KH (2005) Enzymatic synthesis of a new inhibitor of α-amylases: acarviosinyl-isomaltosyl-spiro-thiohydantoin. Carbohydr Res 340:1311–1317. https://doi.org/10.1016/j.carres.2005.03.003
80. Somsák L, Ferrier RJ (1991) Radical-mediated brominations at ring positions of carbohydrates. Adv Carbohydr Chem Biochem 49:37–92. https://doi.org/10.1016/S0065-2318(08)60181-X
81. Somsák L, Czifrák K (2013) Radical-mediated brominations at ring-positions of carbohydrates – 35 years later. Carbohydr Chem 39:1–37. https://doi.org/10.1039/9781849737173-00001
82. Maza S, López O, Martos S, Maya I, Fernández-Bolaños JG (2009) Synthesis of the first selenium-containing acyclic nucleosides and anomeric spironucleosides from carbohydrate precursors. Eur J Org Chem 5239–5246. https://doi.org/10.1002/ejoc.200900793
83. Estevez JC, Ardron H, Wormald MR, Brown D, Fleet GWJ (1994) Spirocyclic peptides at the anomeric position of mannofuranose. Tetrahedron Lett 35:8889–8890. https://doi.org/10.1016/S0040-4039(00)78526-1
84. Estevez JC, Burton JW, Estévez RE, Ardron H, Wormald MR, Dwek RA, Brown D, Fleet GWJ (1998) Spirodiketopiperazines of mannofuranose: carbopeptoid α-amino acid esters at the anomeric position of mannofuranose. Tetrahedron Asymmetry 9:2137–2154. https://doi.org/10.1016/S0957-4166(98)00206-7
85. Estevez JC, Long DD, Wormald MR, Dwek RA, Fleet GWJ (1995) Spirodiketopiperazines at the anomeric position of mannopyranose: novel *N*-linked glycopeptides incorporating an α-amino acid at the anomeric position of mannopyranose. Tetrahedron Lett 36:8287–8290. https://doi.org/10.1016/0040-4039(95)01732-W
86. Krülle TM, Watson KA, Gregoriou M, Johnson LN, Crook S, Watkin DJ, Griffith RC, Nash RJ, Tsitsanou KE, Zographos SE, Oikonomakos NG, Fleet GWJ (1995) Specific inhibition of glycogen phosphorylase by a spirodiketopiperazine at the anomeric position of glucopyranose. Tetrahedron Lett 36:8291–8294. https://doi.org/10.1016/0040-4039(95)01733-X
87. Estevez JC, Smith MD, Lane AL, Crook S, Watkin DJ, Besra GS, Brennan PJ, Nash RJ, Fleet GWJ (1996) Mimics of L-rhamnose: anomeric spirohydantoins and diketopiperazines – approaches to novel *N*-linked glycopeptides of rhamnofuranose. Tetrahedron Asymmetry 7:387–390. https://doi.org/10.1016/0957-4166(96)00013-4
88. Long DD, Tennant-Eyles RJ, Estévez JC, Wormald MR, Dwek RA, Smith MD, Fleet GWJ et al (2001) Carbopeptoids: peptides and diketopiperazines incorporating the anomeric centre of mannopyranose. J Chem Soc Perkin Trans 1:807–813. https://doi.org/10.1039/B009940F
89. Kyogoku Y, Lord RC, Rich A (1968) Specific hydrogen bonding of barbiturates to adenine derivatives. Nature 218:69–72. https://doi.org/10.1038/218069a0
90. Ito T, Suzuki T, Wellman SE, Ho IK (1996) Pharmacology of barbiturate tolerance/dependence: GABAA receptors and molecular aspects. Life Sci 59:169–195. https://doi.org/10.1016/0024-3205(96)00199-3
91. Renard A, Lhomme J, Kotera M (2002) Synthesis and properties of spiro nucleosides containing the barbituric acid moiety. J Org Chem 67:1302–1307. https://doi.org/10.1021/jo016194y
92. Andersch J, Sicker D, Wilde H (1999) Methyl D-arabino-hex-2-ulopyranosonate as a building block for spiro[1,4,-benzoxazine-2,2′-pyrans]. Carbohydr Res 316:85–94. https://doi.org/10.1016/S0008-6215(99)00039-7

93. Andersch J, Sicker D, Wilde H (1999) Synthesis of spiro 1,4-benzothiazine-2,2′-pyrans starting from methyl β-D-arabino-2-hexulopyranosonate. J Heterocyclic Chem 36:457–460. https://doi.org/10.1002/jhet.5570360220
94. Somsák L, Batta G, Farkas I, Parkányi L, Kálmán A, Somogyi Á (1986) Synthesis and stereochemistry of a new spiro glycosylidene heterocycle from the reaction of 1-bromoglycosyl cyanides with S-nucleophiles: complete identification of hydrogen bridges by N.M.R. Methods and X-ray analysis. J Chem Res 12:436–437
95. Ősz E, Szilágyi L, Somsák L, Bényei A (1999) Synthesis of novel glycosylidene-spiro-heterocycles related to hydantocidin. Tetrahedron 55:2419–2430. https://doi.org/10.1016/S0040-4020(99)00034-4
96. Meijer A, Ellervik U (2004) Interhalogens (ICl/IBr_3) and AgOTf in thioglycoside activation; synthesis of bislactam analogues of ganglioside GD3. J Org Chem 69:6249–6256. https://doi.org/10.1021/jo049184g
97. Hossain N, Zapata A, Wilstermann M, Nilsson UJ, Magnusson G (2002) Synthesis of GD3-lactam: a potential ligand for the development of an anti-melanoma vaccine. Carbohydr Res 337:569–580. https://doi.org/10.1016/s0008-6215(02)00036-8
98. Lamberth C, Blarer S (1996) Concise approach to 1-thia-hydantocidin. Synth Commun 26:75–81. https://doi.org/10.1080/00397919608003864
99. Gasch C, Pradera MA, Salameh BAB, Molina JL, Fuentes J (2001) Isothiocyanato derivatives of sugars in the stereoselective synthesis of spironucleosides and spiro-*C*-glycosides. Tetrahedron Asymmetry 12:1267–1277. https://doi.org/10.1016/S0957-4166(01)00223-3
100. Gash C, Pradera MA, Salameh BAB, Molina JL, Fuentes J (2006) Stereocontrolled synthesis of thiohydantoin spiro-nucleosides from sugar spiroacetals. Tetrahedron 62:97–111. https://doi.org/10.1016/j.tet.2005.09.128
101. Silva S, Sánchez-Fernández EM, Ortiz Mellet C, Tatibouët A, Rauter AP, Rollin P (2013) *N*-Thiocarbonyl iminosugars: synthesis and evaluation of castanospermine analogues bearing oxazole-2(3*H*)-thione moieties. Eur J Org Chem 7941–7951. https://doi.org/10.1002/ejoc.201300720
102. Vangala M, Shinde GP (2015) Synthesis of D-fructose-derived spirocyclic 2-substituted-2-oxazoline ribosides. Beilstein J Org Chem 11:2289–2296. https://doi.org/10.3762/bjoc.11.249
103. Kraft J, Golkowski M, Ziegler T (2016) Spiro-fused carbohydrate oxazoline ligands: synthesis and application as enantio-discrimination agents in asymmetric allylic alkylation. Beilstein J Org Chem 12:166–171. https://doi.org/10.3762/bjoc.12.18
104. Soengas RG (2010) A straightforward route to novel α,α-disubstituted tetrahydrofuran β-amino acids and spirodiketopiperazines from sugar lactones. Synlett 2549–2552. https://doi.org/10.1055/s-0030-1258572
105. Taillefumier C, Thielges S, Chapleur Y (2004) Anomeric spiroannelated 1,4-diazepine-2,5-diones from Furano *exo*-glycals: towards a new class of spironucleosides. Tetrahedron 60:2213–2224. https://doi.org/10.1016/j.tet.2004.01.030
106. Li X, Wang R, Wang Y, Chen H, Li Z, Ba C, Zhang J (2008) Stereoselective synthesis and biological activity of novel spirooxazinanone-C-glycosides. Tetrahedron 64:9911–9920. https://doi.org/10.1016/j.tet.2008.08.002
107. Wang B, Wong OA, Zhao M-X, Shi Y (2008) Asymmetric epoxidation of 1,1-disubstituted terminal olefins by chiral dioxirane via a planar-like transition state. J Org Chem 73:9539–9543. https://doi.org/10.1021/jo801576k
108. Wong OA, Wang B, Zhao M-X, Shi Y (2009) Asymmetric epoxidation catalyzed by α,α-dimethylmorpholinone ketone. Methyl group effect on spiro and planar transition states. J Org Chem 74:6335–6338. https://doi.org/10.1021/jo900739q
109. Elek R, Kiss L, Praly JP, Somsák L (2005) β-D-Galactopyranosyl-thiohydroximates and D-galactopyranosylidene-spiro-oxathiazoles: synthesis and enzymatic evaluation against *E. coli* D-galactosidase. Carbohydr Res 340:1397–1402. https://doi.org/10.1016/j.carres.2005.02.021
110. Nagy V, Benltifa M, Vidal S, Berzsényi E, Teilhet C, Czifrák K, Batta G, Docsa T, Gergely P, Somsák L, Praly JP (2009) Glucose-based spiro-heterocycles as potent inhibitors of glycogen phosphorylase. Bioorg Med Chem 17:5696–5707. https://doi.org/10.1016/j.bmc.2009.05.080

111. Somsák L, Nagy V, Vidal S, Czifrák K, Berzsényi E, Praly JP (2008) Novel design principle validated: glucopyranosylidene-spiro-oxathiazole as new nanomolar inhibitor of glycogen phosphorylase, potential antidiabetic agent. Bioorg Med Chem 18:5680–5683. https://doi.org/10.1016/j.bmcl.2008.08.052
112. Szabó KE, Kun S, Mándi A, Kurtán T, Somsák L (2017) Glucopyranosylidene-spiro-thiazolinones: synthetic studies and determination of absolute configuration by TDDFT-ECD calculations. Molecules 22:1760. https://doi.org/10.3390/molecules22101760
113. Czifrák K, Gyóllai V, Kövér KE, Somsák L (2011) Ritter-type reaction of *C*-(1-bromo-1-deoxy-D-glycopyranosyl)formamides and its application for the synthesis of oligopeptides incorporating anomeric α-amino acids. Carbohydr Res 346:2104–2112. https://doi.org/10.1016/j.carres.2011.07.001
114. Somsák L, Bokor É, Czibere B, Czifrák K, Koppány C, Kulcsár L, Kun S, Szilágyi E, Tóth M, Docsa T, Gergely P (2014) Synthesis of *C*-xylopyranosyl- and xylopyranosylidene-spiroheterocycles as potential inhibitors of glycogen phosphorylase. Carbohydr Res 399:38–48. https://doi.org/10.1016/j.carres.2014.05.020
115. Benltifa M, Hayes JM, Vidal S, Gueyrard D, Goekjian PG, Praly JP, Kizilis G, Tiraidis C, Alexacou K, Chrysina ED, Zographos SE, Leonidas DD, Archontis G, Oikonomakos NG (2009) Glucose-based spiro-isoxazolines: a new family of potent glycogen phosphorylase inhibitors. Bioorg Med Chem 17:7368–7380. https://doi.org/10.1016/j.bmc.2009.08.060
116. Goyard D, Kónya B, Chajistamatiou AS, Chrysina ED, Leroy J, Balzarin S, Tournier M, Tousch D, Petit P, Duret C, Maurel P, Somsák L, Docsa T, Gergely P, Praly JP, Azay-Milhau J, Vidal S (2016) Glucose-derived spiro-isoxazolines are anti-hyperglycemic agents against type 2 diabetes through glycogen phosphorylase inhibition. Eur J Med Chem 108:444–454. https://doi.org/10.1016/j.ejmech.2015.12.004
117. Czifrák K, Páhi A, Deák S, Kiss-Szikszai A, Kövér KE, Docsa T, Gergely P, Alexacou KM, Papakonstantinou M, Leonidas DD, Zographos SE, Chrysina ED, Somsák L (2014) Glucopyranosylidene-spiro-iminothiazolidinone, a new bicyclic ring system: synthesis, derivatization, and evaluation for inhibition of glycogen phosphorylase by enzyme kinetic and crystallographic methods. Bioorg Med Chem 22:4028–4041. https://doi.org/10.1016/j.bmc.2014.05.076
118. Chatgilialoglu C, Ferreri C, Gimisis T (1999) Anionically induced formation of anomeric spironucleosides from 1′-C-cyano-2′-deoxyuridine. Tetrahedron Lett 40:2837–2840. https://doi.org/10.1016/S0040-4039(99)00273-7
119. Chatgilialoglu C, Ferreri C, Gimisis T, Roberti M, Balzarini J, de Clercq E (2004) Synthesis and biological evaluation of novel 1′-branched and spironucleoside analogues. Nucleosides Nucleotides Nucleic Acids 23:1565–1581. https://doi.org/10.1081/NCN-200031391
120. Dell'Isola A, McLachlan MMW, Neuman BW, Al-Mullah HMN, Binks AWD, Elvidge W, Shankland K, Cobb JA (2014) Synthesis and antiviral properties of spirocyclic [1,2,3]-triazolooxazine nucleosides. Chem Eur J 20:11685–11689. https://doi.org/10.1002/chem.201403560
121. Tong XG, Zhou LL, Wang YH, Xia C, Wang Y, Liang M, Hou FF, Cheng YX (2011) Acortatarins A and B, two novel antioxidative spiroalkaloids with a naturally unusual morpholine motif from *Acorus tatarinowii*. Org Lett 13:1844–1847. https://doi.org/10.1021/ol100451p
122. Geng HM, Stubbing LA, Chen JL, Furkert DP, Brimble MA (2014) Synthesis of the revised structure of acortatarin A. Eur J Org Chem 6227–6241. https://doi.org/10.1002/ejoc.201403000
123. Cao Z, Li Y, Wang S, Guo X, Wang L, Zhao W (2015) Total synthesis of two pyrrole spiroketal alkaloids: pollenopyrroside A and capparisine B. Synlett 26:921–926. https://doi.org/10.1055/s-0034-1380172
124. Jonckers TH, Lin TI, Buyck C, Lachau-Durand S, Vandyck K, van Hoof S, Vandekerckhove LA, Hu L, Berke JM, Vijgen L, Dillen LL, Cummings MD, de Kock H, Nilsson M, Sund C, Rydegård C, Samuelsson B, Rosenquist A, Fanning G, van Emelen K, Simmen K, Raboisson P (2010) 2′-Deoxy-2′-spirocyclopropylcytidine revisited: a new and selective inhibitor of the

hepatitis C virus NS5B polymerase. J Med Chem 53:8150–8160. https://doi.org/10.1021/jm101050a

125. Shen GH, Hong JH (2012) Synthesis of novel 2′-spirocyclopropyl-5′-deoxyphosphonic acid furanosyl nucleoside analogues as potent antiviral agents. Nucleosides Nucleotides Nucleic Acids 31:503–521. https://doi.org/10.1080/15257770.2012.695839
126. Gosselck J, Schmidt G (1968) A new route to 1,1-disubstituted cyclopropane. Angew Chem Int Ed 7:456–457. https://doi.org/10.1002/anie.196804561
127. Shen GH, Hong JH (2013) Synthesis of novel 2′-spirocyclopropanoid 4′-deoxythreosyl phosphonic acid nucleoside analogues. Bull Kor Chem Soc 34:868–874. https://doi.org/10.5012/bkcs.2013.34.3.868
128. Babu BR, Keinicke L, Petersen M, Nielsen C, Wengel J (2003) 2′-Spiro ribo- and arabinonucleosides: synthesis, molecular modelling and incorporation into oligodeoxynucleotides. Org Biomol Chem 1:3514–3526. https://doi.org/10.1039/B306354B
129. Du J, Chun BK, Mosley RT, Bansal S, Bao H, Espiritu C, Lam AM, Murakami E, Niu C, Micolochick Steuer HM, Furman PA, Sofia JM (2014) Use of 2′-spirocyclic ethers in HCV nucleoside design. J Med Chem 57:1826–1835. https://doi.org/10.1021/jm401224y
130. Versteeg K, Zwilling D, Wang H, Church KM (2010) Synthesis, structure, and sugar dynamics of a 2′-spiroisoxazolidine thymidine analog. Tetrahedron 66:8145–8150. https://doi.org/10.1016/j.tet.2010.08.033
131. Bondada L, Gumina G, Nair R, Ning XH, Schinazi RF, Chu CK (2004) Synthesis of novel spiro[2.3]hexane carbocyclic nucleosides via enzymatic resolution. Org Lett 6:2531–2534. https://doi.org/10.1021/ol0491989
132. Camarasa MJ, Pérez-Pérez MJ, San Félix A, Balzarine J, de Clerq E (1992) 3′-Spiro nucleosides, a new class of specific human immunodeficiency virus type 1 inhibitors: synthesis and antiviral activity of [2′,5′-bis-*O*-(*tert*-butyldimethylsilyl)-β-D-xylo- and- ribofuranose]-3′-spiro-5″-[4″-amino-1″-2″-oxathiole 2″, 2″-dioxide] (TSAO) pyrimidine nucleosides. J Med Chem 35:2721–2727. https://doi.org/10.1021/jm00093a002
133. Balzarini J, Pérez-Pérez MJ, San-Félix A, Schols D, Perno CF, Vandamme AM, Camarasa MJ, De Clercq E (1992) 2′,5′-Bis-O-(tert-butyldimethylsilyl)-3′-spiro-5″-(4″-amino-1″,2″-oxathiole-2″,2″-dioxide)pyrimidine (TSAO) nucleoside analogues: highly selective inhibitors of human immunodeficiency virus type 1 that are targeted at the viral reverse transcriptase. Proc Natl Acad Sci U S A 89:4392–4397
134. Pérez-Pérez MJ, San Félix A, Balzarini J, De Clerq E, Camarasa MJ (1992) TSAO analogs. Stereospecific synthesis and anti-HIV-1 activity of 1-[2′,5′-bis-O-(tert-butyldimethylsilyl)-β-D-ribofuranosyl]-3′-spiro-5″-(4′-amino-1″,2″-oxathiole-2″-2″-dioxide)pyrimidine and pyrimidine-modified nucleosides. J Med Chem 35:2988–2995. https://doi.org/10.1021/jm00094a009
135. Velázquez S, Félix S, Pérez-Pérez MJ, Balzarini J, de Clerq E, Camarasa MJ (1993) TSAO analogs. 3. Synthesis and anti-HIV-1. Activity of [2′-5′-bis-O-(tert-butyldimethylsilyl)-β-D-ribofuranosyl]-3′-spiro-5″-(4′-amino-1″,2″-oxathiole-2″-2″-dioxide)purin and purine-modified nucleosides. J Med Chem 36:3230–3239. https://doi.org/10.1021/jm00074a003
136. San-Félix A, Velázquez S, Pérez-Pérez MJ, Balzarini J, de Clercq E, Camarasa MJ (1994) Novel series of TSAO-T derivatives. Synthesis and anti-HIV-1 activity of 4-, 5-, and 6-substituted pyrimidine analogs. J Med Chem 37:453–460. https://doi.org/10.1021/jm00030a004
137. Alvarez R, Velázquez S, San-Félix A, Aquaro S, de Clercq E, Perno CF, Karlsson A, Balzarini J, Camarasa MJ (1994) 1,2,3-Triazole-[2,5-bis-O-(tert-butyldimethylsilyl)-beta.-D-ribofuranosyl]-3′-spiro-5″-(4′-amino-1″,2″-oxathiole-2″-2″-dioxide) (TSAO) analogs: synthesis and anti-HIV-1 activity. J Med Chem 37:4185–4194. https://doi.org/10.1021/jm00050a015
138. Lobatón E, Velázquez S, San-Félix A, Chamorro C, Tuñón V, Esteban-Gamboa A, de Clercq E, Balzarini J, Camarasa MJ, Pérez-Pérez MJ (1999) Novel TSAO derivatives modified at positions 3″ and 4′ of the spiro moiety. Nucleosides Nucleotides 18:675–676. https://doi.org/10.1080/15257779908041536

139. Lobatón E, Camarasa MJ, Velázquez S (2000) An efficient synthesis of 3′-spiro sultone nucleosides functionalized on the sultone moiety via Pd-catalyzed cross-coupling reaction. Synlett 9:1312–1314. https://doi.org/10.1055/s-2000-7124
140. de Castro S, Lobatón E, Pérez-Pérez MJ, San-Félix A, Cordeiro A, Andrei G, Snoeck R, de Clercq E, Balzarini J, Camarasa MJ, Velázquez S (2005) Novel [2′,5′-bis-O-(tert-butyldimethylsilyl)-α-D-ribofuranosyl]-3′-spiro-5″-(4″-amino-1″,2″-oxathiole-2″,2″-dioxide) derivatives with anti-HIV-1 and anti-human-cytomegalovirus activity. J Med Chem 48:1158–1168. https://doi.org/10.1021/jm040868q
141. Tronchet JM, Kovacs I, Seman M, Dilda P, de Clercq E, Balzarini J (2000) Highly stereoselective synthesis and biological properties of nucleoside analogues bearing a spiro inserted oxirane ring. Nucleosides Nucleotides Nucleic Acids 19:775–779. https://doi.org/10.1080/15257770008035024
142. Gash C, Illagua JM, Merino-Montiel P, Fuentes J (2009) Stereocontrolled synthesis of (5 + 5), (5 + 6) and (6 + 6) 3-spiropseudonucleosides. Tetrahedron 65:4149–4155. https://doi.org/10.1016/j.tet.2009.03.038
143. Turks M, Rodins V, Rolava E, Ostrovskis P, Belyakov S (2013) A practical access to glucose- and allose-based (5 + 5) 3-spiropseudonucleosides from a common intermediate. Carbohydr Res 375:5–15. https://doi.org/10.1016/j.carres.2013.04.008
144. Lipshutz BH, Sharma S, Dimock SH, Behling JR (1992) Preparation of C-4 alkylated dideoxyribosides: potential precursors to a novel series of nucleosides. Synthesis 191–195. https://doi.org/10.1055/s-1992-34158
145. Crich D, Hao X (1999) Asymmetric synthesis of C4′-α-carboxylated 2′-deoxynucleosides. Preparation of oxetanone derivatives and influence of solvent on the stereochemistry of base introduction. J Org Chem 64:4016–4024. https://doi.org/10.1021/jo990046e
146. Paquette LA (2004) Spirocyclic restriction of nucleosides. Aust J Chem 57:7–17. https://doi.org/10.1021/ol010209x
147. Paquette LA, Owen DR, Bibart RT, Seekamp CK, Kahane AL, Lanter JC, Corral MA (2001) 1-Oxaspiro[4.4]nonan-6-ones. Synthetic access via oxonium ion technology, optical resolution, and conversion into enantiopure spirocyclic α,β-butenolides. J Org Chem 66:2828–2834. https://doi.org/10.1021/jo010026o
148. Paquette LA, Bibart RT, Seekamp CK, Kahane AL (2001) Spirocyclic restriction of nucleosides. Synthesis of the first exemplary syn-1-oxaspiro[4.4]nonanyl member. Org Lett 3:4039–4041. https://doi.org/10.1021/ol010209x
149. Paquette LA, Owen DR, Bibart RT, Seekamp CK (2001) Spirocyclic restriction of nucleosides. An analysis of protecting group feasibility while accessing prototype anti-1-oxaspiro [4.4]nonanyl mimics. Org Lett 3:4043–4045. https://doi.org/10.1021/ol010212g
150. Paquette LA, Seekamp CK, Kahane AL (2003) Conformational restriction of nucleosides by spirocyclic annulation at C4′ including synthesis of the complementary dideoxy and didehydrodideoxy analogues. J Org Chem 68:8614–8624. https://doi.org/10.1021/jo0301954
151. Dong S, Paquette LA (2006) Highly stereoselective β-anomeric glycosidation of a 2′-deoxy syn-5′-configured 4′-spironucleoside. J Org Chem 71:1647–1652. https://doi.org/10.1021/jo052324h
152. Maity JK, Achari B, Drew MGB, Mandal SB (2010) Synthesis of a carbohydrate-derived 1-oxaspiro[4.4] nonane skeleton and its conversion into spironucleosides. Synthesis 2533–2542. https://doi.org/10.1055/s-0029-1218828
153. Marquez VE, Lim MI (1986) Carbocyclic nucleosides. Med Res Rev 6:1–40
154. Rodriquez JB, Comin MJ (2003) New progresses in the enantioselective synthesis and biological properties of carbocyclic nucleosides. Mini-Rev Med Chem 3:95–114. https://doi.org/10.2174/1389557033405331
155. Schneller SW (2002) Carbocyclic nucleosides (carbanucleosides) as new therapeutic leads. Curr Top Med Chem 2:1087–1092. https://doi.org/10.2174/1568026023393228
156. Crimmins MT (1998) New developments in the enantioselective synthesis of cyclopentyl carbocyclic nucleosides. Tetrahedron 54:9229–9272. https://doi.org/10.1016/S0040-4020(98)00320-2

157. Huryn DM, Okabe M (1992) AIDS-driven nucleoside chemistry. Chem Rev 92:1745–1768. https://doi.org/10.1021/cr00016a004
158. Borthwick AD, Biggadike K (1992) Synthesis of chiral carbocyclic nucleosides. Tetrahedron 48:571–623. https://doi.org/10.1016/S0040-4020(01)88122-9
159. Paquette LA, Hartung RE, France DJ (2003) Conversion of the enantiomers of spiro[4.4]nonane-1,6-diol into both epimeric carbaspironucleosides having natural C1′ absolute configuration. Org Lett 5:869–871. https://doi.org/10.1021/ol0340033
160. Hartung RE, Paquette LA (2005) Development of a generic stereocontrolled pathway to fully hydroxylated spirocarbocyclic nucleosides as a prelude to RNA targeting. Synthesis 3209–3218. https://doi.org/10.1055/s-2005-918472
161. Dyson MR, Coe PL, Walker RT (1991) The synthesis and antiviral properties of *E*-5-(2-bromovinyl)-4′-thio-2′-deoxyuridine. J Chem Soc Chem Commun 741–742. doi:https://doi.org/10.1039/c39910000741
162. Tiwari KN, Seacrist JA, Montgomery JA (1994) Synthesis and biological activity of 4′-thionucleosides of 2-chloroadenine. Nucleosides Nucleotides 13:1819–1828. https://doi.org/10.1080/15257779408009484
163. Young RJ, Shae-Ponter S, Thomson JB, Miller JA, Cumming JG, Pugh AW, Rider P (1995) Synthesis and antiviral evaluation of enantiomeric 2′,3′-dideoxy- and 2′,3′-didehydro-2′,3′-dideoxy-4′-thionucleosides. Bioorg Med Chem Lett 5:2599–2604. https://doi.org/10.1016/0960-894X(95)00472-6
164. Yoshimura Y, Watanabe M, Satoh H, Ashida N, Ijichi K, Sakata S, Machida H, Matsuda A (1997) A facile, alternative synthesis of 4′-thioarabinonucleosides and their biological activities. J Med Chem 40:2177–2183. https://doi.org/10.1021/jm9701536
165. Satoh H, Yoshimura Y, Watanabe M, Ashida N, Ijichi K, Sakata S, Machida H (1998) Synthesis and antiviral activities of 5-substituted 1-(2-deoxy-2-C-methylene-4-thio-β-D-erythro-pentofuranosyl)uracil. Nucleosides Nucleotides 17:65–79. https://doi.org/10.1080/07328319808005158
166. Paquette LA, Fabris F, Gallou F, Dong S (2003) C4′-Spiroalkylated nucleosides having sulfur incorporated at the apex position. J Org Chem 68:8625–8634. https://doi.org/10.1021/jo030196w
167. Roy A, Achari B, Mandal SB (2006) An easy access to spiroannulated glyco-oxetane, -thietane and -azetane rings: synthesis of spironucleosides. Tetrahedron Lett 47:3875–3879. https://doi.org/10.1016/j.tetlet.2006.03.175
168. Mandal SB (2008) Introduction of vinyl and hydroxymethyl functionalities at C-4 of glucose-derived substrates: synthesis of spirocyclic, bicyclic, and tricyclic nucleosides. J Org Chem 73:4305–4308. https://doi.org/10.1021/jo8002826
169. de Carvalho GSG, Fourrey JL, Dodd RH, Da Silva AD (2009) Synthesis of a 4′,4′-spirothietane-2′, N3-cycloadenosine as a highly constrained analogue of 5′-deoxy-5′-methylthioadenosine (MTA). Tetrahedron Lett 50:463–466. https://doi.org/10.1016/j.tetlet.2008.11.039
170. Tripathi S, Roy BG, Drew MG, Achari B, Mandal SB (2007) Synthesis of oxepane ring containing monocyclic, conformationally restricted bicyclic and spirocyclic nucleosides from D-glucose: a cycloaddition approach. J Org Chem 72:7427–7430. https://doi.org/10.1021/jo070846m
171. Klumpp K, Lévêque V, Le Pogam S, Ma H, Jiang WR, Kang H, Granycome C, Singer M, Laxton C, Hang JQ, Sarma K, Smith DB, Heindl D, Hobbs CJ, Merrett JH, Symons J, Cammack N, Martin JA, Devos R, Nájera I (2006) The novel nucleoside analog R1479 (4′-azidocytidine) is a potent inhibitor of NS5B-dependent RNA synthesis and hepatitis C virus replication in cell culture. J Biol Chem 281:3793–3799. https://doi.org/10.1074/jbc.M510195200
172. Dang Q, Zhang Z, He S, Liu Y, Chen T, Bogen S, Girijavallabhan V, Olsen DB, Meinke PT (2014) Syntheses of 4′-spirocyclic phosphono-nucleosides as potential inhibitors of hepatitis C virus NS5B polymerase. Tetrahedron Lett 55:4407–4409. https://doi.org/10.1016/j.tetlet.2014.06.029

Top Heterocycl Chem (2019) 57: 215–260
DOI: 10.1007/7081_2018_27

Published online: 11 January 2019

Spiroketal Phthalane *C*-Glycosides: Synthesis of Papulacandins and SGLT2 Inhibitors

Yoshihiko Yamamoto

Contents

Abstract Spiroketals are important structural motifs found in diverse natural products, many of which display unique biological activity. Among them, spiroketal phthalane *C*-glycosides, in which a phthalane ring and sugar unit form a spiroketal framework, have garnered enormous attention from wide research areas because such a fascinating spirocycle motif is found in antibiotic natural products, i.e., papulacandins and their relatives. Moreover, recent reports from pharmaceutical researchers have revealed that spiroketal phthalane *C*-glycosides are potent drug candidates for type 2 diabetes. Accordingly, the efficient and selective construction

Y. Yamamoto (✉)
Department of Basic Medicinal Sciences, Graduate School of Pharmaceutical Sciences, Nagoya University, Nagoya, Japan
e-mail: yamamoto-yoshi@ps.nagoya-u.ac.jp

of the spiroketal phthalane *C*-glycoside motif is an important research objective in synthetic organic chemistry. In this chapter, recent advances in the synthesis of spiroketal phthalane *C*-glycosides will be discussed.

Keywords Antibiotics · *C*-arylglycosides · Papulacandins · SGLT2 inhibitors · Spiroketals

1 Introduction

Spiroketal motifs are fascinating structural units as they are found in diverse bioactive natural products such as pinnatifinosides, rubromycins, chaetoquadrins, and papulacandins [1–3]. Among them, papulacandins have garnered enormous attention from wide research areas since they possess attractive structural characteristics. Papulacandins have a common spiroketal *C*-glycoside unit [Although the term "*C*-arylglycoside" has been frequently used, it is a misnomer according to the IUPAC carbohydrate nomenclature ("*C*-glycosyl arene" is correct). However, "*C*-arylglycoside" has been used throughout for the sake of simplicity], in which a phthalane ring and glucose moiety are connected through both C–C and C–O bonds (Fig. 1), and the *C*-glycoside unit is further decorated with a galactose residue and/or a long-chain unsaturated fatty acid [4]. Moreover, papulacandins also display strong activity against *Candida albicans* and several other yeasts [5]. Due to these intriguing features, considerable synthetic efforts have been devoted to the efficient construction of spiroketal phthalane *C*-glycosides relevant to papulacandins. These synthetic efforts have culminated in the total synthesis of papulacandin D [6–10]. Moreover, inspired by the spiroketal phthalane *C*-glycoside motif of papulacandins, pharmaceutical researchers have developed potent drug candidates for type 2 diabetes [11–14]. In this chapter, the synthesis of spiroketal phthalane *C*-glycosides will be discussed. Before discussing synthetic examples, the characteristics of the papulacandin family will be briefly overviewed.

1.1 *Structure and Bioactivity of Papulacandins and Relatives*

Papulacandins A–D were isolated from a strain of *Papularia sphaerosperma* (PERS.) Höhnel by Traxler and coworkers in 1977 (Fig. 2) [15–17]. They have specific activity against yeasts, but little or no activity against filamentous fungi,

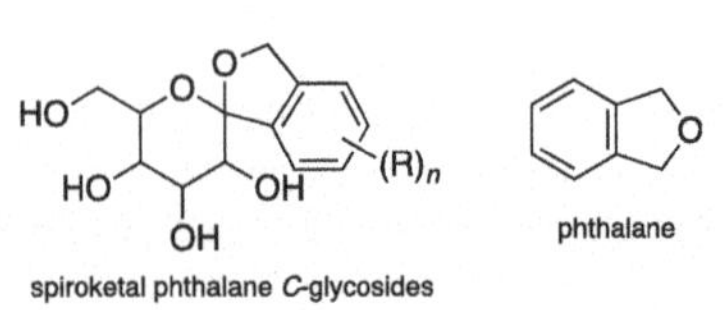

Fig. 1 Spirocyclic phthalane *C*-glycoside motif

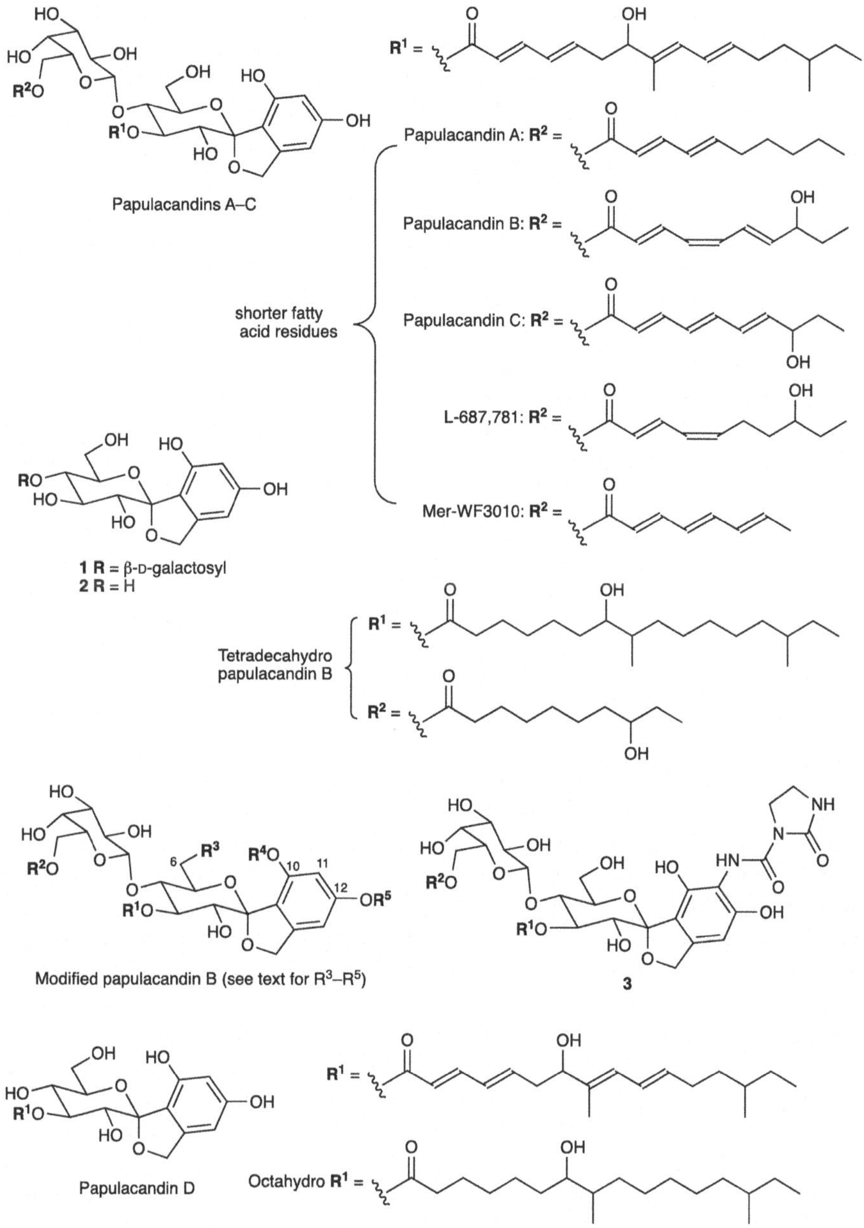

Fig. 2 Structures of papulacandins A–D and derivatives

bacteria, and protozoa [5]. The specific antibiotic activity of papulacandins was ascribed to the inhibition of β-1,3-glucan synthase [18–20]. Since glucan is the main structural component of yeast cell wall, the inhibition of glucan synthase leads to interference of the cell-wall synthesis in yeast. Papulacandin B is the most active among papulacandins, both in terms of growth inhibition of *Candida albicans* and inhibition of glucan biosynthesis. Papulacandins A and C, which have a different shorter fatty acid on the galactose residue, exhibit similar biological activity. Hydrogenation of both the fatty acid residues (tetradecahydro papulacandin B) or the removal of the shorter fatty acid residue has a small impact on the in vitro activity, while the in vivo activity is completely attenuated [21, 22]. Although papulacandin D has no galactose residue connected to the shorter fatty acid, it retains considerable bioactivity. However, hydrogenation of papulacandin D (octahydropapulacandin D) significantly reduces its biological activity. Without the fatty acid residues, spiroketals **1** and **2** exhibit no biological activity. The biological activity was also completely lost when papulacandin B was subjected to exhaustive acetylation (papulacandin B nonaacetate).

In order to improve the biological activity, several derivatives of papulacandin B were synthesized for biological testing (Fig. 2) [21, 22]. However, a significant increase in the biological activity was not achieved. Modifications in the glucose side chain at the 6-position (R^3) did not have a significant impact on the biological activity. On the other hand, polar substituents on the aromatic moiety at the 10-position (OR^4) led to a slight increase in the biological activity in vivo, although substituents OR^5 larger than methoxy at the 12-position led to negative results. Additional substituents at the 11-position of the aromatic ring also reduced the biological activity, although several 11-aminoacyl derivatives presented better biological activity in vivo. Imidazolidinone derivative **3** proved to be five times more active than papulacandin B.

Several other natural spiroketal *C*-arylglycosides have also been identified. L-687,781 and Mer-WF3010 are papulacandin analogs, which bear a different shorter fatty acid on the galactose residue (Fig. 2) [23–27]. There are other papulacandin analogs, in which both fatty acid residues are different from those of papulacandins (Fig. 3). BU-4794F has a pentaene-type fatty acid residue with no methyl branches [28], while BE-29602 and saricandin share a common tetraene-type fatty acid on the glucose unit [29, 30]. Saricandin is the sole papulacandin analog bearing a cinnamate substituent on the galactose residue. There are four members in the F-10748 family, in which A_1 has no shorter fatty acid and B_1–D_1 have a saturated acid of different length on the galactose residue [31]. These natural products exhibit biological activities similar to those of papulacandins and inhibit 1,3-β-D-glucan synthase. *Pneumonocystis carinii* pneumonia is a major cause of death in AIDS patients. Since the wall of the cyst form of *P. carinii* contains high levels of 1,3-β-glucan, papulacandins are potent drugs that inhibit the formation of *P. carinii* cysts. In fact, it was reported that L-687,781 was effective on a *P. carinii* pneumonia rat model [32].

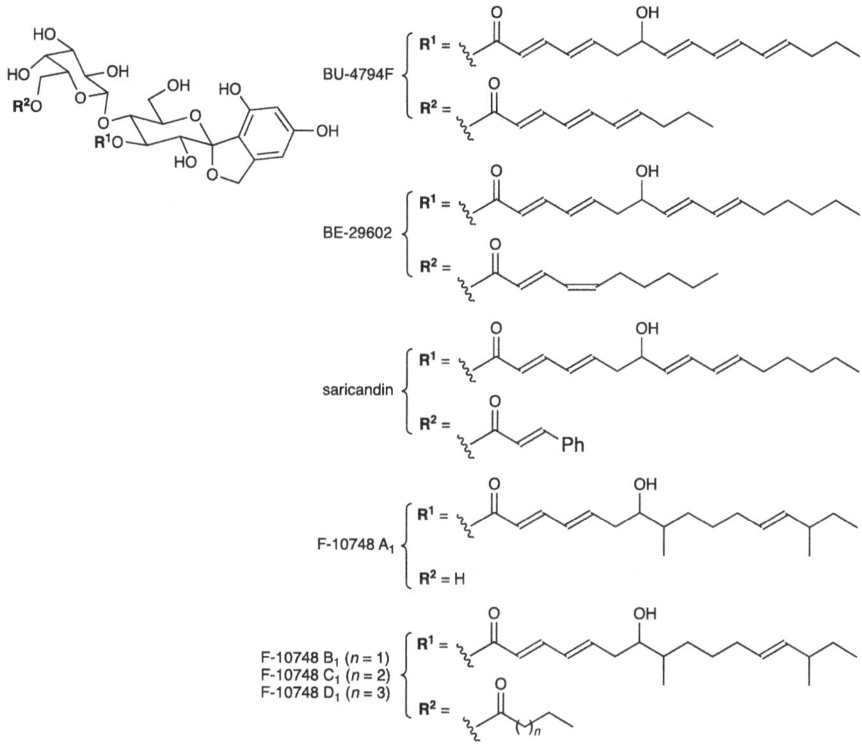

Fig. 3 Structures of papulacandin-type natural products

1.2 Synthetic Strategies for the Spiroketal Phthalane C-Glycoside Motif

C-Arylglycosides, which have an aromatic ring directly connected to the anomeric carbon of a sugar through a C–C bond, have attracted considerable attention owing to their presence in diverse natural products and their biological relevance as robust analogs of *O*-arylglycosides [33–36]. However, the efficient and stereoselective C–C bond formation at the anomeric position of complex sugar substrates has been a formidable challenge [37–39]. This is also true for the construction of spiroketal *C*-arylglycosides, in which a spirocyclic framework is formed through C–C and C–O bonds at the anomeric carbon. Thus, a wide variety of methods have been devised, whose central strategy can be categorized into four major types; i.e., synthesis *via* (a) the coupling of a sugar electrophile with an aryllithium reagent, (b) coupling of a metalated glucal with an aryl electrophile, (c) de novo construction of a sugar framework, and (d) construction of an aromatic ring (Fig. 4).

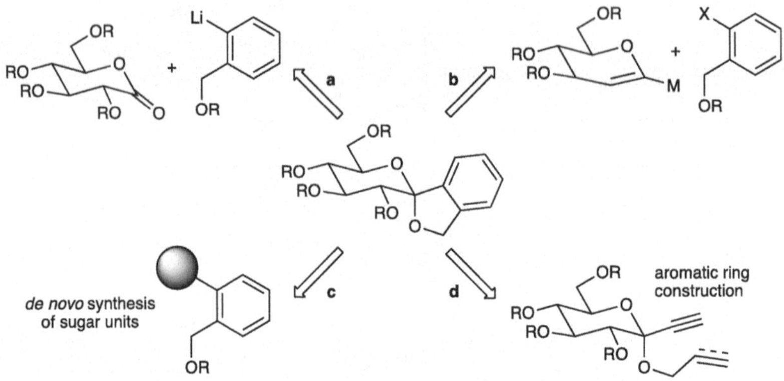

Fig. 4 Synthetic strategies for the spiroketal phthalane *C*-glycoside motif

In the following sections, the synthesis of spiroketal phthalane *C*-glycosides will be discussed primarily according to the compound class (papulacandins, SGLT2 inhibitors, and others). The next section for the synthesis of papulacandins will be divided into subsections according to the employed synthetic strategy.

2 Synthesis of Spiroketal Phthalane *C*-Glycosides Relevant to Papulacandins

In this section, the synthesis of the papulacandin spiroketal core will be overviewed following the categories for the strategies employed (Fig 4).

2.1 Synthesis of the Papulacandin Spiroketal Core via Coupling of a Sugar Electrophile with an Aryllithium Reagent

The earliest example in this category was reported by Schmidt and Frick, who employed D-glucose-derived aldehyde **4** as a sugar electrophile (Scheme 1) [40]. Protected aryl bromides **6** underwent bromine–lithium exchange upon treatment with ^{n}BuLi in tetrahydrofuran (THF) at –15°C. The resultant aryllithium reagents were allowed to react with aldehyde **4** at the same temperature to afford alcohols **7** as 1:1 diastereomeric mixtures in moderate yields. The subsequent dimethylsulfoxide (DMSO) oxidation of **7** produced ketones **8** in high yields. Finally, debenzylation followed by exhaustive acetylation converted **8** into protected papulacandin spiroketal core **9** in 45–62% yields. The addition of the aryllithium to ester **5**, which was prepared from **4** in two steps, directly afforded **8**, albeit in a moderate yield.

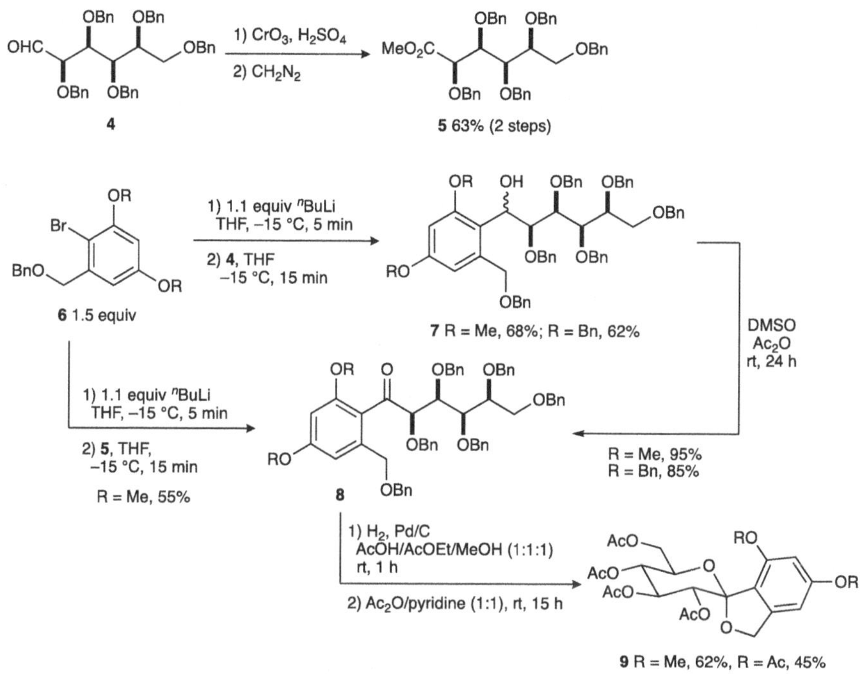

Scheme 1 Synthesis of spiroketal *C*-glycosides **9** from glucose-derived aldehyde **4**

Although Schmidt and Frick described that the addition of the aryllithium to 2,3,4,6-tetra-*O*-benzyl-D-gluconolactone (**10**) was not effective [40], later, Rosenblum and Bihovsky tried to utilize **10** as the electrophile (Scheme 2) [41]. The aryllithium reagent derived from partially protected aryl bromide **11** reacted with **10** at low temperature to produce spiroketal anomers α/β-**12** in low yields. This inefficiency is ascribed to the formation of undesired β-elimination byproduct **13**. Thus, the fully protected aryllithium reagent derived from **14** was used, and hydroxyketone **15** was obtained in 42% yield. After desilylation of **15** using tetra-*n*-butylammonium fluoride (TBAF), the obtained diol **16** was subjected to hydrogenolysis of the benzyl protecting groups to afford spiroketal **17** in a high yield. The final removal of methoxymethyl (MOM) groups using Dowex 50 × 4 resin afforded the unprotected papulacandin spiroketal core **2** in 77% yield.

They further investigated the introduction of a fatty acid to the papulacandin spiroketal core (Scheme 3) [41]. After regioselective benzylidenation of **17**, the obtained **18** was subjected to acylation using palmitoyl chloride under phase-transfer conditions. This uncontrolled acylation produced mono-acylated products **19** and **20** along with bis-acylated product **21** with low selectivity. Nevertheless, global deprotection of **19** using Dowex 50 × 4 resin afforded the unnatural papulacandin D analog **22** in 70% yield. A similar method using D-gluconolactone **10** was also reported by Czernecki and Perlat [42]. In their study, ^{s}BuLi was used for the bromine–lithium exchange of a differently protected aryl bromide to generate the corresponding aryllithium reagent.

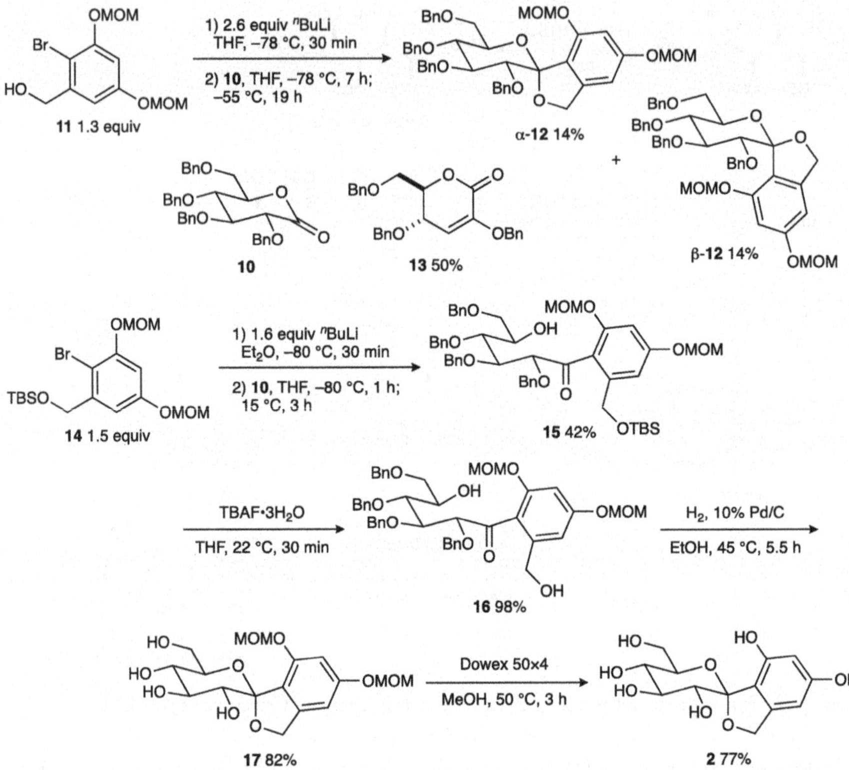

Scheme 2 Synthesis of papulacandin spiroketal core **2** from gluconolactone **10**

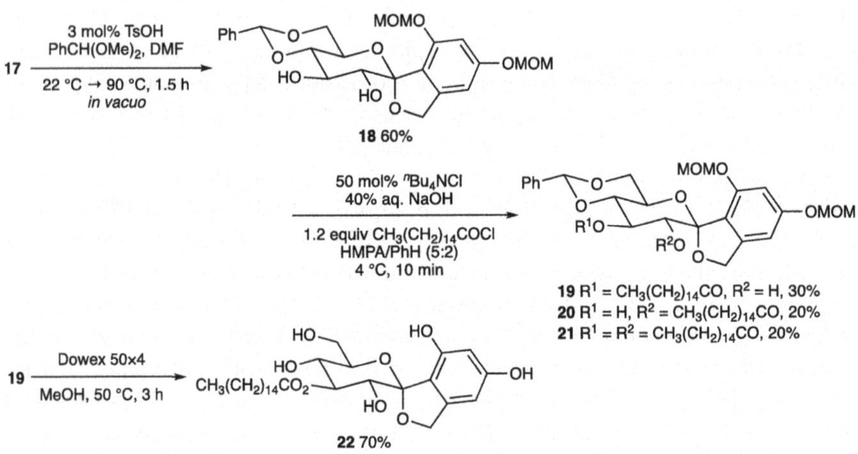

Scheme 3 Synthesis of papulacandin D analog **22**

The first total synthesis of papulacandin D was accomplished *via* coupling of a sugar electrophile with an aryllithium reagent by the Barrett group [6–8]. In their study, trimethylsilyl (TMS)-protected D-gluconolactone **23** and triisopropylsilyl (TIPS)/*tert*-butyldimethylsilyl (TBS)-protected aryl bromide **24** were used for the construction of a spiroketal unit (Scheme 4). The bromine–lithium exchange of **24** with ^{t}BuLi in Et_2O at –78°C generated the corresponding aryllithium reagent, which reacted with gluconolactone **23** at the same temperature. The resultant crude hemiketal **25** was directly treated with an ion-exchange resin (Amberlist 120) to produce the desired spiroketal as a single anomer. After exhaustive acetylation, the protected papulacandin spiroketal core **26** was obtained in 33% overall yield.

Although the stereochemistry of the papulacandin spiroketal core was established [4], both the relative and absolute configurations of the chiral centers in the unsaturated fatty acid residue were unknown. Since L-isoleucine was presumed to be the biosynthetic precursor for the fatty acid, its synthesis started with known (*S*)-3-methylpentan-1-ol (**27**) derived from L-isoleucine (Scheme 5). Alcohol **27** was converted into nitrile **28**, which was subjected to diisobutylaluminum hydride (DIBAL-H) reduction, followed by Wittig olefination, affording enoate **29**. A similar sequence was applied to **29** to obtain dienoate **30**, which was then converted to dienal **31**. Propargylation of **31** afforded a diastereomeric mixture of homopropargylic alcohol **32**. The kinetic resolution of diastereomers **32** was performed by Sharpless asymmetric epoxidation to afford the desired optically active alcohol **33** with the *S* configuration at the secondary alcohol moiety and epoxide **34** in 45% and 14% yields, respectively. After protection of the hydroxyl group of **33**, the obtained silyl ether was subjected to hydrozirconation, followed by cross coupling with methyl 3-bromo-*E*-acrylate to produce tetraenoyl ester **35**. The hydroxy ester obtained by desilylation of **35** was found to be identical to the authentic sample derived from the degradation of natural papulacandins. It was finally confirmed that the natural fatty acid has the absolute configurations 7*S* and 14*S*.

To achieve coupling with the fatty acid component, spiroketal **36**, derived from **26**, was protected with a ^{t}Bu$_2$Si group to afford **37** in 85% yield (Scheme 6). Upon treatment of fatty acid **38** with 2,4,6-trichlorobenzoyl chloride, the resultant

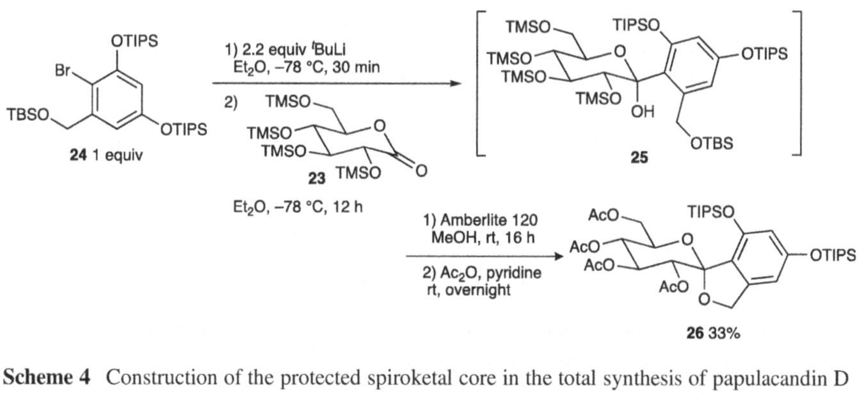

Scheme 4 Construction of the protected spiroketal core in the total synthesis of papulacandin D

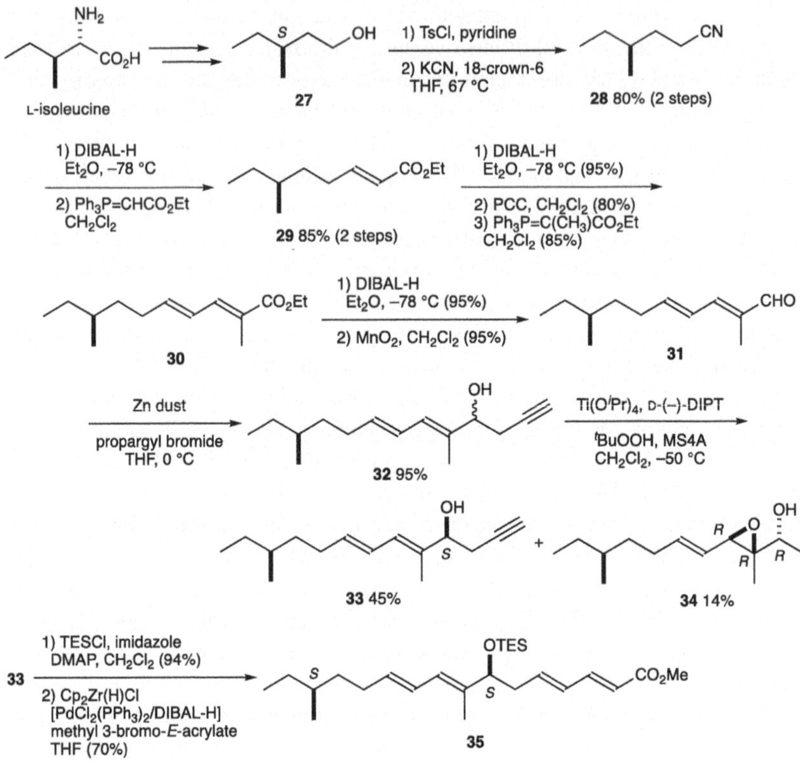

Scheme 5 Preparation of optically active fatty acid ester **35** required for the total synthesis of papulacandin D

Yamaguchi mixed anhydride was subjected to coupling with spiroketal **37** to afford protected papulacandin D (**39**), along with undesired byproduct **40** in 70% overall yield with a **39/40** ratio of 4:1. The final global deprotection of **39** using tris (dimethylamino)sulfonium difluorotrimethylsilicate (TASF) in THF at 0°C completed the first total synthesis of (+)-papulacandin D. Following the report from the Barrett group, a papulacandin D analog with a fatty acid residue identical to that of saricandin was synthesized using a similar procedure [43, 44]. The antifungal activity of the obtained analog was evaluated in enzyme and cell-based assays, but no significant activity was observed. A modified procedure for selective acylation was also reported [44].

2.2 *Synthesis of the Papulacandin Spiroketal Core via Coupling of a Metalated Glucal with an Aryl Electrophile*

The construction of a spiroketal glycoside unit was also achieved using the reverse polarity strategy (Scheme 7) [45]. Protected D-glucal **41** was treated with ^{t}BuLi and

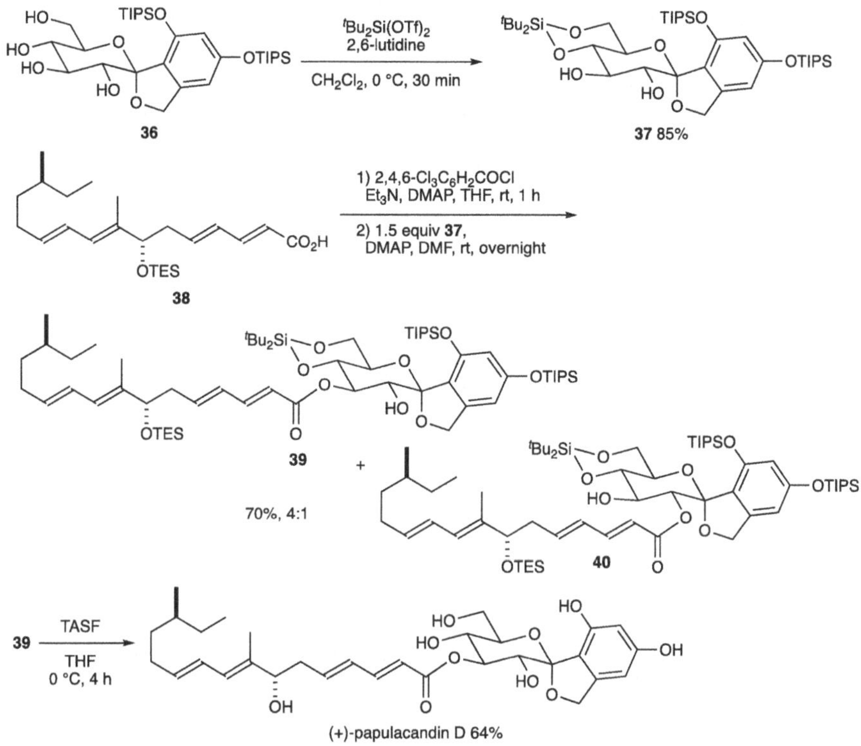

Scheme 6 Completion of the total synthesis of (+)-papulacandin D

the resultant lithiated species **42** was allowed to react with quinone **43** in the presence of $Et_2O{\cdot}BF_3$ at –78°C. The expected adduct **44** was obtained, albeit in a low yield, along with unreacted **43**. Then, **44** was aromatized to afford 1-arylglucal **45** after benzylation of the phenolic hydroxyl groups. The obtained glucal **45** was treated with *m*-chloroperbenzoic acid (mCPBA) in MeOH/THF to afford *C*-arylglycosides **46** as an anomeric mixture, which were converted into protected papulacandin spiroketal core **47** *via* removal of the benzyl protecting groups, followed by silylation of the phenolic hydroxyl groups.

As discussed in the former section, the C–C bond formation strategies using aryllithium reagents afford relatively low yields. The reverse polarity method introduced by Parker and Georges also needs further improvement. Hereinafter, alternative methods based on the transition-metal-catalyzed coupling of metalated glucals with aryl electrophiles will be overviewed.

The Friesen group and Beau group independently developed the synthesis of spirocyclic phthalane *C*-glycosides *via* Stille coupling. In the study by Friesen and Sturino, Stille coupling of 3,4,6-tri-*O*-(*tert*-butyldimethylsilyl)-1-(tributylstannyl)-D-glucal (**48**) with aryl bromide **49** using 5 mol% $PdCl_2(PPh_3)_2$ produced *C*-arylglucal **50** (Scheme 8) [46]. Since the undesired self-dimerization of **48** occurred, the reaction was conducted using 2.1 equiv of **48** in toluene under reflux, affording **50** in 85% yield based on **49**. After reductive removal of the acetyl group, the obtained

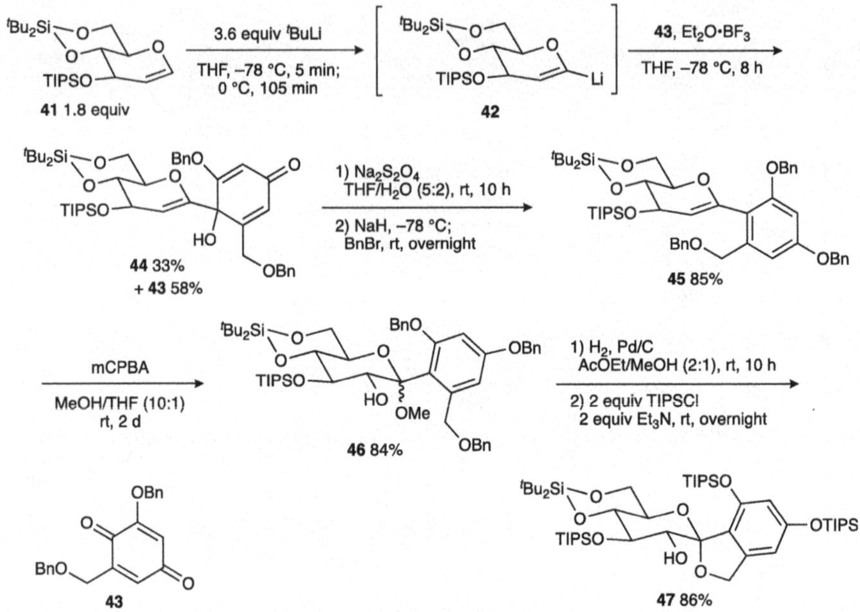

Scheme 7 Synthesis of the protected papulacandin spiroketal core using the reverse polarity strategy

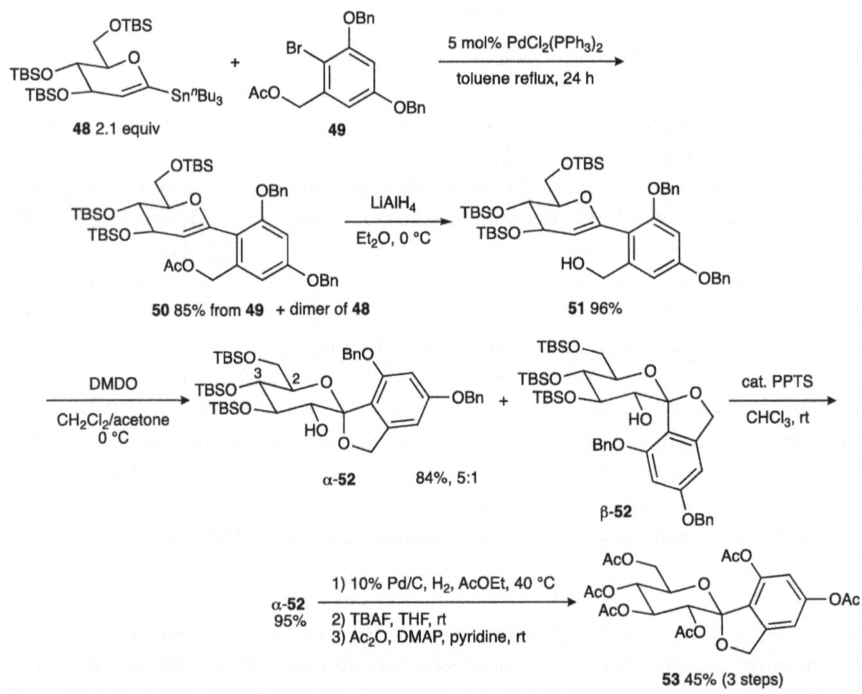

Scheme 8 Synthesis of peracetylated papulacandin core **53** *via* Stille coupling of **48** with **49**

benzylic alcohol **51** was treated with dimethyldioxirane (DMDO) in CH_2Cl_2/acetone at 0°C to induce spiroketalization. Two anomers were formed in 84% yield with an α-**52**/β-**52** ratio of 5:1. The stereochemistry of α-**52** was established by the H2 and H3 proton resonances, which displayed coupling constants of $J_{2,3} = 9.3$ Hz and $J_{3,4} = 8.4$ Hz. Such large coupling constants are the results of the *trans*-diaxial relationships between H2 and H3, as well as H3 and H4 in a $^4C_1$ conformation. In contrast, smaller coupling constants of $J_{2,3} = 3.3$ Hz and $J_{3,4} = 4$ Hz were observed for anomer β-**52**, which prefers a twisted conformation. The obtained anomeric mixture was further treated with catalytic amounts of pyridinium *p*-toluenesulfonate (PPTS) in $CHCl_3$ at room temperature to afford α-**52** as the sole product in 95% yield, as a result of the acid-catalyzed epimerization of β-**52**. Further deprotection, followed by exhaustive acetylation of the hydroxyl groups, afforded the protected papulacandin spiroketal core **53** in 45% yield from α-**52**.

In the study of Dubois and Beau, Stille coupling of 1-stannyl-D-glucal **54** with aryl bromide **55** bearing an unprotected benzylic alcohol was conducted using 5 mol% $Pd(PPh_3)_4$ to afford deoxy spiroketal **56** in 72% yield along with self-dimer **57** in 15% yield (Scheme 9) [47, 48]. It was considered that the desired coupling product **58** directly underwent acid-promoted cyclization under the

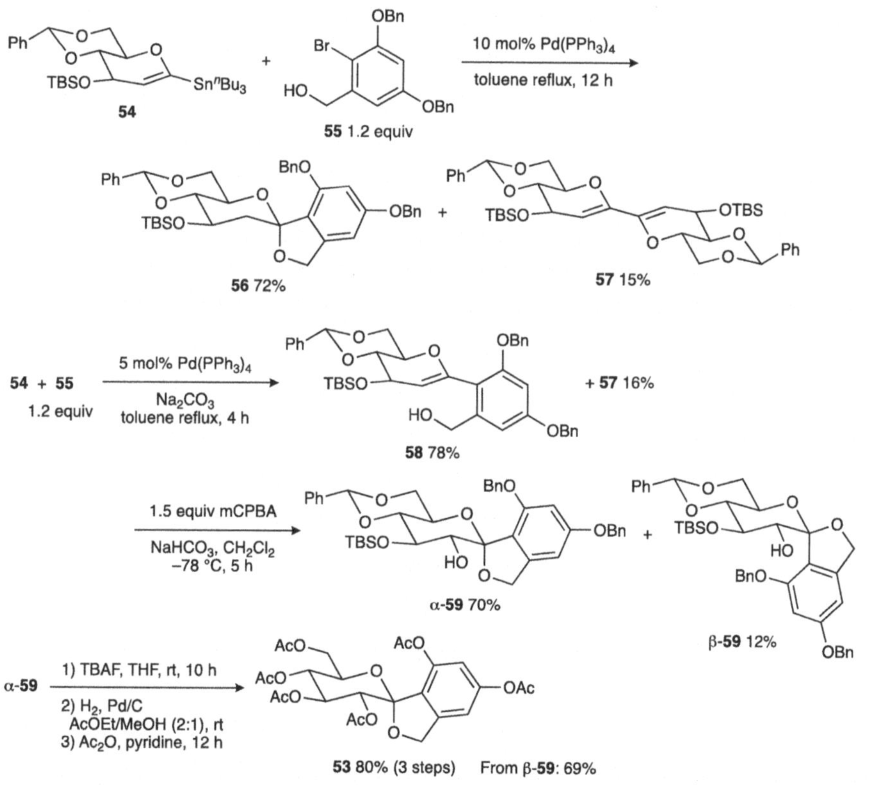

Scheme 9 Synthesis of peracetylated papulacandin core **53** *via* Stille coupling of **54** with **55**

employed reaction conditions. Thus, the reaction was repeated in the presence of Na_2CO_3 as the base to produce **58** in 78% yield along with **57**. Subsequent treatment of **58** with mCPBA at low temperature afforded spiroketals α-**59** and β-**59** in 70% and 12% yields, respectively. The stereochemistry of α-**59** was confirmed by the coupling constants of $J_{2,3} = J_{3,4} = 9.0$ Hz and $J_{4,5} = 10$ Hz, which indicate H2, H3, H4, and H5 to be axial and the conformation to be $^4C_1$. In addition, β-**59** also displayed similar coupling constants ($J_{2,3} = 7.5$ Hz, $J_{3,4} = 8.5$ Hz and $J_{4,5} = 9.8$ Hz), even though it is highly strained. This was ascribed to the conformational rigidity imposed by the benzylidene ring moiety. Each obtained anomer was then subjected to deprotection followed by exhaustive acetylation, affording the acetylated papulacandin spiroketal core **53** as a single product, even though the acid-promoted epimerization of β-**59** was not observed. Thus, these results suggest that epimerization should occur after deprotection.

The selective introduction of a fatty acid was also investigated (Scheme 10). Upon treatment with NaH, α-**59** underwent silyl migration to afford **60**, which was then subjected to acylation using stearic acid to afford **61** in 73% yield. The final deprotection steps, followed by exhaustive acetylation, gave protected papulacandin D analog **62** in 61% yield over three steps.

In the above studies, the initial Stille coupling proceeded efficiently, while the subsequent spirocyclization of the resultant arylated glucal produced a mixture of anomers with stereoselectivity of 5:1–6:1 in favor of the α-anomers. Tan and coworkers further investigated the stereoselectivity in spiroketalizations in detail [49, 50]. As examples closely relevant to the papulacandin synthesis, *o*-hydroxymethylphenyl-substituted glucals *cis*-**65** and *trans*-**65** were subjected to several cyclization conditions, affording the corresponding spiroketal *C*-arylglycosides (Scheme 11). Substrate *cis*-**65**, which was prepared by the Stille coupling of D-1-stannylglucal *cis*-**63** with aryl bromide **64** and subsequent deacetylation, was treated with DMDO in CH_2Cl_2/acetone at –78°C for 10 min, and then the produced epoxide intermediate **66** was treated with $Ti(O^iPr)_4$ at the

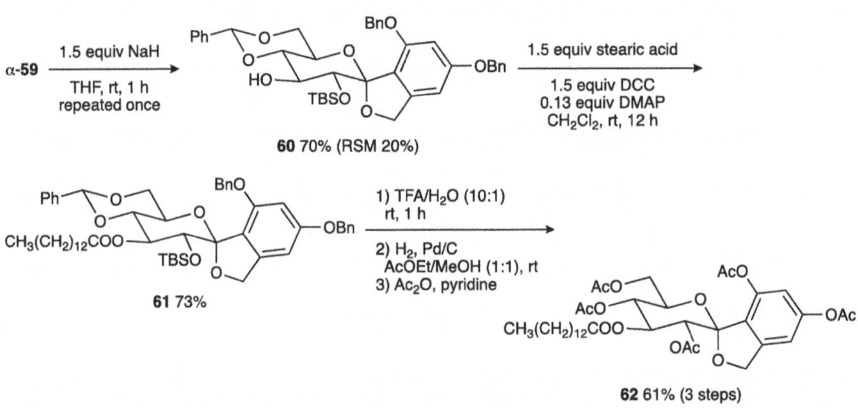

Scheme 10 Synthesis of papulacandin D analog *via* silyl migration and selective acylation

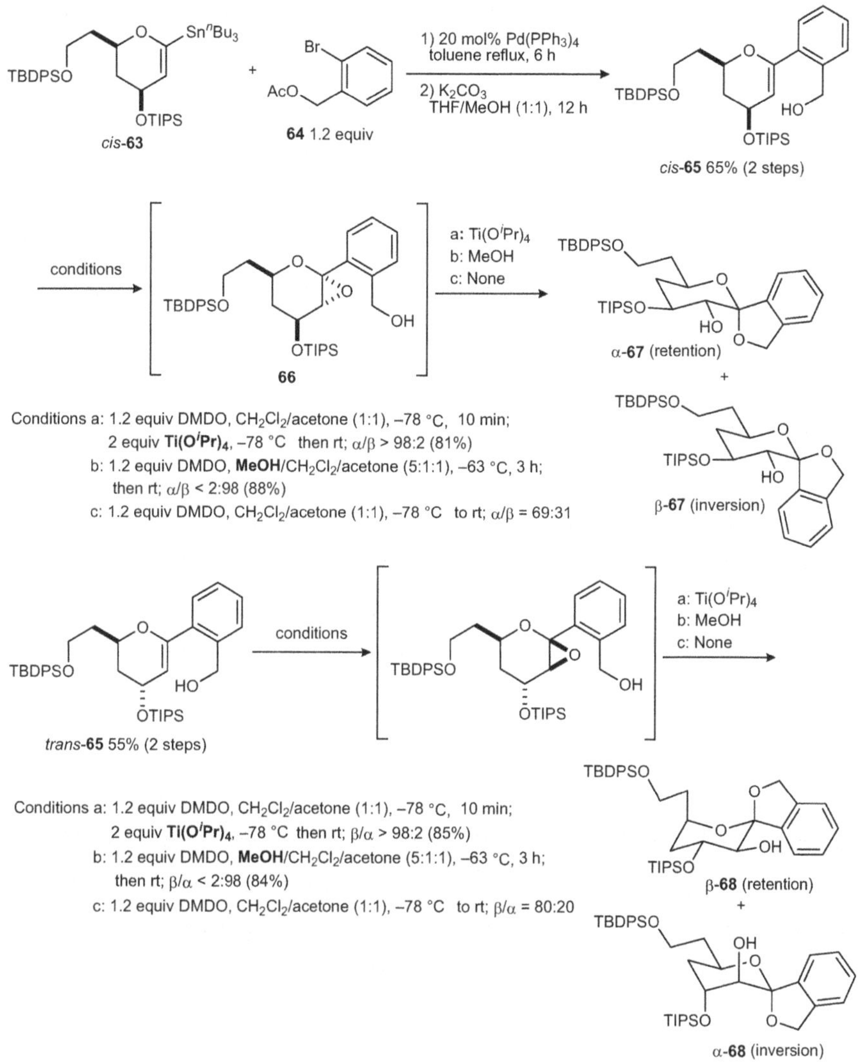

Scheme 11 Stereo-divergent spiroketalization of *C*-arylglucals *cis*/*trans*-**65**

same temperature (conditions a). As a result, cyclization occurred with retention of the anomeric configuration to produce the expected spiroketal α-**67** as an almost exclusive anomer in a high yield. On the other hand, the epoxidation of *cis*-**65** was conducted in the presence of excess MeOH at –63°C for 3 h to afford the opposite anomer β-**67** in a high yield with inversion of the anomeric configuration (conditions b). Such a stereo-controlled formation of one of the two anomers α/β-**67** is in striking contrast to the much lower α/β ratio (69:31) observed for the epoxidation reaction without additives (conditions c). Similar results were obtained when the epimer *trans*-**65** was used as the substrate. NOESY analysis and coupling constants indicated that α-**67**, β-**67**, and α-**68** adapt $^{4}C_1$ chair conformations, while β-**68** adapts an alternative $^{1}C_4$ chair conformation.

The electronic influence of the C1 aryl group (R in **69**) on the spiroketalization stereoselectivity was investigated in the presence or absence of MeOH (Scheme 12 and Table 1). In the absence of MeOH (conditions a), stereoselectivity influenced by the substituent R was observed: substrates bearing an electron-donating substituent tended to afford C1-retention products α-**70**, while the ratio of C1-inversion products β-**70** increased with the increase in electron-withdrawing nature of R. In contrast, in the presence of methanol (conditions b), C1-inversion products β-**70** became predominant independently of the substituent R, although they are less favored for the electron-rich substrates than for electron-deficient ones. From these results, it was inferred that C1-retention products α-**70** were formed *via* oxocarbenium intermediates (S_N1 mechanism), which are stabilized by electron-donating substituents. The formation of C1-inversion products β-**70** can be ascribed to S_N2-type cyclization without the involvement of oxocarbenium intermediates. The observation of trace cyclization (R = H) by low temperature ^{1}H NMR spectroscopy revealed that the epoxide intermediate is stable at –63°C and that cyclization occurred at –35°C in the presence of MeOH-d_4 (11.9 and 17.8 M) to produce β-**70** with high stereoselectivity (>98:2). On the other hand, no cyclization was observed in the absence of MeOH-d_4 at –35°C. Moreover, treatment of C1-inversion products β-**70** with TsOH (10 mol% in $CDCl_3$) at room temperature caused epimerization leading

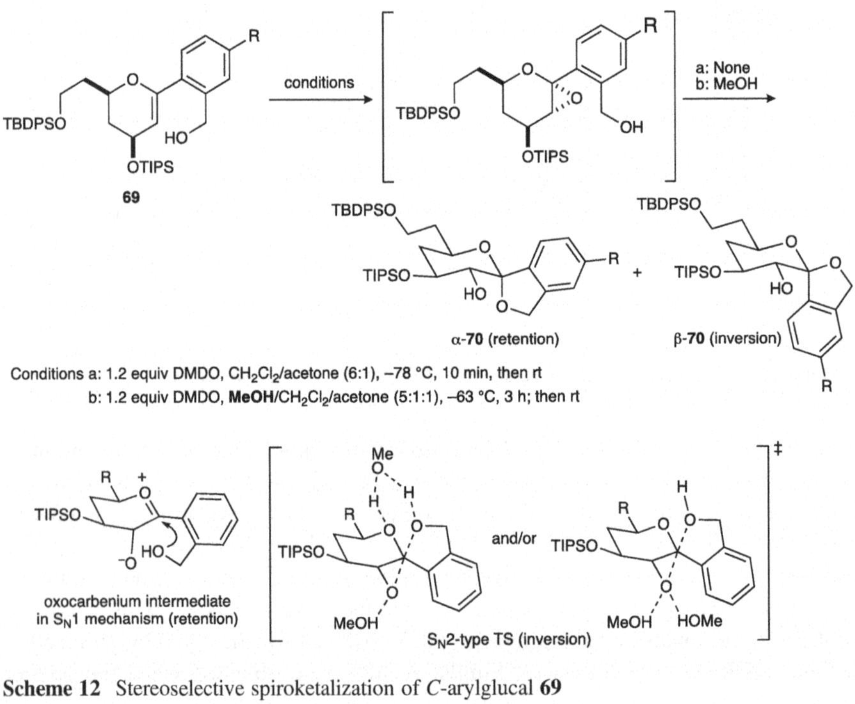

Scheme 12 Stereoselective spiroketalization of *C*-arylglucal **69**

Table 1 α/β ratio depending on the R substituent for the spiroketalization of C-arylglucal **69** (for conditions a and b, see Scheme 12)

	α/β ratio					
R	OMe	Me	H	Cl	CF_3	NO_2
Conditions a	>98:2	87:13	69:31	68:32	73:27	33:67
Conditions b	12:88	5:95	<2:98	<2:98	<2:98	<2:98

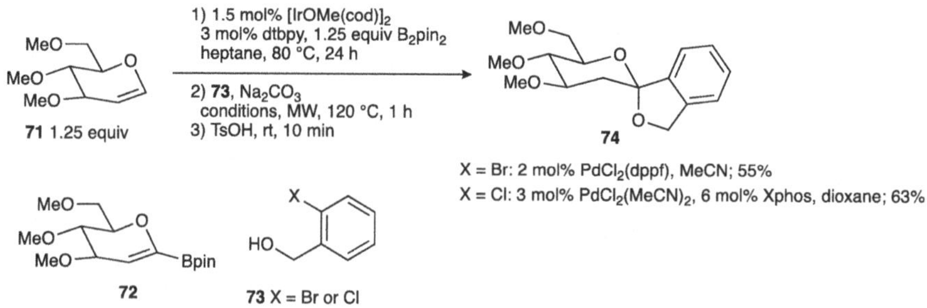

Scheme 13 Synthesis of 2-deoxy spiroketal C-arylglycoside **74** *via* Suzuki–Miyaura coupling

to α-**70**. The rate of conversion increased with electron-rich substrates owing to the stabilization of the oxocarbenium intermediates. Accordingly, the rate of conversion followed the order R = MeO > Me > H > Cl >> CF_3, and no conversion was observed for the *p*-nitro-substituted substrate over 72 h. The ρ value obtained from the plot of $\log(k_R/k_H)$ *vs.* Hammet constants σ was –5.1, which is a typical value for S_N1 mechanisms. In contrast, a smaller ρ value of –1.3 was obtained for spiroketalization in $CD_3OD/CDCl_3$ at –35°C. The kinetic study also revealed that S_N2-type spiroketalization is second-order dependent on MeOH. Thus, two molecules of MeOH are involved in the activation of the alcohol nucleophile and/or the epoxide.

The Suzuki–Miyaura coupling of a borylated glucal was employed for the synthesis of spiroketal phthalane C-glycoside by Cossy and coworkers (Scheme 13) [51]. The iridium-catalyzed C–H borylation of D-glucal **71** using bis(pinacolato) diboron (B_2pin_2) produced 1-borylglucal **72**, which was directly subjected to subsequent Suzuki–Miyaura coupling with aryl halides **73** under microwave (MW)-irradiation conditions using different palladium catalysts depending on the halogen substituent (X). After cyclization under acidic conditions, the corresponding 2-deoxy spiroketal C-arylglycoside **74** was obtained as a single anomer in 55–63% yields over three steps. This study demonstrated the utility of C–H borylation to circumvent the need for strong bases during the preparation of 1-metalated glucals.

Subsequently, the Kotora group used Suzuki–Miyaura coupling for the synthesis of the papulacandin spiroketal core (Scheme 14) [52]. 1-Borylglucal **75** was prepared from a protected D-glucal according to standard lithiation/borylation procedures and subjected to the Suzuki–Miyaura coupling with aryl iodide **76** to afford 1-arylglucal **77** in 87% yield. Owing to the self-dimerization of **75**, **76** was used as the limiting reagent. Product **77** was then treated with excess

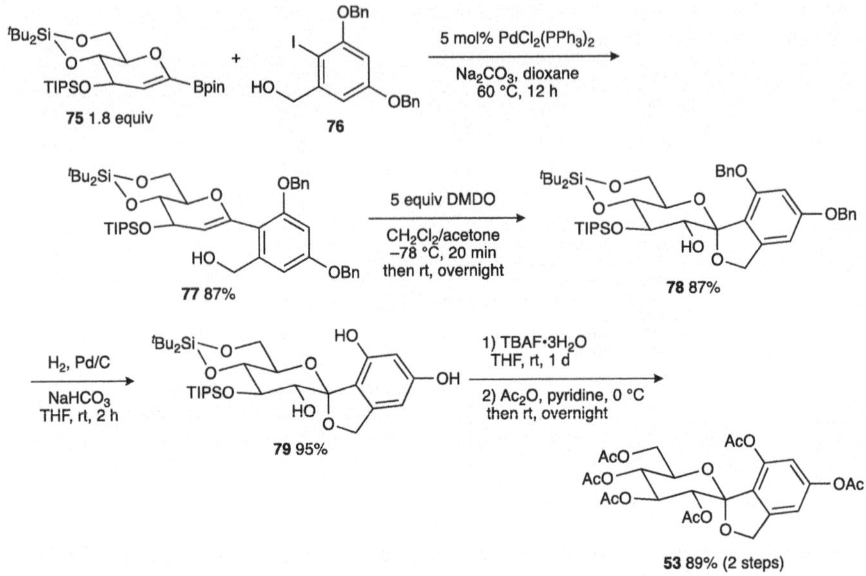

Scheme 14 Synthesis of peracetylated papulacandin spiroketal core **53** *via* Suzuki–Miyaura coupling

DMDO at –78°C to achieve the spiroketalization, stereoselectively affording the desired anomer **78** in 87% yield. Subsequently, sequential deprotection followed by exhaustive acetylation afforded the known acetyl-protected papulacandin spiroketal core **53**.

The second total synthesis of (+)-papulacandin D was accomplished *via* Hiyama–Denmark coupling by the Denmark group [9, 10]. The required glucal **81** bearing a 1-silanol moiety was prepared from protected D-glucal **80** *via* lithiation/silylation, followed by ruthenium-catalyzed oxidative hydrolysis (Scheme 15). The obtained silanol **81** was subjected to coupling with aryl iodide **82** under palladium-catalyzed conditions to afford the desired 1-arylglucal **83** in 82% yield. After reductive removal of the pivaloyl group, the obtained benzyl alcohol **84** was treated with mCPBA at 0°C to produce spiroketal anomers α-**85** and β-**85** in 77% and 15% yields, respectively. Nevertheless, the undesired anomer β-**85** could be converted into α-**85** upon treatment with HCl.

The enantioselective synthesis of required fatty acid **90** is outlined in Scheme 16 [9, 10]. (*S*)-Citronellol, which can be prepared by asymmetric hydrogenation of geraniol, was transformed into dienyl ester **86** *via* removal of the hydroxyl group, ozonolysis, and the Horner–Wadsworth–Emmons reaction of the resultant aldehyde. The ester group of **86** was transformed into an aldehyde, which was subjected to asymmetric allylation using allyltrichlorosilane and chiral bisphosphoramide (*R*,*R*)-**87**. The obtained homoallylic alcohol **88** was then subjected to the cross metathesis with acrolein using Grubbs 2G catalyst to afford unsaturated aldehyde **89**. Finally, **89** was converted to **90** *via* triethylsilyl (TES) protection of the hydroxyl group, Wittig olefination, and the saponification of the resultant ester with KOTMS.

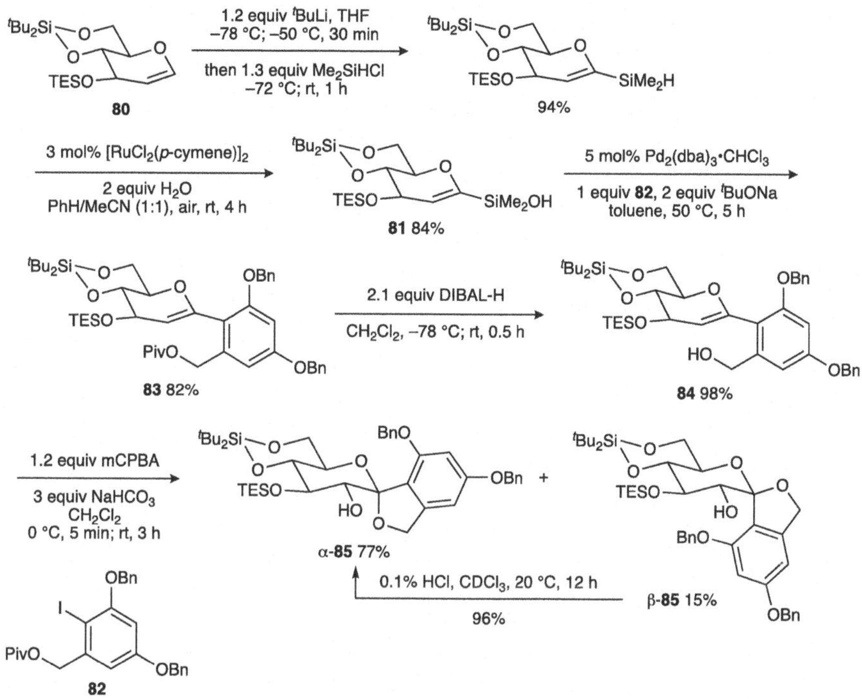

Scheme 15 Synthesis of protected papulacandin spiroketal core **85** *via* Hiyama–Denmark coupling

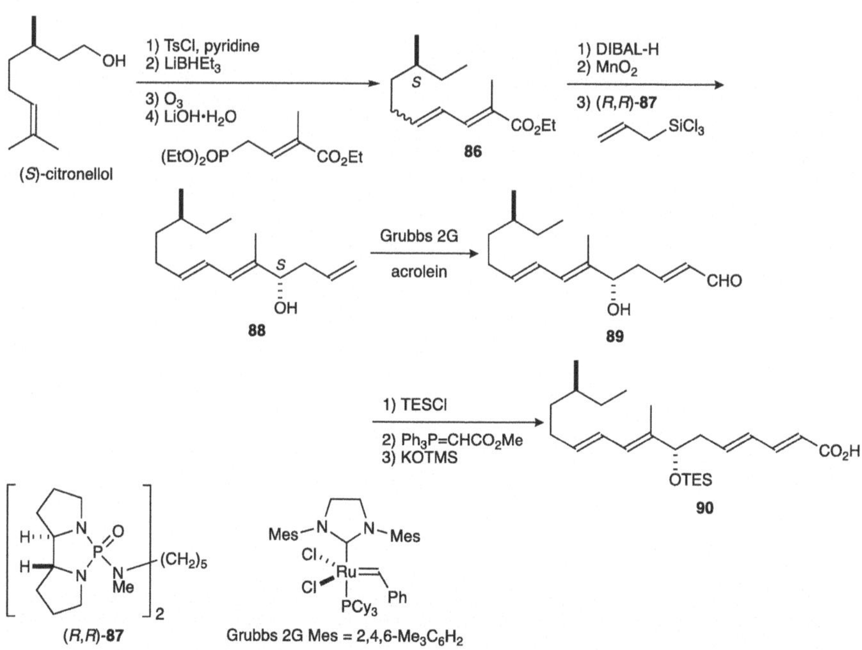

Scheme 16 Enantioselective synthesis of protected fatty acid **90**

Although orthogonally protected papulacandin spiroketal core α-**85** was successfully obtained as shown in Scheme 15, the removal of the protecting groups was found to be problematic. Thus, α-**85** was transformed into **91** over four steps, prior to the introduction of the unsaturated fatty acid residue (Scheme 17) [9, 10]. The coupling of fatty acid **90** with spiroketal **91** was implemented using the Yamaguchi mixed anhydride method to afford **92** in 87% yield. The global deprotection of **92** was performed upon treatment with HF·NEt_3 to ultimately complete the total synthesis of (+)-papulacandin D. This convergent strategy afforded the final natural product in 9.2% yield over 31 steps from commercial triacetyl-D-glucal and geraniol.

Following the report from the Denmark group, unnatural papulacandin D analogs were synthesized [53]. As shown in Scheme 18, the reactions of protected papulacandin core **93** with sorbic acid, palmitic acid, linoleic acid, or *trans*-retinoic acid using the Yamaguchi protocol afforded the corresponding esters **94a–d** in high yields. The subsequent removal of the bis(*tert*-butyl)silyl protecting groups was conducted using $^{n}Bu_3N$·HF to deliver diols **95a–d** in varying yields: **95b** and **95c** were obtained in high yields, whereas **95a** and **95d** with acyl side chains including conjugated olefins gave lower yields of 25–39%. The final removal of MOM groups afforded unnatural papulacandin D analogs **96a–d**, albeit in low yields. The biological activity of **96a–d** was tested for inhibition of *Candida albicans*. Sorbate **96a** and *trans*-retinoate **96d** appeared to be totally ineffective, while palmitate **96b** and linoleate **96c** showed some inhibitory activity, although much lower than that of natural papulacandin D.

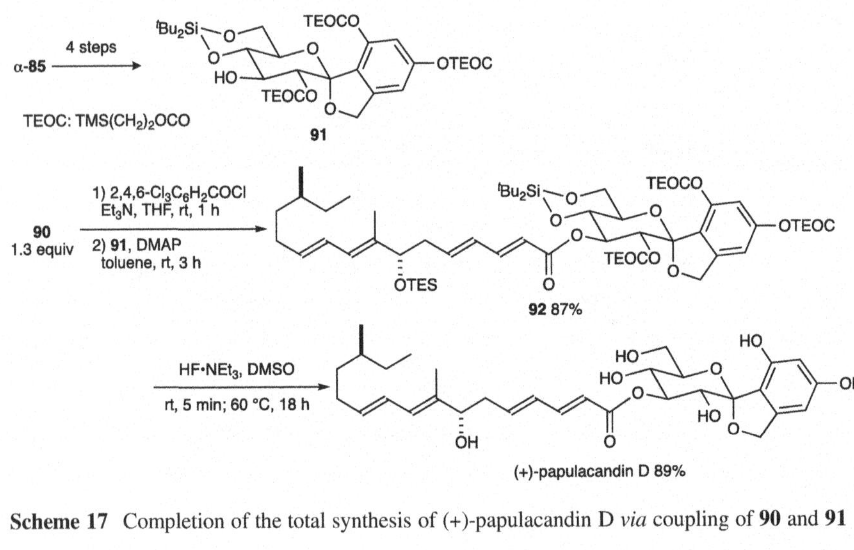

Scheme 17 Completion of the total synthesis of (+)-papulacandin D *via* coupling of **90** and **91**

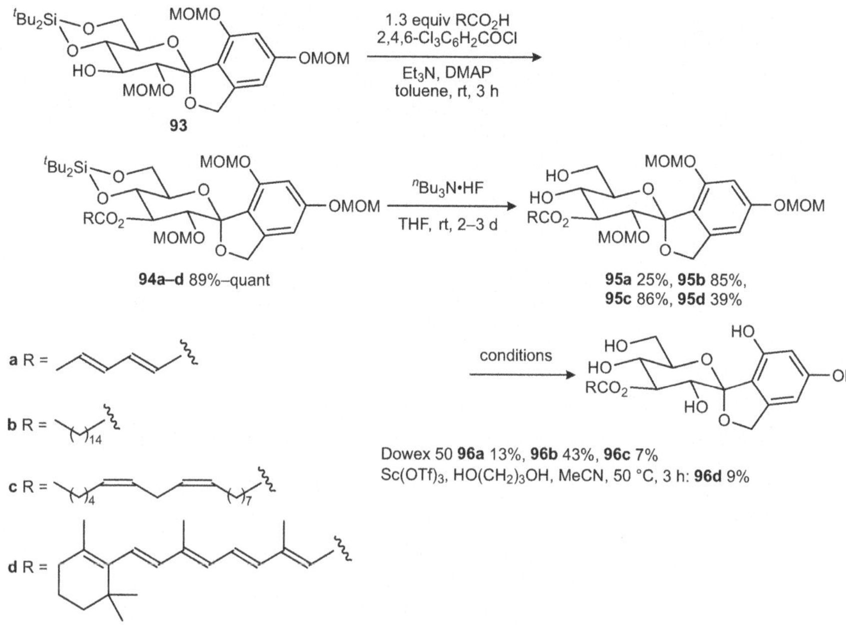

Scheme 18 Synthesis of unnatural papulacandin D analogs **96a–d**

2.3 *Synthesis of the Papulacandin Spiroketal Core via De Novo Construction of the Sugar Framework*

The previous sections outlined the synthetic strategies based on the coupling of sugar and aromatic precursors. In this section, completely different approaches to papulacandin-type spiroketal *C*-aryglycosides will be discussed. The first example of this category was the racemic synthesis of papulacandin spiroketal core (±)-**53** reported by the Danishefsky group (Scheme 19) [54]. They used the hetero-Diels–Alder reaction of Danishefsky's diene **97** and substituted benzaldehyde **98**. The desired 2-aryl-3,4-dihydro-2*H*-pyran-4-one (±)-**99** was obtained in 92% yield. Subsequently, 1,4-addition of a vinyl Grignard reagent to (±)-**99** in the presence of CuI produced (±)-**100**. The oxidative cleavage of the vinyl group of (±)-**100** afforded the corresponding aldehyde, which was reduced with $LiAlH(OCEt_3)_3$; subsequent benzoylation of the resultant hydroxy group afforded (±)-**101**. The enolization and silylation of (±)-**101** produced regioisomeric silyl enol ethers, which were treated with mCPBA to afford the desired oxidation product. After benzoylation of the introduced hydroxyl group, (±)-**102** was obtained, albeit in a low yield. The Ito–Saegusa oxidation of (±)-**102** delivered enone (±)-**103**, which was subjected to stereoselective DIBAL-H reduction and subsequent acetylation to yield glucal (±)-**104**. The treatment of (±)-**104** with mCPBA in MeOH/THF

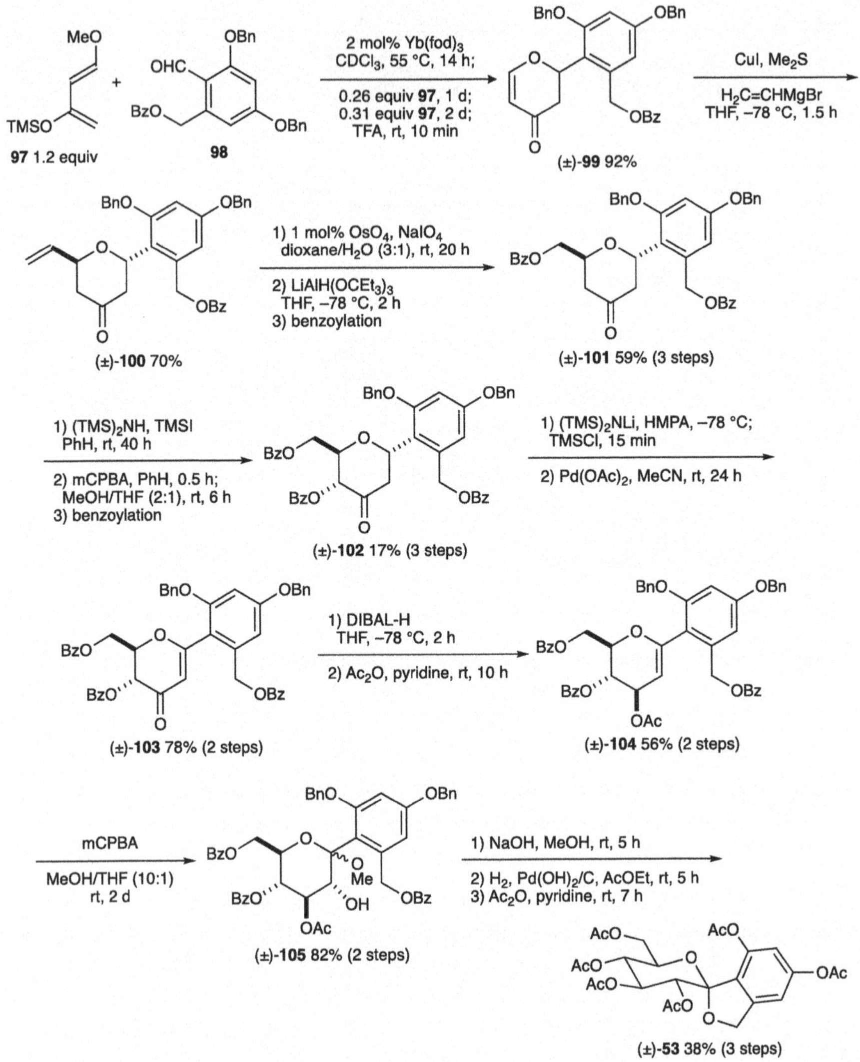

Scheme 19 Synthesis of racemic papulacandin spiroketal core (±)-**53** *via* Diels–Alder reaction

produced glycosides (±)-**105** as an anomeric mixture. Finally, global deprotection of (±)-**105** induced spiroketalization and the acetylation of the resultant unprotected product afforded (±)-**53**. Although many synthetic sequences were required for this process, the first synthesis of the papulacandin spiroketal core was accomplished, albeit in racemic form.

Later, Balachari and O'Doherty reported a de novo approach to the *manno*- and *allo*-isomers of the papulacandin spiroketal core (Scheme 20) [55]. The enantioselective construction of spiroketals was enabled by Sharpless asymmetric

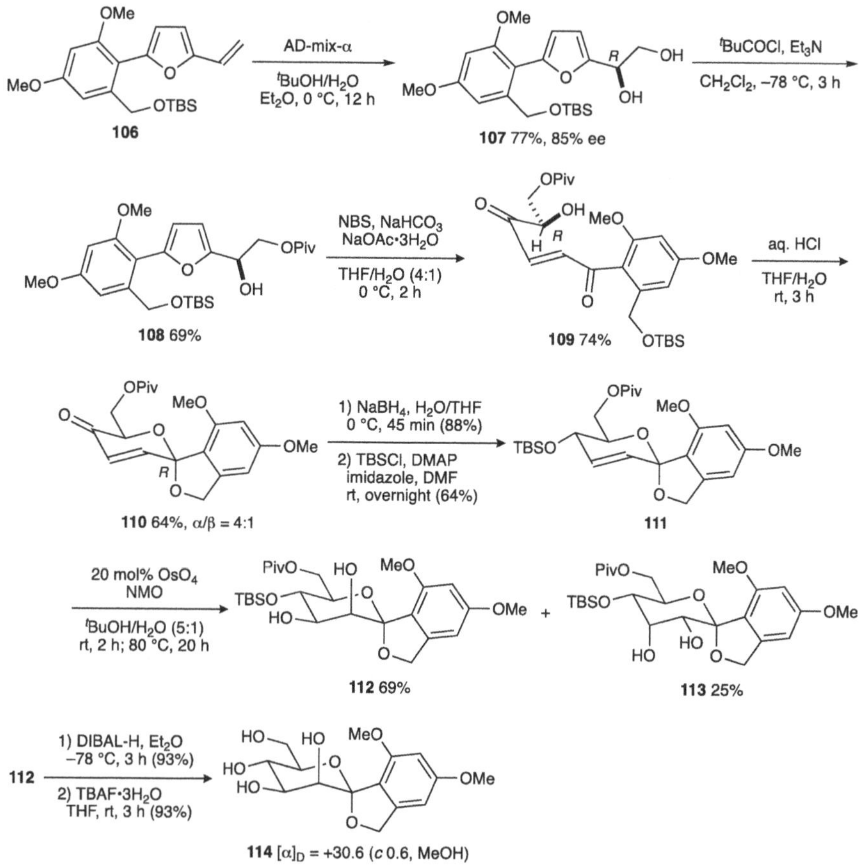

Scheme 20 De novo approach to D-*manno* analog of the papulacandin spiroketal core

dihydroxylation and subsequent Achmatowicz oxidative ring expansion of a 5-aryl-2-vinylfuran precursor. Thus, vinylfuran **106** was subjected to asymmetric dihydroxylation using commercial AD-mix-α, affording diol **107** in 77% yield with 85% ee. After selective pivaloylation of the primary alcohol of **107**, the obtained hydroxyl ester **108** was subjected to Achmatowicz/spiroketalization sequence to afford cyclic enone **110** *via* **109** in a moderate overall yield as a 4:1 anomeric mixture. Subsequent $NaBH_4$ reduction, followed by TBS protection, converted **110** into spiro-pseudoglucal **111**. Dihydroxylation of **111** was performed under forcing conditions (cat. OsO_4, *N*-methylmorpholine-*N*-oxide (NMO), $^tBuOH/H_2O$, 80°C, 20 h) to produce protected *manno*-isomer **112** in 69% yield along with protected *allo*-isomer **113** in 25% yield. The removal of pivaloyl and TBS groups from **112** afforded mannose derivative **114**. In this de novo synthesis, the D/L ratio of **114** is dependent on the enantioselectivity of the asymmetric dihydroxylation of **106**. Since (*R*)-**107** was obtained as the major enantiomer, D-**114** is considered to be the major enantiomer, even though its enantiopurity was not provided.

To synthesize the natural papulacandin spiroketal core, inversion of the C2 axial hydroxy group of protected D-*manno*-isomer **112** was also attempted (Scheme 21) [56]. Selective TBS protection of the C3 hydroxy group afforded a high yield of **115**, which was subjected to oxidation using Dess–Martin periodinane (DMP) to give ketone **116** in 98% yield. The treatment of **116** with DIBAL-H led to the selective removal of the pivaloyl group, affording **117** in 98% yield. Further reduction of the ketone carbonyl group was achieved using $LiAlH_4$ to afford the desired protected papulacandin spiroketal core **118** in 88% yield. Similarly, protected *allo*-isomer **119** obtained from **113** was subjected to Swern oxidation to afford ketone **120**; however,

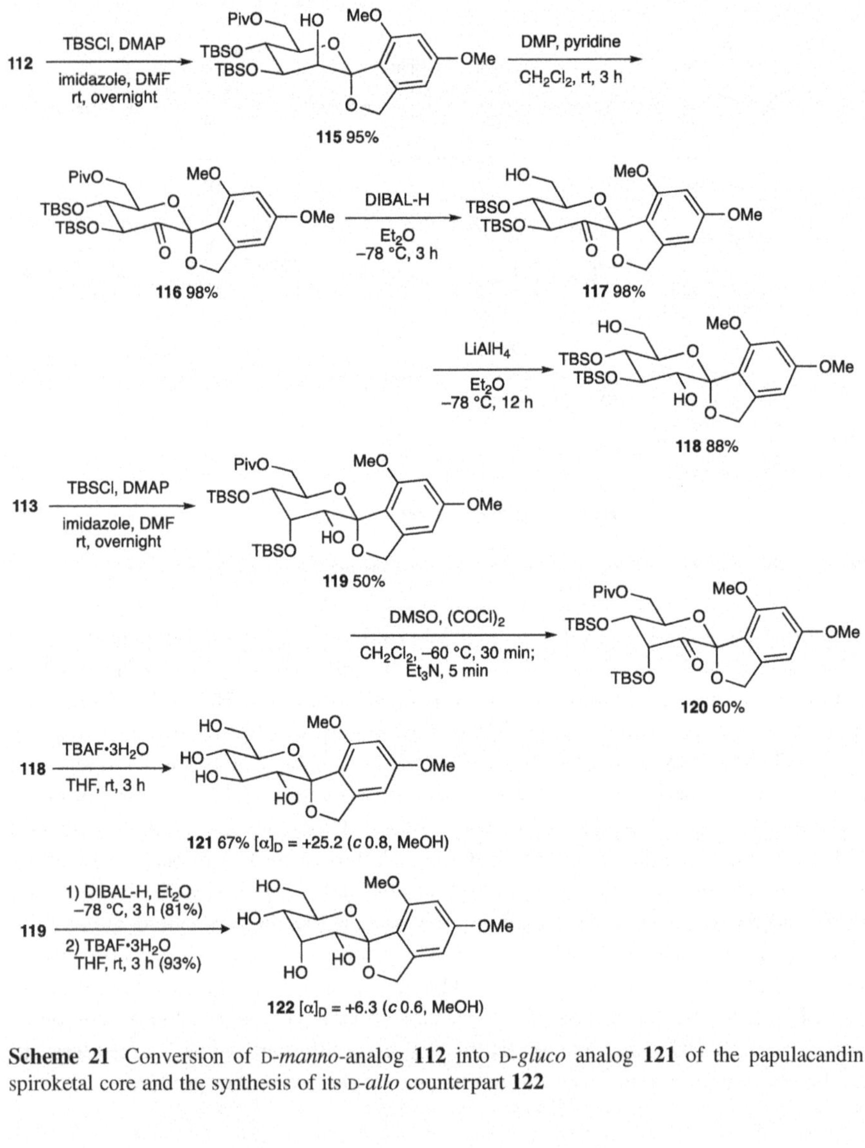

Scheme 21 Conversion of D-*manno*-analog **112** into D-*gluco* analog **121** of the papulacandin spiroketal core and the synthesis of its D-*allo* counterpart **122**

the subsequent ketone reduction was unsuccessful. Finally, deprotection of **118** and **119** afforded D-glucose and D-allose derivatives **121** and **122**, respectively. The optical rotation of **121** ($[\alpha]_D$ = +25.2, *c* 0.8, MeOH) is similar to that reported for D-**121** derived from benzyl-protected D-gluconolactone **10** ($[\alpha]_D$ = +48.1, *c* 0.43, $CHCl_3$) [42].

The de novo synthesis of a D-*galacto*-spiroketal core was also investigated (Scheme 22) [57]. The required dienone **124** was prepared from aryl iodide **123** in three steps. The key Sharpless asymmetric dihydroxylation of **124** was performed using AD-mix-β to regioselectively afford diol **125** in 60% yield with 90% ee. Subsequent diastereoselective dihydroxylation, followed by acetylation, converted **125** into tetracetate **126** in 70% yield with a 4:1 diastereomer ratio. Removal of the TBS group from **126** under acidic conditions in MeOH produced ketal **127** in 90% yield. Further deacetylation of **127** under basic conditions followed by acidification produced a 2:1 mixture of the expected *galacto*-pyrano spiroketal **128** and *galacto*-furano spiroketal **129** in 80% combined yield. The 2:1 isomeric ratio was found to be thermodynamic; **128** or **129** were separately treated with TsOH·pyridine in MeOH for 36 h to produce the same 2:1 mixture of **128** and **129** in 95% yield.

The de novo synthesis of D-*allo*- and D-*altro*-isomers of the papulacandin spiroketal core was investigated using organocatalytic asymmetric aldol reactions (Scheme 23) [58]. The asymmetric self-aldol reaction of the known aldehyde **130** using D-proline afforded β-hydroxy aldehyde **131** in 70% yield with 9:1 de and 95% ee. The key coupling of **131** and acetophenone derivative **132** was

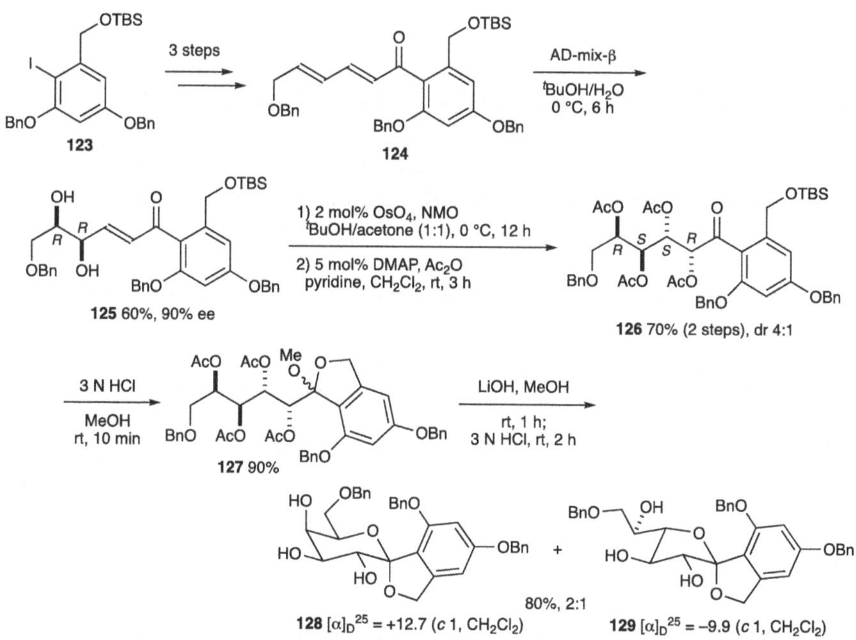

Scheme 22 Preparation of D-*galacto*-spiroketals **128** and **129**

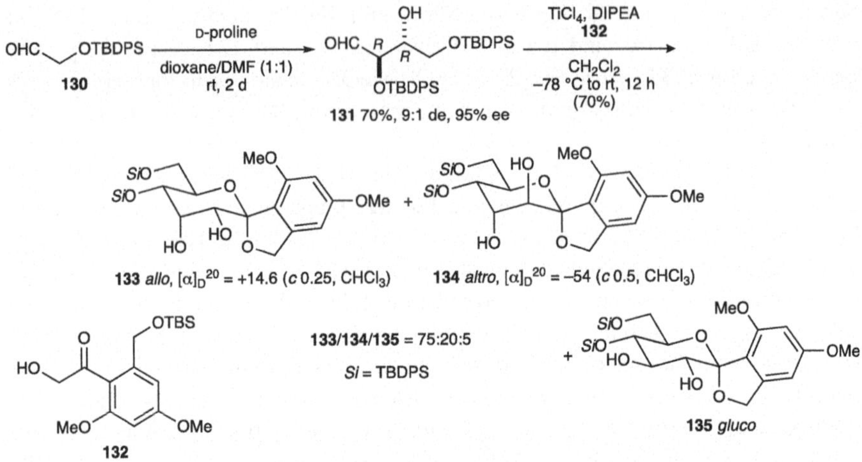

Scheme 23 Synthesis of the D-*allo/altro/gluco*-isomers of the papulacandin spiroketal core

conducted upon treatment with $TiCl_4$ and *N*,*N*-diisopropylethylamine (DIPEA), affording a 75:20:5 mixture of *allo-*, *altro-*, and *gluco*-spiroketals (**133**, **134**, and **135**, respectively) in 70% combined yield.

In contrast to D-glucose, D-mannose, and D-galactose, D-altrose is not found in nature. Thus, the selective synthesis of an *altro*-isomer of papulacandins is interesting in terms of its biological activity. To this aim, the known hydroxy aldehyde **136**, which can be obtained by organocatalytic asymmetric dimerization, was used as the starting material (Scheme 24) [58]. TBS protection of **136** afforded **137**, which was used for aldol condensation with acetophenone derivative **132**. The aimed condensation was achieved in three steps: **132** and **137** were treated with $TiCl_4$ and DIPEA to produce β-hydroxy ketone **138** as a 9:1 diastereomeric mixture in 89% yield, and subsequent mesylation of the hydroxyl group was followed by β-elimination using 1,8-diazabicyclo[5.4.0]undec-7-ene (DBU), affording enone **139** in 90% yield. Then, diastereoselective dihydroxylation of **139** afforded **140** in 80% yield. After acetonide protection, the obtained product **141** was treated with TBAF in THF to induce desilylation of the benzylic TBS ether. The resultant hemiketal **142** was then subjected to acidic conditions in MeOH to afford D-β-*altro*-spiroketal **143** with concomitant removal of both the TBS and acetonide groups. NOESY experiments and a coupling constant ($J_{2,3} = 10.6$ Hz) of the diacetylated derivative **144** suggested that this compound adopts a $^1C_4$ rather than a $^4C_1$ conformation. Although no NOE was observed between OMe and the sugar ring protons, the spiro configuration was assumed to be α based on literature precedence and coupling constants observed for the sugar ring protons.

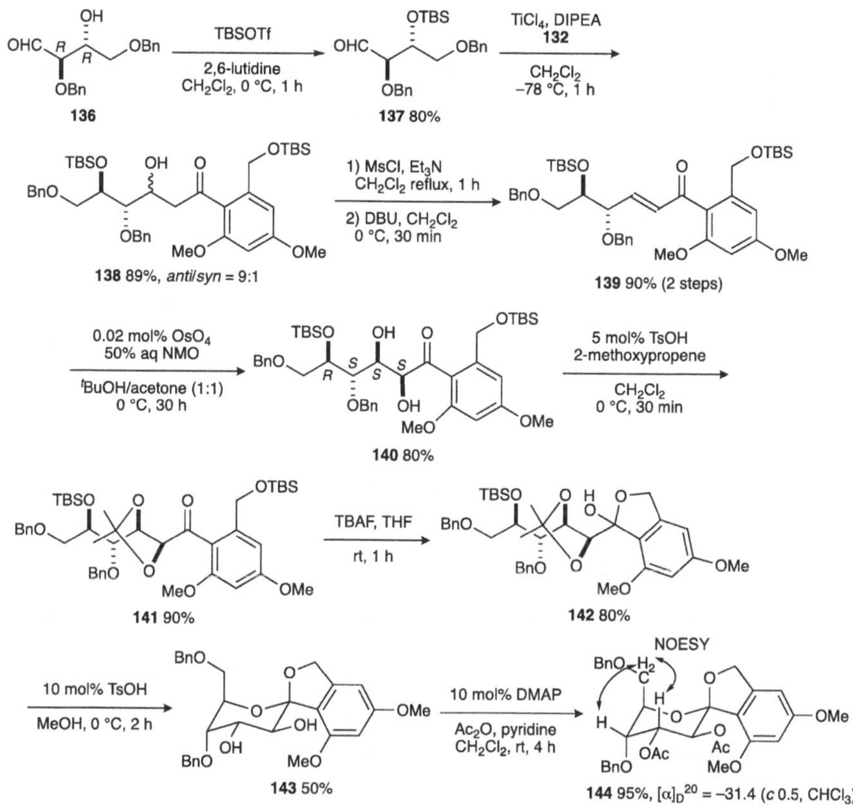

Scheme 24 Synthesis of the D-*altro*-isomer of the papulacandin spiroketal core

3 Synthesis of Sodium Glucose Cotransporter 2 Inhibitors

Sodium glucose cotransporter 2 (SGLT2) plays a major role in the renal tubular reabsorption of glucose from urine. Therefore, SGLT2-selective inhibitors increase the excretion of glucose into urine. Since this effect is independent of the insulin secretion or insulin sensitivity, SGLT2 inhibitors are potential drugs for type 2 diabetes [59]. Moreover, SGLT2 inhibitors are also expected to reduce the body weight through the improved excretion of glucose. The first SGLT inhibitor phlorizin is a natural *O*-glycoside with a dihydrochalcone aglycone (Fig. 5). However, phlorizin has limited applicability due to low SGLT1/SGLT2 selectivity [60]. For instance, phlorizin can cause adverse gastrointestinal effects because it blocks SGLT1 highly expressed in the intestine. Thus, more selective phlorizin analogs have been developed, but they displayed high metabolic instability due to their readily hydrolysable *O*-glycosidic bonds.

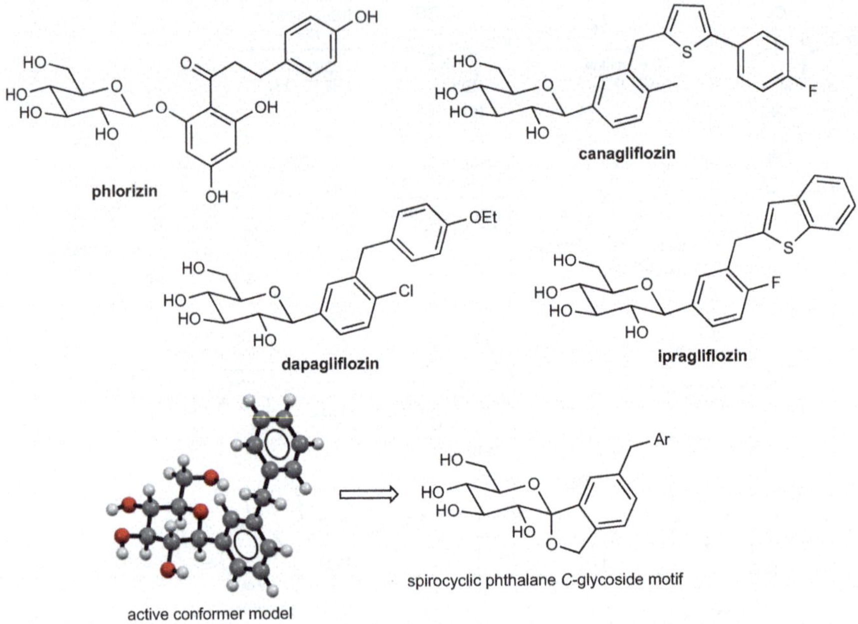

Fig. 5 Selected SGLT inhibitors and spirocyclic phthalane *C*-glycoside motif

Therefore, *C*-glycoside-type gliflozins have been developed as potent and selective SGLT2 inhibitors (Fig. 5) [11–14, 37]. As a common structural feature, they have a *C*-arylglycoside motif, in which a diarylmethane is directly connected to a glycosyl core *via* a C–C bond. Canagliflozin is the first-in-class SGLT2 inhibitor developed by Johnson & Johnson. Various gliflozins, including dapagliflozin and ipragliflozin, have been developed by modifying the diarylmethane moiety. The structural modeling of potential SGLT2 inhibitors suggested that the aryl aglycone is almost perpendicular to the carbohydrate moiety in the lowest energy conformation [11–14]. Therefore, to obtain new potent SGLT2-selective inhibitors, the spiroketal phthalane *C*-glycoside motif was introduced since such a spiroketal framework conveys the desired perpendicular relationship between the aryl and sugar moieties. In this section, the syntheses of spirocyclic SGLT2 inhibitors will be overviewed.

3.1 Syntheses of Spirocyclic SGLT2 Inhibitors

Various spiroketal phthalane *C*-glycosides were synthesized starting from benzyl-protected D-gluconolactone **10** and aryl bromide **145** by the Chugai Pharmaceutical Co. research group (Scheme 25) [61]. The addition of an aryllithium reagent derived from **145** to **10** produced hemiketal **146**, which underwent detritylation upon

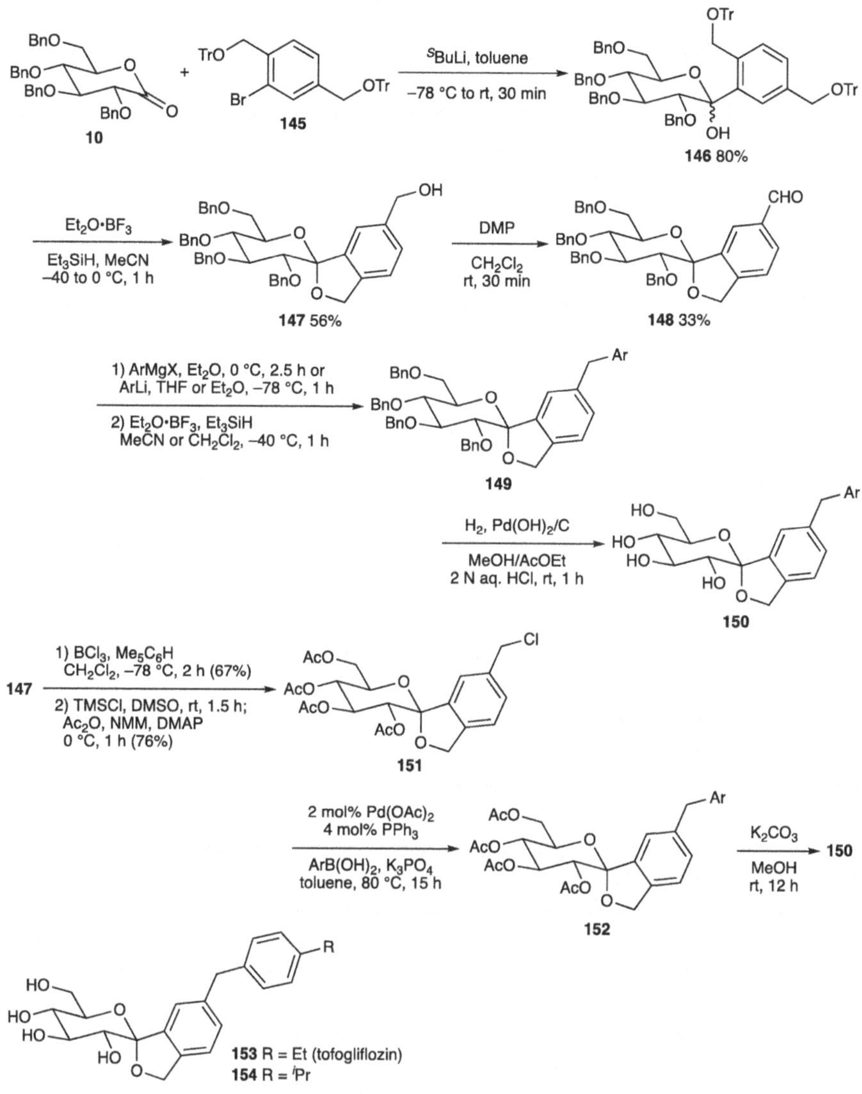

Scheme 25 Synthesis of benzyl-substituted spiroketal phthalane *C*-glycosides **149**

treatment with $Et_2O{\cdot}BF_3$ and Et_3SiH to afford spiroketal **147** bearing a benzylic alcohol moiety in 56% yield. The Dess–Martin oxidation of **147** generated aldehyde **148** in 33% yield. The addition of aryl nucleophiles followed by triethylsilane reduction produced **149** bearing a diarylmethane moiety, which were then subjected to debenzylation conditions to afford unprotected spiroketal *C*-glycosides **150**. Alternatively, **147** was converted to acetyl-protected spiroketal **151** bearing a benzylic chloride moiety, which was subjected to Suzuki–Miyaura coupling with arylboronic acids and subsequent deacetylation of the resultant products **152** to afford **150**. Among the synthesized compounds, ethyl derivative **153** (tofogliflozin) or isopropyl derivative **154** were found to have significant potency for the inhibition of hSGLT2 in vitro (IC_{50} <5 nM) and led to a significant increase in renal glucose excretion and decrease in the blood glucose level in vivo. In addition, the oral bioavailability of **153** was higher (75% in mice and 85% in monkeys) than that of **154**. It has been experimentally shown that tofogliflozin (**153**) is the most specific SGLT2 inhibitor with potent antidiabetic effects by increasing the urinary glucose excretion without the risk of hypoglycemia and unprecedented side effects due to interactions with glucose-related reactions and off-target reactions [62, 63]. Tofogliflozin hydrate has been approved for the treatment of type 2 diabetes mellitus (T2DM) in Japan [64].

Spiroketal phthalane *C*-glycosides relevant to dapagliflozin and tofogliflozin were synthesized by Xu and coworkers (Scheme 26) [65–67]. The target compounds (e.g., **155**) were synthesized by the addition of aryllithium reagents to TMS-protected D-gluconolactone **23**, followed by acid-promoted global deprotection/spiroketalization. Similarly, spiroketal *C*-glycoside **161** with a diaryl ketone moiety was obtained from acetal **160** in 68% yield over two steps. Moreover, **161** was transformed into **162** bearing a difluoromethylene moiety *via* a hydrazone. The biological activity of the obtained compounds was compared to that of dapagliflozin in a cell-based SGLT functional assay and hSGLT inhibitory activity (IC_{50}) at a screening concentration of 1 μM. Although compound **155** without a chlorine substituent on the phthalane ring was much less effective, compounds **157**, **158a**, and **162** proved to be similarly potent to dapagliflozin in terms of the hSGLT2 inhibitory activity. No hSGLT2 inhibitory activity was found for **156**, which is a regioisomer of tofogliflozin. On the other hand, compounds **158b** and **159** were found to be more potent and selective (hSGLT2 *vs.* hSGLT1) than dapagliflozin. Further improvement of the selectivity was observed when the phthalane ring of **157** was replaced with an indane ring [67].

The modification of 6′-hydroxy group of **157** produced compounds **163–167** bearing fluoro, alkoxy, azide, and amino substituents (Scheme 27) [66]. The 6′-fluoro derivative **163** showed moderate in vitro SGLT2 inhibitory activity and in vivo blood glucose-lowering effect. The Chugai group also developed a divergent synthesis of protected tofogliflozin derivatives using 4-chloro analog **168** as the platform (Scheme 28) [61]. Derivatizations of **168** were conducted using palladium-mediated cross-coupling reactions. Ultimately, methyl, ethynyl, and phenyl derivatives **169–171** were obtained after deprotection.

3.2 *Practical Syntheses of Tofogliflozin*

The medicinal chemistry synthetic route to tofogliflozin (**153**) presents challenges for the development of a large-scale synthetic process: crystalline intermediates are necessary for easy quality control. Thus, a modified synthetic method was developed, as outlined in Scheme 29 [68]. Diol **172** was protected as bis(1-methoxy-1-methylethyl) ether (**173**), which was converted into an aryllithium reagent and

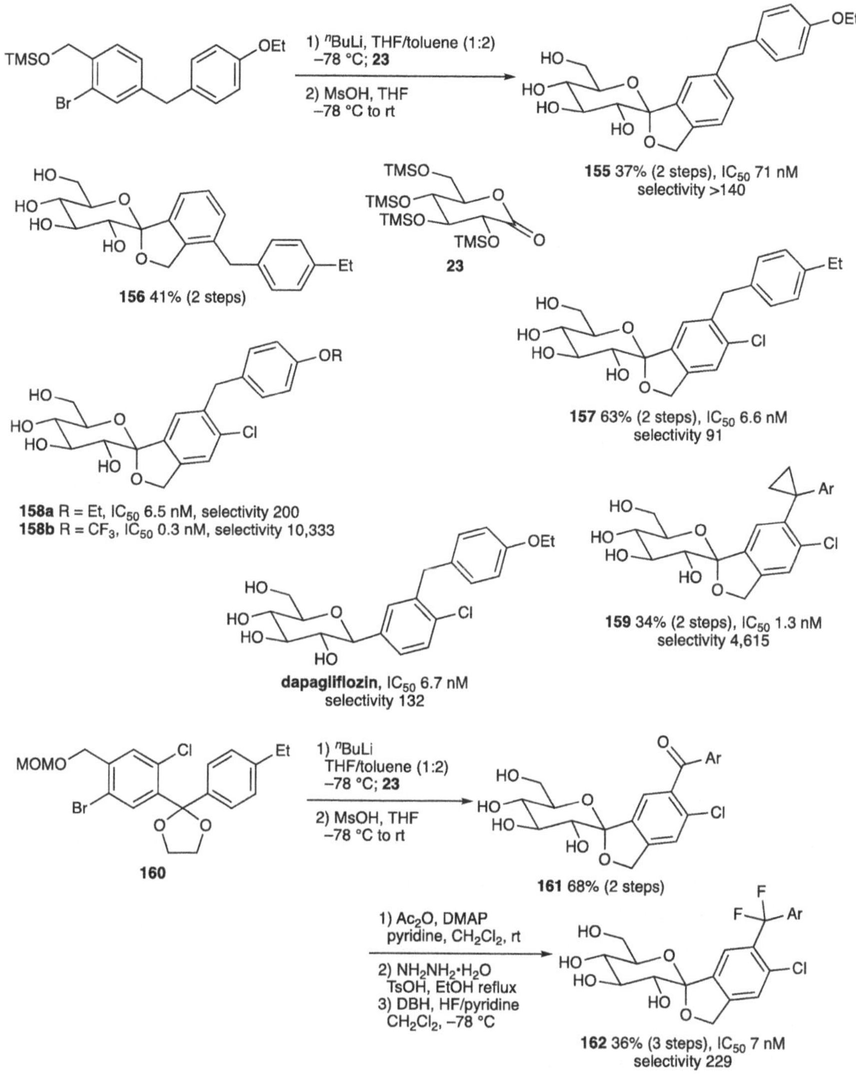

Scheme 26 Synthesis and biological evaluation of spiroketal phthalane C-glycosides relevant to dapagliflozin and tofogliflozin

reacted with TMS-protected D-gluconolactone **23**. The generated crude hemiacetal was directly treated with TsOH in THF/MeOH at room temperature to afford the desired spiroketal phthalane C-glycoside **174**, which was isolated by trituration with methyl *tert*-butyl ether (MTBE) as a stable crystalline product in 60% yield from **172**. After exhaustive methoxycarboxylation, the obtained carbonate-protected product **175** was subjected to Suzuki–Miyaura coupling with *p*-ethylphenylboronic acid to produce diarylmethane derivative **176** in 88% yield. In this step, the residual

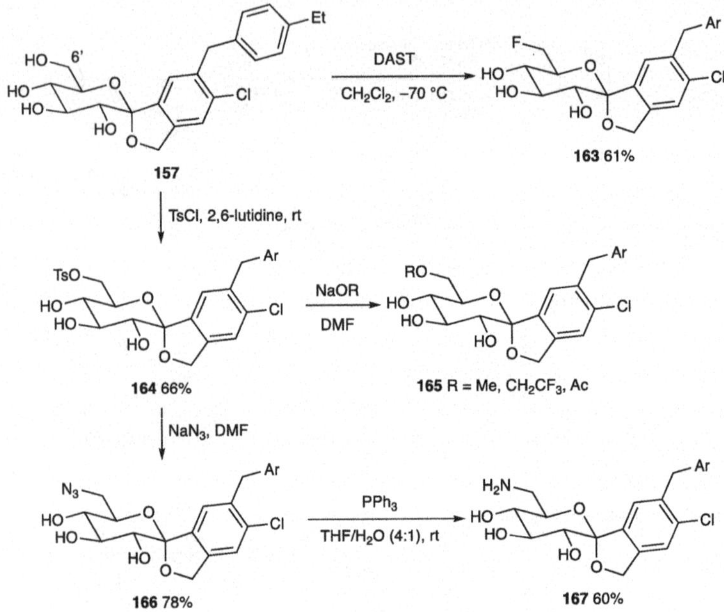

Scheme 27 Modification of potent SGLT2 inhibitor **157** at the 6′ position

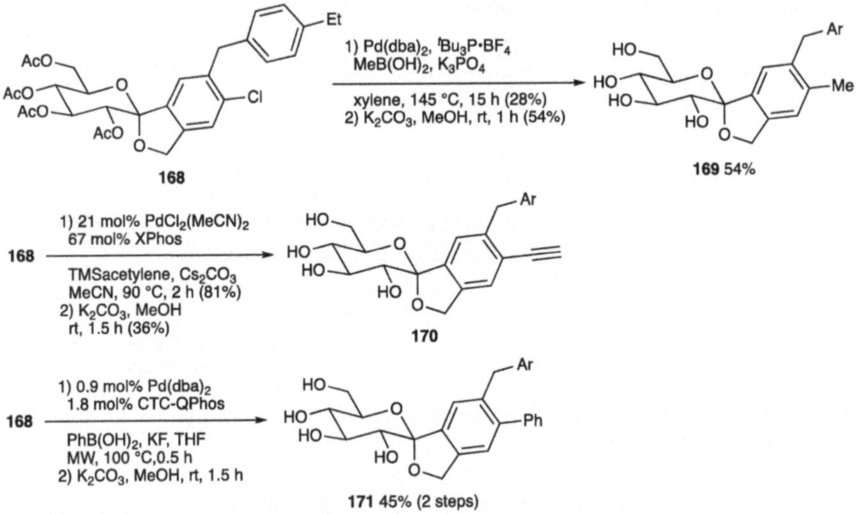

Scheme 28 Divergent synthesis of tofogliflozin derivatives using 4-chloro derivative **168** as the platform

palladium was efficiently removed with filtration of the precipitates formed upon adding 20 mol% of *N*-acetyl-L-cysteine (NAC) to the reaction mixture. Moreover, treatment of the crude mixture with EtOH and H_2O containing NAC afforded

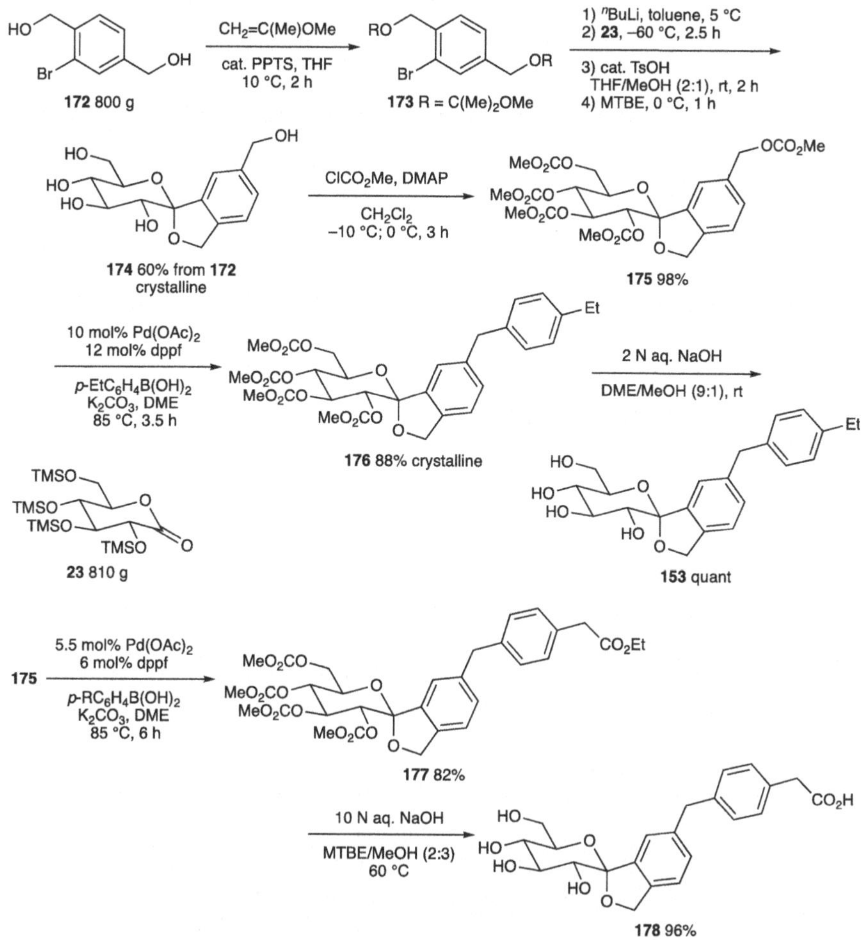

Scheme 29 Large-scale synthesis of tofogliflozin (**153**)

crystalline **176** with >99% purity and <1 ppm of residual palladium. After deprotection, tofogliflozin (**153**) was quantitatively obtained as a colorless amorphous solid with >99% purity by extraction and concentration. In a similar manner, the coupling of intermediate **175** with *p*-(ethoxycarbonylmethyl)phenylboronic acid generated ester **177**, which was then subjected to basic deprotection conditions to afford **178** in 96% yield. The obtained carboxylic acid derivative **178** is the major metabolite of tofogliflozin [69].

A simplified synthetic route to tofogliflozin was further developed by sequential metalation–functionalization of a 2,4-dibromobenzyl alcohol derivative (Scheme 30) [70]. 2,4-Dibromobenzyl alcohol **179** protected with a 1-methoxy-1-methylethyl group was carefully metalated *via* halogen–metal exchange to selectively generate mono-lithiated species **180** with an *ortho*/*para* ratio of 53:1. The addition of **180**

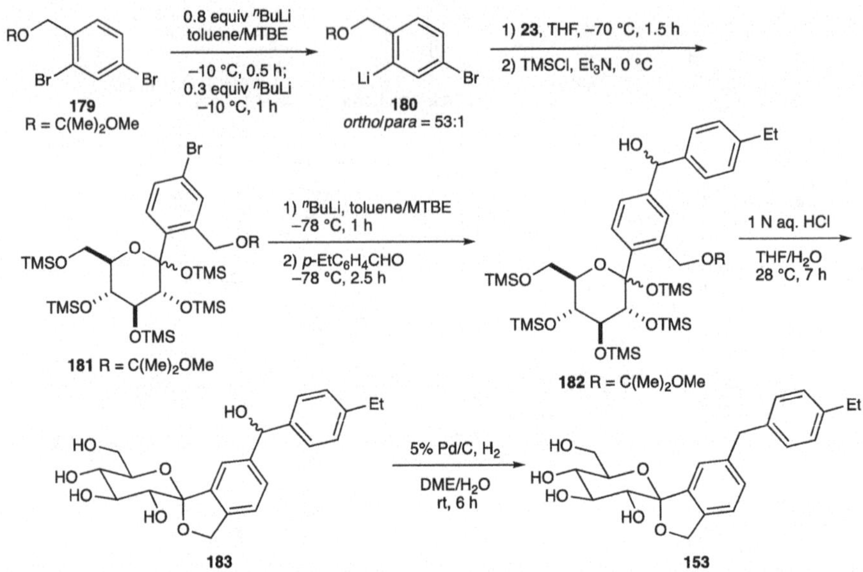

Scheme 30 Simplified synthetic process for tofogliflozin (**153**)

to TMS-protected D-gluconolactone **23** was followed by silylation of the resultant hemiacetal to afford **181**. Crude **181** was then subjected to a second halogen–metal exchange, followed by the trapping of the resultant aryllithium species with *p*-ethylbenzaldehyde, affording diarylmethanol derivative **182**. Upon treatment of **182** with 1N HCl, stereoselective spirocyclization occurred to afford spiroketal **183**, which was then subjected to hydrogenation conditions to afford tofogliflozin (**153**). The crude product was ultimately purified by exhaustive methoxycarbonylation, followed by deprotection of the recrystallized carbonate. This process eliminates the need for palladium catalysts for the cross-coupling reactions. The above-mentioned syntheses depend on bromine–lithium exchange at low temperatures. Thus, an improved method involving iodine–magnesium exchange under milder conditions was developed (Scheme 31) [71]. An aryl Grignard reagent derived from aryl iodide **184** with ^{i}PrMgBr·LiCl was allowed to react with TMS-protected D-gluconolactone **23** at 0–5°C. The resultant adduct was treated with $MeSO_3H$ in MeOH/THF to afford crude tofogliflozin (**153**). According to the established method, exhaustive methoxycarbonylation of **153** afforded carbonate-protected intermediate **176** in 60% yield from **184**. Then, deprotection under basic conditions, followed by treatment with acetone/H_2O, afforded pure tofogliflozin hydrate (**153**·H_2O). A large-scale synthetic process for the spirocyclic dapagliflozin analog **157** was also developed [72, 73].

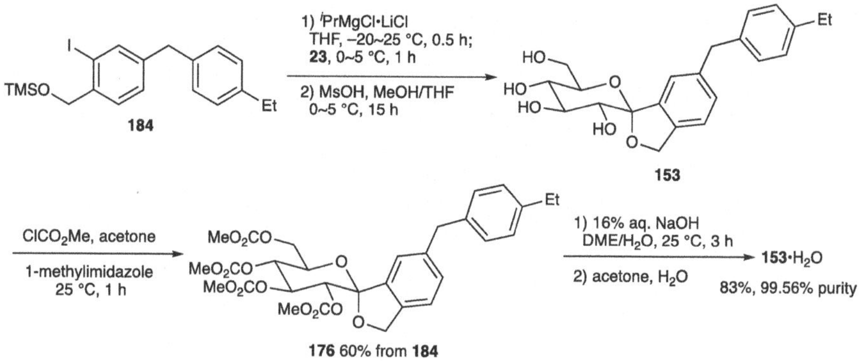

Scheme 31 Modified synthetic process for tofogliflozin hydrate (**153**·H_2O)

4 Syntheses of Miscellaneous Spiroketal Phthalane *C*-Glycosides

In this section, a series of methods to assemble unnatural spiroketal phthalane *C*-glycosides will be overviewed. In particular, spirocyclic *C*-arylribosides, in which the pyranose ring of the papulacandin-type compound is replaced by a ribose, are fascinating as unnatural analogs. Transition-metal catalysis is featured in the synthesis of highly functionalized unnatural spiroketal *C*-glycosides.

A transition-metal-catalyzed [2+2+2] cycloaddition reaction was employed by McDonald and coworkers for the first time for the synthesis of a spiroketal phthalane *C*-glycoside (Scheme 32) [74]. The addition of an acetylide derived from (trimethylsilyl)acetylene to benzyl-protected D-gluconolactone **10** produced hemiketal **185** as a 1:1 mixture of anomers. After acetylation of **185**, the resultant acetate **186** reacted with propargyl TMS ether in the presence of 10 mol% $SnCl_4/AgClO_4$ to afford the expected diyne, albeit in a modest yield. Subsequent desilylation under phase-transfer conditions afforded the desired diyne substrate **187** as a 2.2:1 mixture of anomers. Finally, [2+2+2] cycloaddition of **187** with acetylene was conducted using 10 mol% Wilkinson catalyst ($RhCl(PPh_3)_3$) in EtOH at room temperature for 16 h to afford spiroketal phthalane *C*-arylglycoside **188** in a high yield, although the stereochemistry was not assigned.

By improving the McDonald's procedure, sugar diynes were prepared in shorter routes and were subjected to ruthenium-catalyzed [2+2+2] cycloadditions (Scheme 33) [75, 76]. The addition of propynyllithium to methyl-protected D-gluconolactone **189**, followed by the reaction with *C*-TMS propargyl alcohol in the presence of montmorillonite K10 clay, produced the expected diyne as a mixture of anomers. The subsequent desilylation under phase-transfer conditions afforded diyne α-**190** as the major anomer along with the minor anomer β-**190**. The combined yield was 73% over three steps. Then, the major anomer α-**190** was subjected to

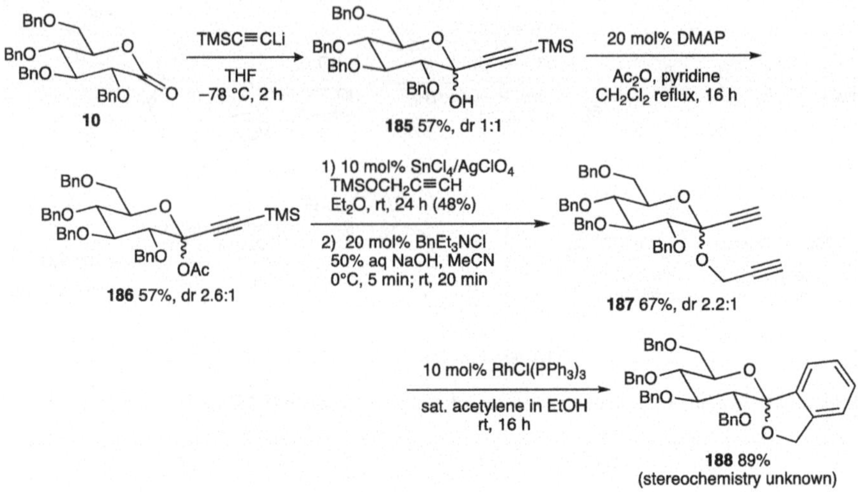

Scheme 32 Synthesis of spiroketal phthalane *C*-glycoside **188** *via* rhodium-catalyzed [2+2+2] cycloaddition

ruthenium-catalyzed [2+2+2] cycloaddition with 1-hexyne at room temperature to regioselectively afford spiroketal phthalane *C*-glycoside **191** in a good yield. Moreover, a similar reaction with chloroacetonitrile produced spiroketal **192**, a pyridine-analog of papulacandin, in 83% yield. The structure of **192** was unambiguously confirmed by single crystal X-ray diffraction analysis. Similarly, ribose-derived diyne **194** was prepared from D-γ-ribonolactone **193**. In this case, the β-anomer was selectively produced, but was inseparable from the minor α-anomer. Nevertheless, **194** could be converted into interesting ribose-analogs of papulacandins **195–197** in good yields *via* ruthenium-catalyzed [2+2+2] cycloadditions with 1-hexyne, chloroacetonitrile, or *N*-propyl isocyanate. The anomeric configuration of β-**194** was determined by comparison of its ^{1}H NMR data with those of a closely related compound, and was later confirmed by the X-ray analysis of its cycloadduct (see below).

The diversified synthesis of spiroketal phthalane *C*-glycosides was accomplished *via* ruthenium-catalyzed [2+2+2] cycloaddition of the corresponding iododiynes. The required iododiynes were readily synthesized by silver-catalyzed iodination of the terminal alkyne moieties of α-**190** and **194** (Scheme 34) [75, 76]. The resultant iododiynes **198** and **200** underwent cycloaddition with acetylene to afford iodinated spiroketal phthalane *C*-glycoside **199** and *C*-riboside **201**, respectively, in good yields. Their palladium-catalyzed cross-coupling reactions were conducted as shown in Scheme 35. The Sonogashira reaction with phenylacetylene, Heck reaction with styrene, and Suzuki–Miyaura coupling with (*p*-methoxyphenyl)boronic acid converted **199** or **201** into the corresponding products **202–207** in good yields. Moreover, the Suzuki–Miyaura coupling of **199** with the known boronate **208** derived from tyrosine successfully afforded spirocyclic *C*-arylglycoside–amino acid hybrid molecule **209** in 75% yield. The copper-catalyzed C–N coupling of

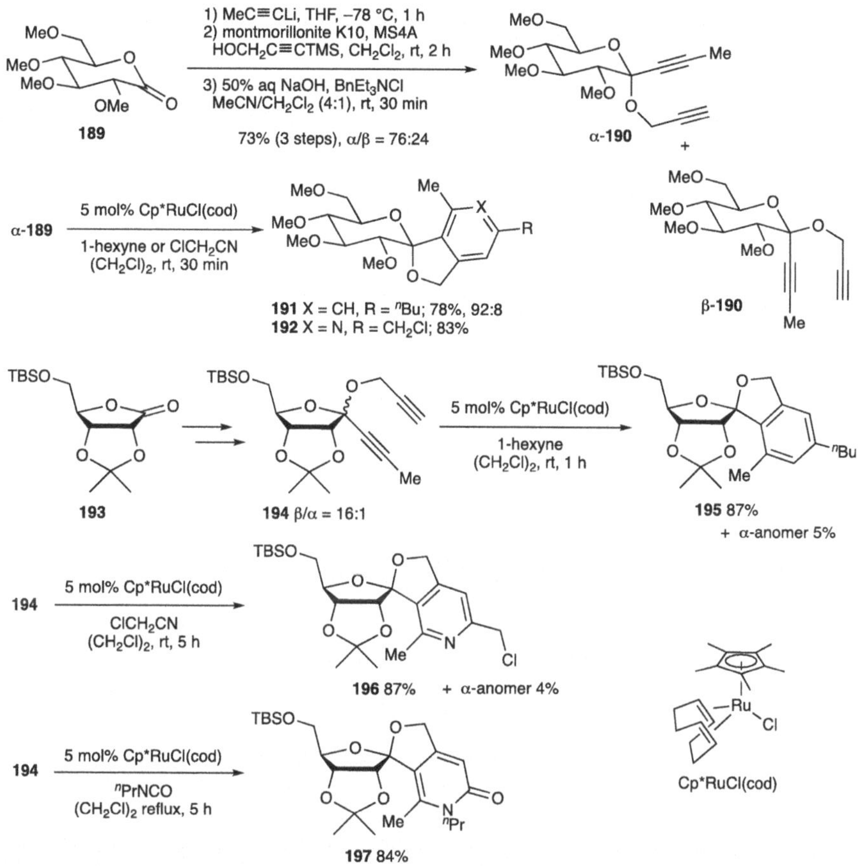

Scheme 33 Regioselective synthesis of spiroketal phthalane *C*-glycosides/ribosides *via* ruthenium-catalyzed [2+2+2] cycloadditions

201 with imidazole also produced spirocyclic *C*-riboside **210** with an *N*-heterocycle-substituted phthalane in 89% yield. An X-ray crystallographic analysis unambiguously proved the depicted structure of **210**.

The synthesis of unnatural derivatives using an enyne metathesis/Diels–Alder sequence was reported by Kaliappan and coworkers (Scheme 36) [77]. The addition of (trimethylsilyl)acetylide to D-mannose-derived lactone **211**, followed by desilylation using TBAF, afforded hemiketal **212**. In the presence of 10-camphorsulfonic acid (CSA), **212** reacted with allyl alcohol to produce 1,6-enyne **213** as a single anomer in 61% yield. After acetylation of the hydroxyl groups, the resultant product **214** was subjected to enyne methathesis using Grubbs 1G catalyst, affording spiroketal **215** bearing a 1,3-diene moiety in 75% yield. The anomeric configuration was later established by NOESY analysis of cycloadduct **218**. Similarly, ribose-derivative **217** was synthesized from D-γ-ribonolactone **193**. The anomeric configuration was determined by comparison of intermediate **216**

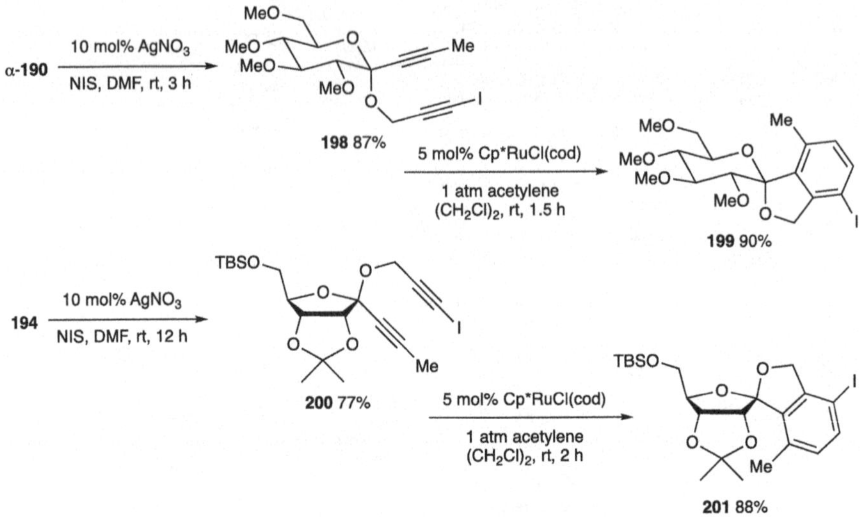

Scheme 34 Synthesis of iodinated spiroketal phthalane *C*-glycoside and riboside

with related diyne **194** (see Scheme 33). Then, the prepared dienes were used for the Diels–Alder reaction with naphthoquinone or dimethyl acetylenedicarboxylate (DMAD). Diene **215** reacted with naphthoquinone or DMAD in toluene at 80°C to generate the corresponding cycloadducts, which were directly subjected to aromatization to afford *manno*-spiroketal *C*-arylglycosides **218** and **219** in 62% and 20% yields, respectively. Analogously, *ribo*-analogs **220** and **221** were obtained in 60% and 58% yields, respectively, from diene **217**.

The de novo synthesis of *ribo*-spiroketals has also been reported [78]. Partially protected tetraol **221**, which was derived from 2-butene-1,4-diol, was subjected to kinetic resolution using lipase AK to obtain monoacetate **222** in 90% yield with >98% ee (Scheme 37). After exchange of the protecting groups, the remaining alcohol moiety of **223** was oxidized to an aldehyde, which underwent Horner–Wadsworth–Emmons reaction to afford enoate **224** in 94% yield over two steps. After DIBAL-H reduction of the ester moiety of **224**, tetrahydropyran (THP)-protection of the resultant hydroxy group and the removal of the TBS group produced primary alcohol **225**. The subsequent oxidation of **225** using 2,2,6,6-tetramethylpiperidine 1-oxyl and condensation of the resultant carboxylic acid with HN(OMe)Me afforded Weinreb amide **226** in 71% yield over two steps.

The reaction of Weinreb amide **226** with aryllithium **227** produced hemiacetal **228** in 96% yield (Scheme 38). The key spiroketalization of **228** was conducted using 20 mol% $PdCl_2(PhCN)_2$ in THF at room temperature, affording a 12.6:4.6:1 mixture of (1*R*,4*R*)-**229**, (1*R*,4*S*)-**229**, and (1*S*,4*R*)-**229** in 91% combined yield. The subsequent oxidative cleavage of the alkene moiety of the major isomer (1*R*,4*R*)-**229** was followed by $NaBH_4$ reduction to generate D-riboside β-**230** in 76% yield over two steps. The X-ray analysis of β-**230** established its anomeric configuration. After benzoylation, the resultant ester β-**231** was treated with aqueous trifluoroacetic acid

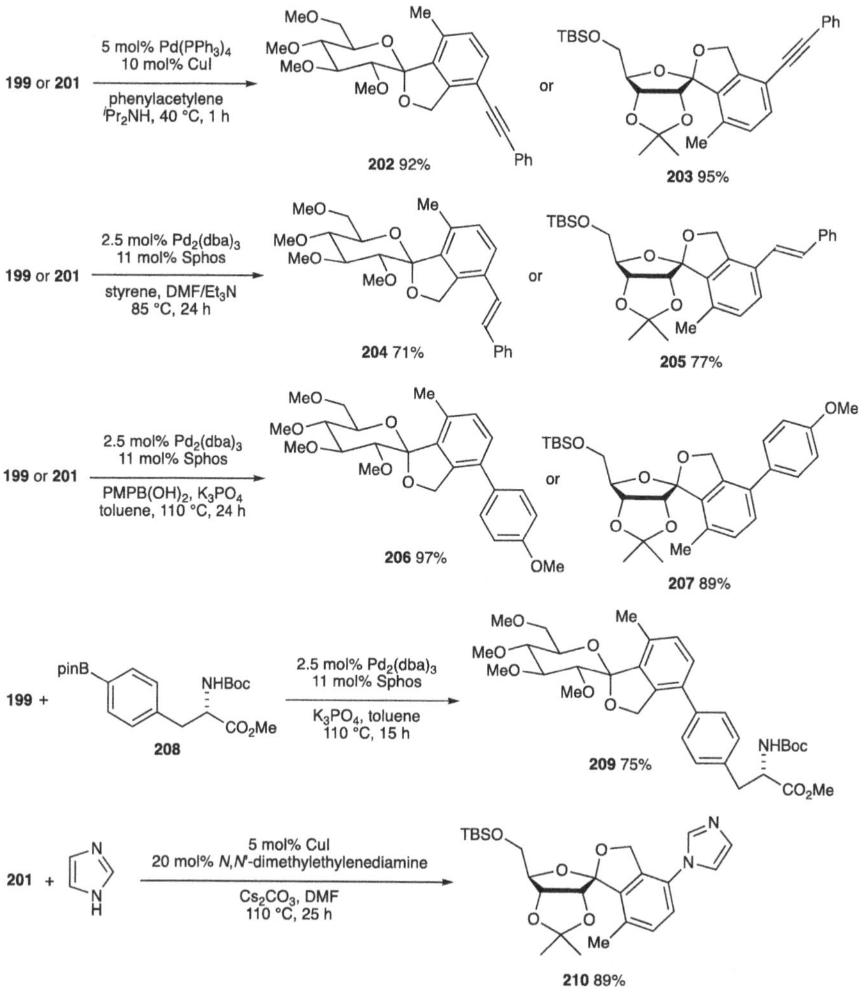

Scheme 35 Palladium- and copper-catalyzed coupling reactions of **199** and **201**

(TFA) for the removal of the acetonide protecting group, affording a 1:10 mixture of anomers α/β-**232** in 98% combined yield. Epimerization of the spiroketal moiety occurred to produce α-**232** as the major product. Then, removal of the benzoyl group from α-**232** afforded unprotected spiroketal phthalane *C*-riboside α-**233** in 87% yield.

An inseparable mixture of (1*R*,4*S*)- and (1*S*,4*R*)-**229** was subjected to oxidative cleavage of the alkene moiety and subsequent $NaBH_4$ reduction to generate L-lyxofuranoside α-**234** and D-ribofuranoside α-**233** in 87% combined yield over two steps (Scheme 39). Subsequently, their 4-bromobenzoylation afforded the corresponding esters α-**235** and α-**236** in 68% combined yield. At this stage, these products could be separated and analyzed by X-ray crystallography to confirm their structures.

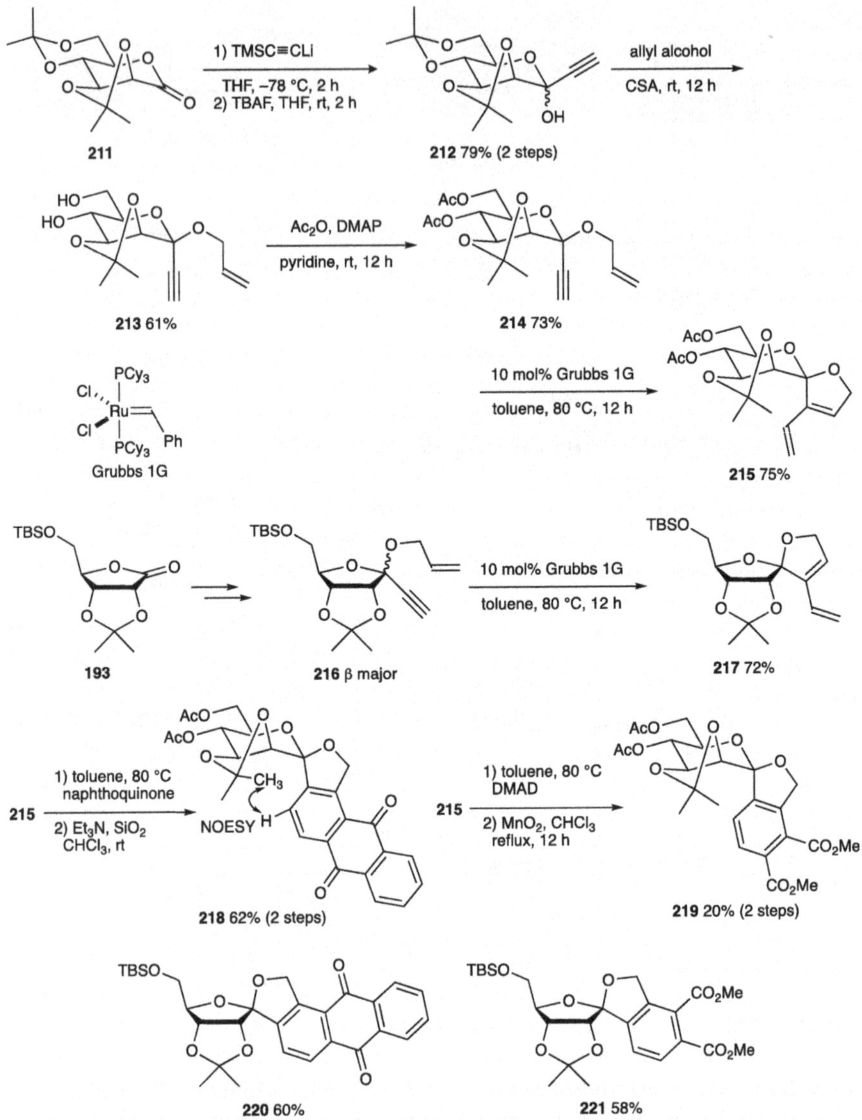

Scheme 36 Synthesis of spiroketals *via* the enyne metathesis/Diels–Alder sequence

The introduction of fluorine atoms into bioactive molecules has been extensively employed in drug discovery, because the highly electronegative fluorine can alter the molecular properties of the parent compounds such as lipophilicity, interactions with receptors, and metabolic stability [79–82]. In this respect, a spiroketal phthalane *C*-glycoside, in which a hydroxyl group is replaced by a fluorine atom, is a fascinating unnatural analog of papulacandins. Sadurní and Gilmour achieved the synthesis of a 2-fluorinated papulacandin core analog as shown in Scheme 40 [83].

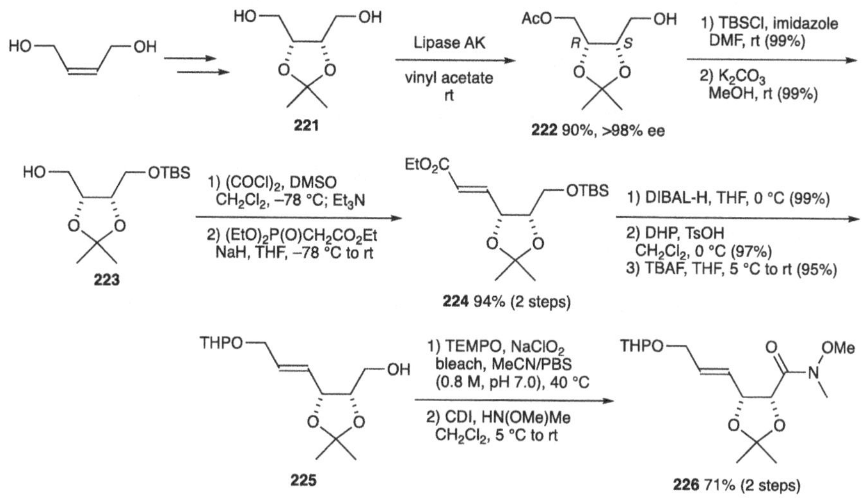

Scheme 37 Preparation of optically active Weinreb amide **226**

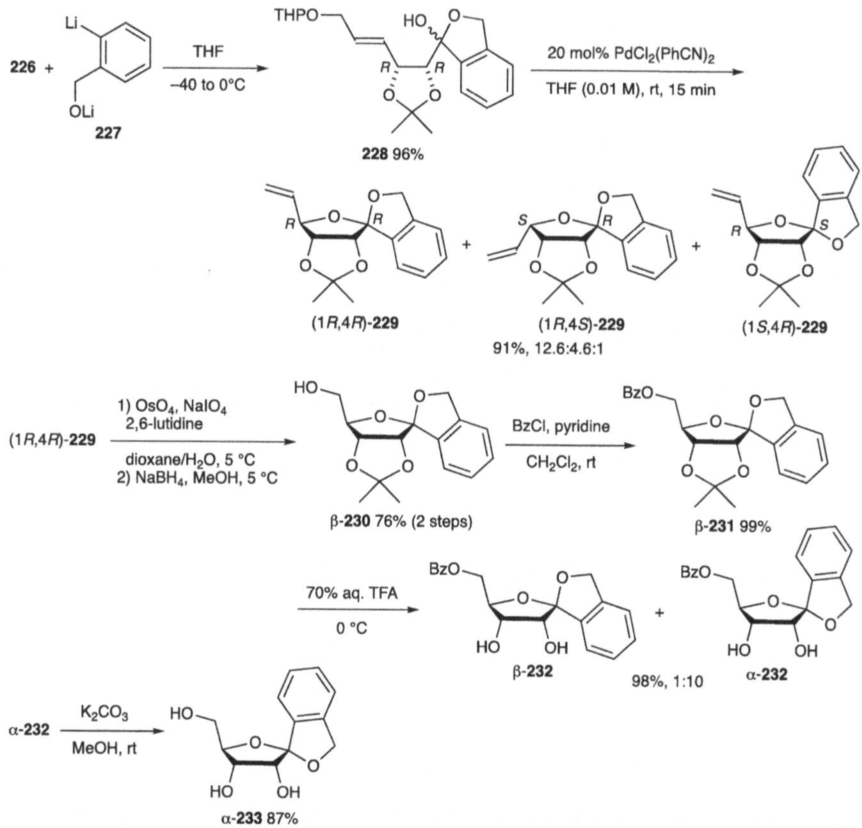

Scheme 38 Synthesis of spiroketal phthalane *C*-riboside α-**233** from Weinreb amide **226**

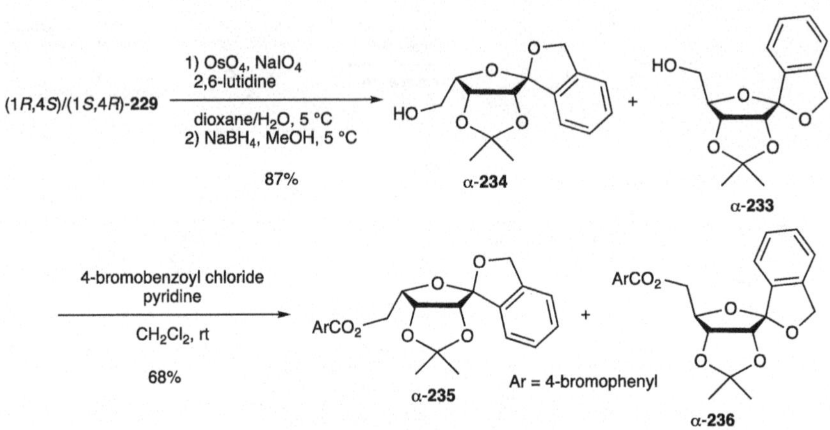

Scheme 39 Transformations of (1*R*,4*S*)- and (1*S*,4*R*)-**229** into α-**235** and α-**236**

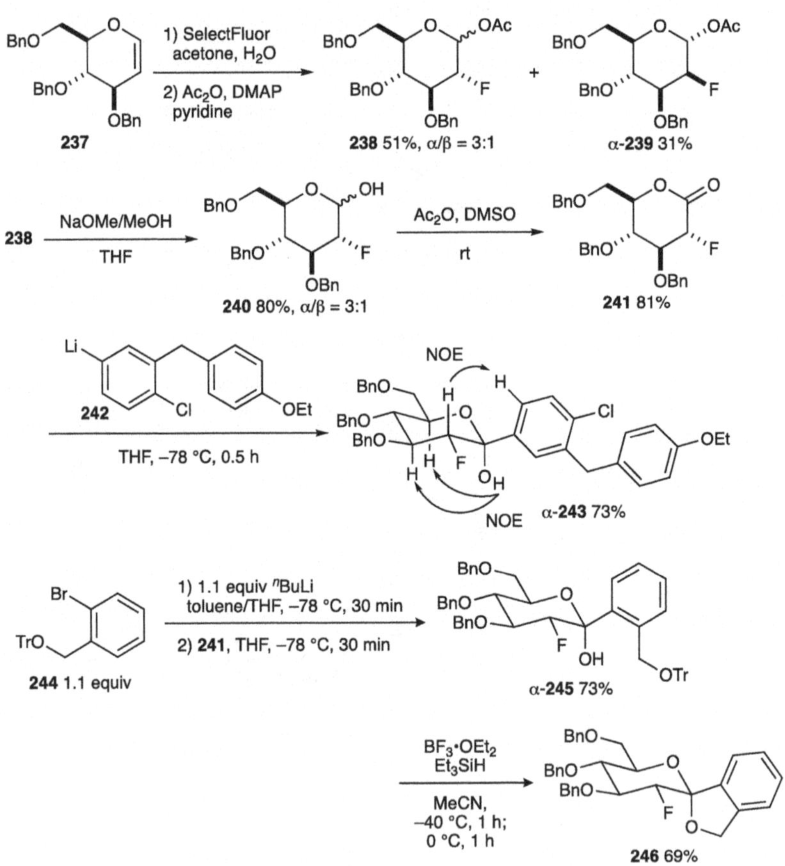

Scheme 40 Synthesis of 2-fluorinated papulacandin core analog **246**

The Gilmour group prepared 2-fluoro-D-gluconolactone **241** from the known benzyl-protected D-glucal **237** [84, 85]. The reaction of **237** with SelectFluor in acetone/H_2O afforded α/β-2-fluoroglucose derivative **238** and α-2-fluoromannose derivative **239** in 51% and 31% yields, respectively, after acetylation. After deacetylation of **238**, the obtained **240** was oxidized to give the desired lactone **241**. The addition reaction of aryllithium reagent **242** to **241** selectively afforded α-anomer **243** in 73% yield. The anomeric configuration of **243** was confirmed by NOE experiments. Similarly, the addition of an aryllithium reagent derived from trityl-protected *o*-bromobenzyl alcohol **244** to **241** selectively afforded α-**245** in 73% yield. The subsequent removal of the trityl group under reduction conditions using $BF_3{\cdot}OEt_2$ and Et_3SiH produced the desired 2-fluorinated papulacandin core analog **246** in 69% yield.

5 Summary

Inspired by the fascinating natural antibiotic papulacandins, diverse synthetic methodologies have been developed for the efficient construction of spiroketal phthalane *C*-glycosides. Early investigations aimed at the natural product synthesis based on the coupling of gluconolactones with aryllithium reagents, albeit with moderate efficiency. To improve the efficiency of the coupling between the sugar and arene moieties, the palladium-catalyzed cross coupling of metalated glucals with arene electrophiles was employed in later studies. To date, two total syntheses of (+)-papulacandin D along these lines have been independently reported by the Barrett and Denmark groups. Moreover, de novo approaches to papulacandin analogs have also been developed utilizing asymmetric transformations such as Sharpless asymmetric dihydroxylation and organocatalytic asymmetric aldol reactions.

In addition to target-oriented investigations, diversity-oriented synthetic approaches have also been developed to obtain unnatural spiroketal phthalane *C*-glycosides. Pharmaceutical company researchers synthesized a variety of spiroketal phthalane *C*-glycosides to identify effective SGLT2 inhibitors. As a result, tofogliflozin was obtained as a highly selective SGLT2 inhibitor with potent antidiabetic effects. Tofogliflozin hydrate has been approved for the treatment of type 2 diabetes mellitus (T2DM). Transition-metal-catalyzed reactions such as [2+2+2] cycloadditions and enyne metathesis have been successfully applied to the assembly of diverse unnatural spiroketal phthalane *C*-glycosides and *C*-ribosides.

In the future, new creative strategies may be expected to be devised to improve the efficiency of the assembly of spiroketal phthalane *C*-glycosides. Those methods to be developed will enable finding of novel drug candidates through drug discovery studies.

References

1. Aho JE, Pihko PM, Rissa TK (2005). Chem Rev 105:4406–4440
2. Sperry J, Wilson ZE, Rathwell DCK, Brimble MA (2010). Nat Prod Rep 27:1117–1137
3. Atkinson DJ, Brimble MA (2015). Nat Prod Rep 32:811–840
4. Rihs G, Traxler P (1981). Helv Chim Acta 64:1533–1539
5. Traxler P, Fritz H, Fuhrer H, Richter WJ (1980). J Antibiot 33:967–978
6. Barrett AGM, Peña M, Willardsen JA (1995). J Chem Soc Chem Commun:1145–1146
7. Barrett AGM, Peña M, Willardsen JA (1995). J Chem Soc Chem Commun:1147–1148
8. Barrett AGM, Peña M, Willardsen JA (1996). J Org Chem 61:1082–1100
9. Denmark SE, Regens CS, Kobayashi T (2007). J Am Chem Soc 129:2774–2776
10. Denmark SE, Kobayashi T, Regens CS (2010). Tetrahedron 66:4745–4759
11. Somsák L, Bokor É, Czifrák K, Juhász L, Tóth M (2011) Carbohydrate derivatives and glycomimetic compounds in established and investigational therapies of type 2 diabetes mellitus. In: Zimering MB (ed) Topics in the prevention, treatment and complications of type 2 diabetes. InTech, Rijeka, pp 103–126
12. Washburn WN (2012) SGLT2 inhibitors in development. In: Jones RM (ed) New therapeutic strategies for type 2 diabetes: small molecule approaches. The Royal Society of Chemistry, Cambridge, pp 29–87
13. Zhang Y, Liu Z-P (2016). Curr Med Chem 23:832–849
14. Aguillón AR, Mascarello A, Segretti ND, de Azevedo HFZ, Guimaraes CRW, Miranda LSM, de Souza ROMA (2018). Org Process Res Dev 22:467–488
15. Traxler P, Gruner J, Auden JAL (1977). J Antibiot 30:289–296
16. Traxler P, Fritz H, Richter WJ (1977). Helv Chim Acta 60:578–584
17. Pérez P, García-Acha I, Durán A (1983). J Gen Microbiol 129:245–250
18. Baguley BC, Römmele G, Gruner J, Wehrli W (1979). Eur J Biochem 97:345–351
19. Pérez P, Varona R, Garcia-Acha I, Durán A (1981). FEBS Lett 129:249–252
20. Varona R, Pérez P, Durán A (1983). FEMS Microbiol Lett 20:243–247
21. Römmele G, Traxler P, Wefrli W (1983). J Antibiot 36:1539–1542
22. Traxler P, Tosch W, Zak O (1987). J Antibiot 40:1146–1164
23. VanMiddlesworth F, Omstead MN, Schmatz D, Bartizal K, Fromtling R, Bills G, Nollstadt K, Honeycutt S, Zweerink M, Garrity G, Wilson K (1991). J Antibiot 44:45–51
24. VanMiddlesworth F, Dufresne C, Smith J, Wilson KE (1991). Tetrahedron 47:7563–7568
25. Bartizal K, Abruzzo G, Trainor C, Krupa D, Nollstadt K, Schmatz D, Schmartz R, Hammond M, Balkovec J, VanMiddlesworth F (1992). Antimicrob Agents Chemother 36:1648–1657
26. Kaneto R, Chiba H, Agematu H, Shibamoto N, Yoshioka T, Nishida H, Okamoto R (1993). J Antibiot 46:247–250
27. Chiba H, Kaneto R, Agematu H, Yoshioka T, Nishida H, Okamoto R (1993). J Antibiot 46:356–358
28. Aoki M, Andoh T, Ueki T, Masuyoshi S, Sugawara K, Oki T (1993). J Antibiot 46:952–960
29. Okada H, Nagashima M, Suzuki H, Nakajima S, Kojiri K, Suda H (1996). J Antibiot 49:103–106
30. Chen RH, Tennant S, Frost D, O'Beirne MJ, Karwowski JP, Humphrey PE, Malmberg L-H, Choi W, Brandt KD, West P, Kadam SK, Clement JJ, McAlpine JB (1996). J Antibiot 49:596–598
31. Ohyama T, Iwadate-Kurihara Y, Hosoya T, Ishikawa T, Miyakoshi S, Hamano K, Inukai M (2002). J Antibiot 55:758–763
32. Schmatz DM, Romancheck MA, Pittarelli LA, Schwartz RE, Fromtling RA, Nollstadt KH, Vanmiddlesworth FL, Wilson KE, Turner MJ (1990). Proc Natl Acad Sci U S A 87:5950–5954
33. Jaramillo C, Knapp S (1994). Synthesis:1–20
34. Bililign T, Griffith BR, Thorson JS (2005). Nat Prod Rep 22:742–760
35. Wellington KW, Benner SA (2006). Nuclos Nucleot Nucleic Acids 25:1309–1333

36. Kitamura K, Ando Y, Matsumoto T, Suzuki K (2018). Chem Rev 118:1495–1598
37. Bokor É, Kun S, Goyard D, Tóth M, Praly J-P, Vidal S, Somsák L (2017). Chem Rev 117:1687–1764
38. Yang Y, Yu B (2017). Chem Rev 117:12281–12356
39. Liao H, Ma J, Yao H, Liu X-W (2018). Org Biomol Chem 16:1791–1806
40. Schmidt RR, Frick W (1988). Tetrahedron 44:7163–7169
41. Rosenblum SB, Bihovsky R (1990). J Am Chem Soc 112:2746–2748
42. Czernecki S, Perlat M-C (1991). J Org Chem 56:6289–6292
43. Hamdouchi C, Sanchez-Martinez C (2001). Synthesis:833–840
44. Hamdouchi C, Jaramillo C, Lopez-Prados J, Rubio A (2002). Tetrahedron Lett 43:3875–3878
45. Parker KA, Georges AT (2000). Org Lett 2:497–499
46. Friesen RW, Sturino CF (1990). J Org Chem 55:5808–5810
47. Dubois E, Beau J-M (1990). Tetrahedron Lett 31:5165–5168
48. Dubois E, Beau J-M (1992). Carbohydr Res 223:157–167
49. Liu G, Wurst JM, Tan DS (2009). Org Lett 11:3670–3673
50. Wurst JM, Liu G, Tan DS (2011). J Am Chem Soc 133:7916–7925
51. Butkevich AN, Corbu A, Meerpoel L, Stanfield I, Angibaud P, Bonnet P, Cossy J (2012). Org Lett 14:4998–5001
52. Parkan K, Pohl R, Kotora M (2014). Chem A Eur J 20:4414–4419
53. van der Kaaden M, Breukink E, Pieters RJ (2012). Beilstein J Org Chem 8:732–737
54. Danishefsky S, Phillips G, Ciufolini M (1987). Carbohydr Res 171:317–327
55. Balachari D, O'Doherty GA (2000). Org Lett 2:863–866
56. Balachari D, O'Doherty GA (2000). Org Lett 2:4033–4036
57. Ahmed MM, O'Doherty GA (2005). Tetrahedron Lett 46:4151–4155
58. Mainkar PS, Johny K, Rao TP, Chandrasekhar S (2012). J Org Chem 77:2519–2525
59. DeFronzo RA, Norton L, Abdul-Ghani M (2017). Nat Rev Nephrol 13:11–26
60. Ehrenkranz JRL, Lewis NG, Kahn CR, Roth J (2005). Diabetes Metab Res Rev 21:31–38
61. Ohtake Y, Sato T, Kobayashi T, Nishimoto M, Taka N, Takano K, Yamamoto K, Ohmori M, Yamaguchi M, Takami K, Yeu S-Y, Ahn K-H, Matsuoka H, Morikawa K, Suzuki M, Hagita H, Ozawa K, Yamaguchi K, Kato M, Ikeda S (2012). J Med Chem 55:7828–7840
62. Suzuki M, Honda K, Fukazawa M, Ozawa K, Hagita H, Kawai T, Takeda M, Yata T, Kawai M, Fukuzawa T, Kobayashi T, Sato T, Kawabe Y, Ikeda S (2012). J Pharmacol Exp Ther 341:692–701
63. Ikeda S, Takano Y, Cynshi O, Tanaka R, Christ AD, Boerlin V, Beyer U, Beck A, Ciorciaro C, Meyer M, Kadowaki T (2015). Diabetes Obes Exp Metab 17:984–993
64. Poole RM, Prossler JE (2014). Drugs 74:939–944
65. Xu B, Lv B, Feng Y, Xu G, Du J, Welihinda A, Sheng Z, Seed B, Chen Y (2009). Bioorg Med Chem Lett 19:5632–5635
66. Lv B, Xu B, Feng Y, Peng K, Xu G, Du J, Zhang L, Zhang W, Zhang T, Zhu L, Ding H, Sheng Z, Welihinda A, Seed B, Chen Y (2009). Bioorg Med Chem Lett 19:6877–6881
67. Lv B, Feng Y, Dong J, Xu M, Xu B, Zhang W, Sheng Z, Welihinda A, Seed B, Chen Y (2010). ChemMedChem 5:827–831
68. Ohtake Y, Emura T, Nishimoto M, Takano K, Yamamoto K, Tsuchiya S, Yeu S-Y, Kito Y, Kimura N, Takeda S, Tsukazaki M, Murakata M, Sato T (2016). J Org Chem 81:2148–2153
69. Yamane M, Kawashima K, Yamaguchi K, Nagao S, Sato M, Suzuki M, Honda K, Hagita H, Kuhlmann O, Polirier A, Fowler S, Funk C, Simon S, Aso Y, Ikeda S, Ishigai M (2015). Xenobiotica 45:230–238
70. Murakata M, Ikeda T, Kimura N, Kawase A, Nagase M, Kimura M, Maeda K, Honma A, Shimizu H (2017). Tetrahedron 73:655–660
71. Yang X-D, Pan Z-X, Li D-J, Wang G, Liu M, Wu R-G, Wu Y-H, Gao Y-C (2016). Org Process Res Dev 20:1821–1827
72. Liu Y-H, Fu T-M, Ou C-Y, Fan W-L, Peng G-P (2013). Chin Chem Lett 24:131–133
73. Liu Y, Fu T, Chen Z, Ou C (2015). Monatsh Chem 146:1715–1721

74. McDonald FE, Zhu HYH, Holmquist CR (1995). J Am Chem Soc 117:6605–6606
75. Yamamoto Y, Hashimoto T, Hattori K, Kikuchi M, Nishiyama H (2006). Org Lett 8:3565–3568
76. Yamamoto Y, Yamashita K, Hotta T, Hashimoto T, Kikuchi M, Nishiyama H (2007). Chem Asian J 2:1388–1399
77. Subrahmanyam AV, Palanichamy K, Kaliappan KP (2010). Chem Eur J 16:8545–8556
78. Awasaguchi K, Miyazawa M, Uoya I, Inoue K, Nakamura K, Yokoyama H, Kakuda H, Hirai Y (2010). Synlett:2392–2396
79. Wang J, Sánchez-Roselló M, Aceña JL, del Pozo C, Sorochinsky AE, Fustero S, Soloshonok VA, Liu H (2014). Chem Rev 114:2432–2506
80. Gillis EP, Eastman KJ, Hill MD, Donnelly DJ, Meanwell NA (2015). J Med Chem 58:8315–8359
81. Zhou Y, Wang J, Gu Z, Wang S, Zhu W, Aceña JL, Soloshonok VA, Izawa K, Liu H (2016). Chem Rev 116:422–518
82. Yerien DE, Bonesi S, Postigo A (2016). Org Biomol Chem 14:8398–8427
83. Sadurní A, Gilmour R (2018). Eur J Org Chem 2018:3684–3687
84. Bucher C, Gilmour R (2010). Angew Chem Int Ed 49:8724–8728
85. Sadurní A, Kehr G, Ahlqvist M, Wernevik J, Sjögren HP, Kankkonen C, Knerr L, Gilmour R (2018). Chem Eur J 24:2832–2836

Top Heterocycl Chem (2019) 57: 261–290
DOI: 10.1007/7081_2019_29

Published online: 16 February 2019

Spiro Iminosugars: Structural Diversity and Synthetic Strategies

Damien Hazelard, Raphaël Hensienne, Jean-Bernard Behr, and Philippe Compain

Contents

Abstract From their discovery in the late 1960s, iminosugars have undergone an expansion from an area of science limited to a few researchers to a field that now attracts the interest of members of the whole synthetic organic chemistry community. Indeed, many tasks concern structural modifications of standard iminosugars in order to improve their biological and pharmacological properties. In this way, the introduction of an adjoining *spiro*cycle afforded unprecedented polyhydroxy-azaspiranes, the structures and syntheses of which are presented in this chapter.

D. Hazelard, R. Hensienne, and P. Compain (✉)
Laboratoire d'Innovation Moléculaire et Applications, Univ. de Strasbourg, Univ. de Haute-Alsace, CNRS, LIMA (UMR 7042), Equipe de Synthèse Organique et Molécules Bioactives (SYBIO), ECPM, Strasbourg, France
e-mail: philippe.compain@unistra.fr

J.-B. Behr
Institut de Chimie Moléculaire de Reims, Univ. Reims Champagne Ardenne, CNRS (UMR 7312), Méthodologie en Synthèse Organique (MSO), Reims Cedex 2, France
e-mail: jb.behr@univ-reims.fr

Special attention is paid to the key steps involved in the generation of the pivotal quaternary spiro atom.

Keywords Conformational constraint · Glycochemistry · Glycomimetics · Glycosidase inhibitors · Iminosugars · Spiro compounds

1 Introduction

Azaspirocycles represent structural motifs found in a diversity of natural products and biologically active compounds including alkaloids, such as the neurotoxin histrionicotoxins, or spirocyclic nucleosides related to hydantocidin [1–7]. The list includes also stemonamine [5], nankakurine A [6], halichlorine, and cephalotaxine [1–7]. Inspired by nature, chemists have recently designed original glycomimetics based on spiranic frameworks. In view of the biological importance of carbohydrate mimics with nitrogen atom replacing the ring oxygen, increasing efforts are directed toward the synthesis of "spiro iminosugars." Historically known as potent glycosidase inhibitors, iminosugars have been shown to inhibit an ever-growing number of therapeutically relevant carbohydrate-processing enzymes such as glycogen phosphorylases and, most recently also, enzymes that act on non-sugar substrates [8–14]. Since the discovery of their potential as antidiabetic agents in the 1970s, the interest of chemists and clinical researchers for iminosugars has been steadily strengthened by important breakthroughs. The list includes the development of the pharmacological chaperone concept that culminated in 2016 with the approval of Galafold™ as the first oral drug for the treatment of Fabry disease [15] or the discovery of large multivalent effects in glycosidase inhibition in the late 2000s [16–18] (Fig. 1). In this review, a spiro iminosugar is defined as a glycomimetic based on a polyhydroxylated spiroazacycle. Such structures combine the advantages of classical iminosugars and of spirocycles. The ability of endocyclic amines to become protonated at physiological pH may induce effective biological activity by mimicking the structure of the enzymatic oxocarbenium ion-like transition state and/or by promoting strong electrostatic interactions with carboxylate residues in the enzyme active site [8, 19, 20]. The introduction of a spiranic center is expected to provide conformational rigidity and give access to a diversity of conformations other than the traditional chair or boat of six-membered rings. The original distributions of

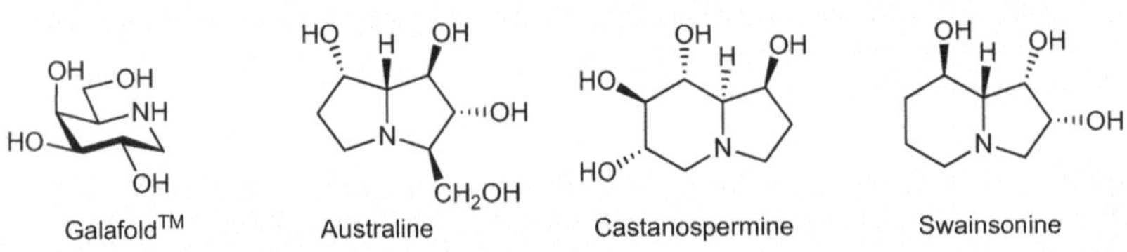

Fig. 1 Some representative mono- and bicyclic iminosugars

hydroxyl groups thus obtained are thought to be of likely significance for receptor recognition purposes. Constraining a given structure in orientations that fit the enzymatic active site or binding pocket of the biological target is a strategy widely used in drug discovery ([21], for a review on conformationally restricted glycoside derivatives, see [22]). Within the context of glycomimetic-based inhibitors/ligands, holding hydroxyl groups in precisely defined arrangements may be a promising tool to gain specificity for a particular protein. In addition, this approach allows the exploration of unfrequented regions of chemical and intellectual property spaces. In contrast to fused bicyclic iminosugars based on pyrrolizidine or indolizidine skeletons such as castanospermine, swainsonine, or australine (Fig. 1) [23], the field of spiro iminosugars has emerged only recently. Although the first description of polyhydroxylated spiroazacycles dates back to almost 50 years [24], increased scientific interest for such structures has emerged only in the early 2000s. This may be explained in part by the synthetic difficulties associated with spiranic structures including the stereocontrolled formation of the stereogenic quaternary spirocyclic center and the functionalization of the newly formed ring system [25, 26].

The purpose of this review is to give an overview of the synthetic strategies designed to address the multiple challenges posed by such structures. Our intent is to focus on the key steps leading to the formation of spiro iminosugars. The review will be organized in three sections: cyclitol-based azaspiranes, glycoside-based azaspiranes, and iminosugars spiro-linked with carba-, oxa-, or azacycles (Fig. 2).

We hope that this review will stimulate further research in the area by providing a description of the different types of spiro iminosugars that have been reported so far. Data on their biological activity will be also given whenever such data are available.

2 Cyclitol-Based Azaspiranes

A limited number of polyhydroxylated spiroazacycles in which the hydroxyl groups are positioned on a cycloalkane ring have been described in the literature so far. In 2003, Vasella et al. reported the synthesis of spiroaziridines **5** (Scheme 1) [27].

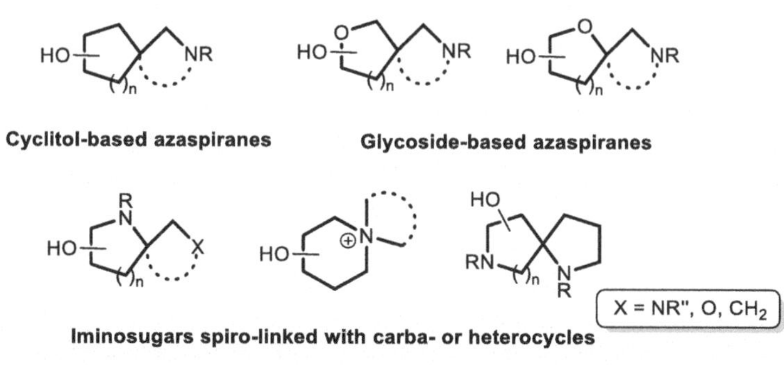

Fig. 2 Spiro iminosugar structures

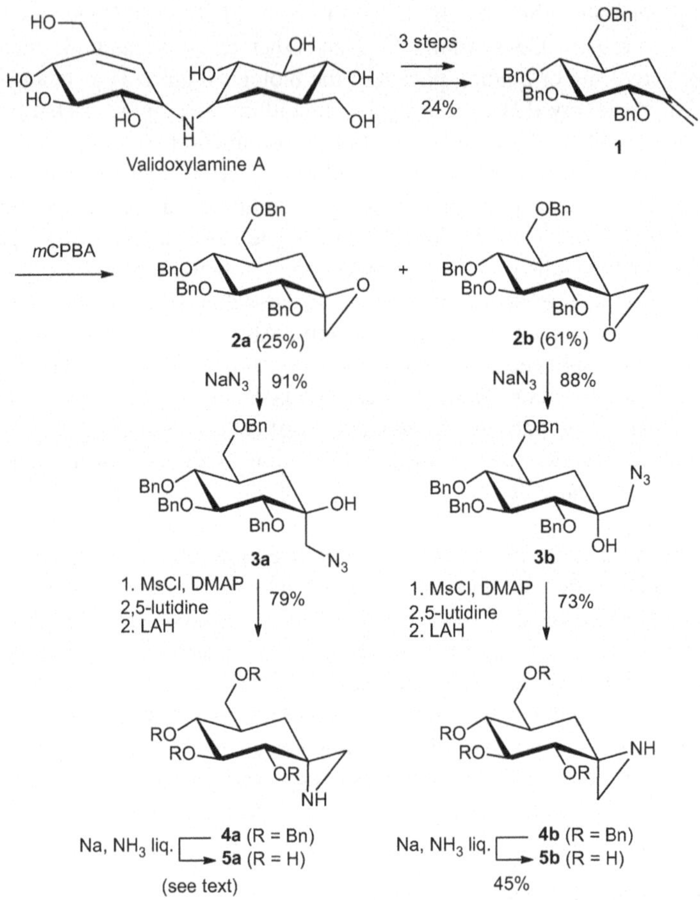

Scheme 1 Synthesis of spiroaziridines **5**

These compounds were designed as mechanistic probes to study the impact of the position of the basic nitrogen atom on the inhibition profile toward a panel of glycosidases. The key alkene intermediate **1** was prepared in three steps from validoxylamine A (Scheme 1). The targeted aziridines were synthesized via the corresponding epoxides **2** obtained as a mixture of diastereomers after treatment of **1** with *m*CPBA (86% yield, dr 2.4:1 in favor of **2b**). Attempts to convert exocyclic alkene **1** into the corresponding *N*-tosyl aziridine using chloramine-T and phenyl (trimethyl)ammonium tribromide led to poor yields. After separation by HPLC, epoxides **2a** and **2b** were treated with NaN_3. The resulting ring-opening products **3** were converted to the corresponding azido methanesulfonate intermediates which were transformed into aziridines **4** by treatment with $LiAlH_4$ in THF. Debenzylation of the spiroaziridines under Birch conditions provided the expected deprotected aziridines **5**. Spiroaziridine **5a** was found to be particularly unstable and was obtained

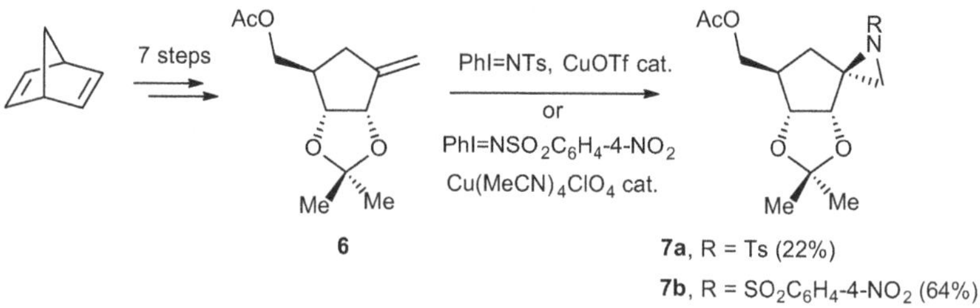

Scheme 2 Synthesis of spiroaziridines **7** via copper-catalyzed aziridination

with a small amount of an unidentified by-product after purification by Sephadex chromatography. Evaluation of azaspiranes **5** toward a panel of three glycosidases indicated that **5b** was a weak irreversible inhibitor of the β-glucosidase from *Caldocellum saccharolyticum* and a weak reversible inhibitor of the β-glucosidase from yeast. No inhibition was observed for β-glucosidase from sweet almonds. Spiroaziridine **5a** was a poor inhibitor of the three enzymes.

Phosphoribosyl transferase is an attractive target for antiprotozoal chemotherapy. Within the context of the synthesis of carbocyclic phosphoribosyl transferase transition state analogues, Borhani and co-workers reported the synthesis of protected spiroaziridines **7** based on a five-membered carbasugar [28]. Interestingly, direct copper-catalyzed aziridination of exocyclic alkene **6** using PhI=NTs provided aziridine **7a** as a single diastereomer, albeit in low yields (Scheme 2). The structure of **7a** was unequivocally confirmed by X-ray analysis of the corresponding ring-opening product obtained after treatment with Li_2NiBr_4. The yield of the aziridination reaction could be significantly improved using [(4-nitrobenzenesulfonyl)-imino]-phenyliodinane and $Cu(MeCN)_4ClO_4$ as the catalyst.

In 1987, the group of Harrisson reported a model study toward the total synthesis of enantiomerically pure (−)-histrionicotoxins (HTX) from D-mannose (Scheme 3) [29]. Their strategy was to take advantage of the chirality of carbohydrates to access enantiopure natural products and related compounds. The 1-azaspiro[5.5]undecane ring system was synthesized by way of successive inter- and intramolecular Henry reaction. KOH-catalyzed addition of protected 5-nitro-pentan-1-ol **14** onto aldehyde **8** obtained in three steps from D-mannose afforded nitro alcohols **9** as a complex mixture of diastereomers. Treatment of the 1-*O*-silyl protected furanose **9** with TBAF liberated the anomeric aldehyde group giving rise to an intramolecular Henry reaction. This process afforded the formation of the key quaternary C–N bond of the 1-azaspiro[5.5]undecane skeleton. The nitrocyclitols **10** were however obtained as a mixture of diastereomers (61% yield). It is noteworthy that, quite remarkably, the related two-step sequence afforded only a single diastereomer in 70% yield when nitroethane is used instead of **14**. As a prelude to the formation of the piperidine ring by way of intramolecular S_N2, the tetrahydropyranyl group was deprotected, and the corresponding alcohol was tosylated to afford **11**. At this stage, after purification on silica gel, the synthesis was performed on the major diastereomer. Acetylation of diastereomerically pure triol **11** with acetic anhydride in the

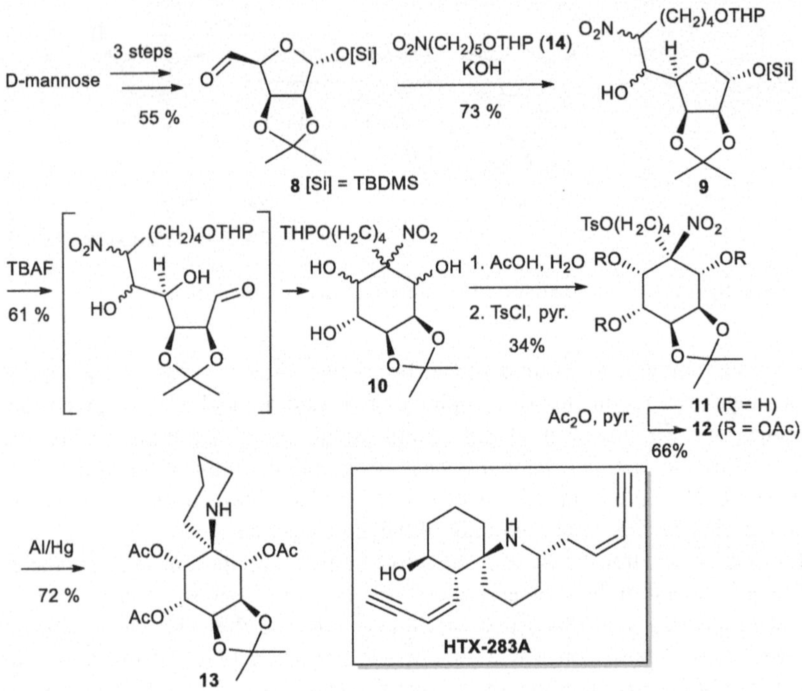

Scheme 3 Synthesis of polyoxygenated 1-azaspiro[5.5]undecane **13** by way of inter- and intramolecular Henry reactions

presence of pyridine provided the nitro derivative **12** which was reduced with aluminum amalgam. Spontaneous ring closing then afforded the expected polyoxygenated azaspirane **13** in 72% yield.

Other examples of spiro iminosugars with a nitrogen atom directly connected to the quaternary spiro carbon atom were recently described by the group of Compain [30, 31]. The key step of the synthesis was the stereocontrolled formation of the pivotal C–N bond via nitrene insertion into a C–H bond at the spiro carbon atom (Scheme 4). This process which is catalyzed by rhodium(II) complexes occurs with retention of configuration. However, applying such a reaction to polyoxygenated substrates represents a challenge in terms of regioselectivity since insertion is known to be favored in α-ethereal C–H bonds. The first part of the synthesis is the preparation of the functionalized square carbasugar **16** ([32], for a review on square sugars, see [33]). This key intermediate was obtained in 67% yield by way of SmI_2-mediated intramolecular coupling reaction of γ,δ-unsaturated aldehyde **15** prepared in six steps from vitamin C [32]. To achieve a complete regiocontrol in the pivotal C–H amination step, a strategy using a combination of activating and electron-withdrawing groups has to be followed. The mere introduction of a vinylic group was indeed not sufficient to reach high regioselectivity at the spiro carbon atom; electron-withdrawing protecting groups were also required to reduce the electron density at the undesired C–H insertion site in α-position to the carbamate group.

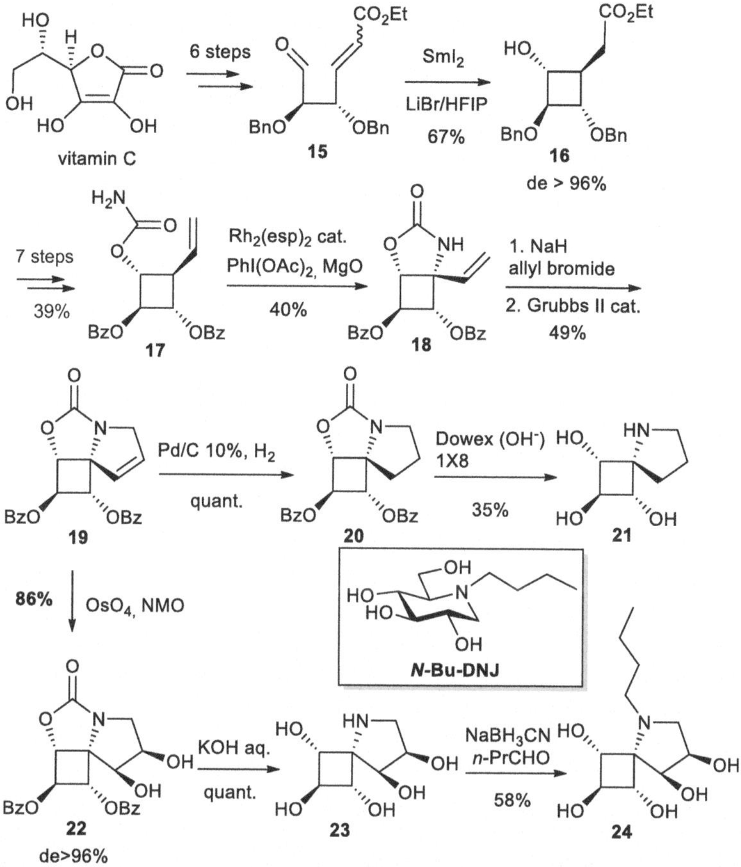

Scheme 4 Synthesis of four-membered cyclitol-based spiro iminosugars

After various attempts, the best C–H amination substrate was found to be carbamate **17** obtained in seven steps from the cyclization product **16**. After having generated the pivotal C–N bond of the final targets with retention of configuration and high regiocontrol, the final key step of the synthesis entailed the formation of the azacycle by ring-closing metathesis. Despite the additional ring strain generated by the five-membered ring closure, the expected tricyclic spirocycle derivative **19** was obtained in high yield using 5 mol% of Grubbs II catalyst. Unprecedented constrained iminosugars **21** and **23–24** were then obtained in two or three steps from the common intermediate **19** thus generated (Scheme 4). Preliminary biological evaluations were performed and led to the identification of a new class of correctors of defective F508del-CFTR gating involved in cystic fibrosis (CFTR, cystic fibrosis transmembrane conductance regulator). The best corrector of the spiranic series, iminosugar **24**, displayed a F508del-CFTR activity rescue not significantly different

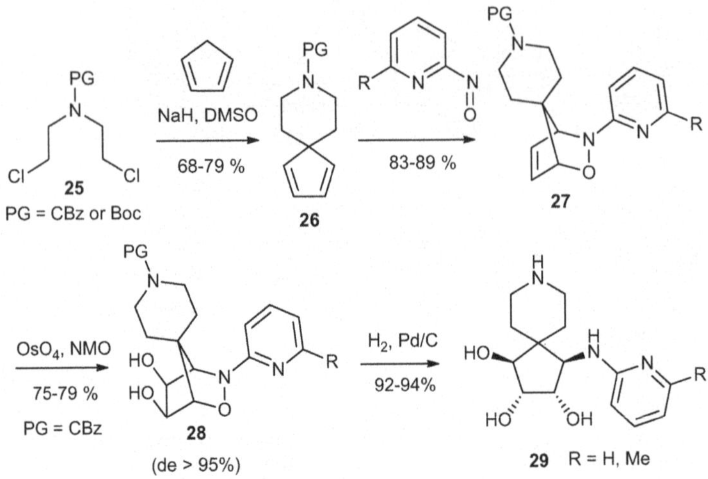

Scheme 5 Synthesis of 4-amino-8-azaspiro[4.5]decanols **29**

from the *N*-Bu DNJ-induced one [31]. *N*-Bu DNJ – also named miglustat – is a clinical candidate for the treatment of cystic fibrosis.

Cyclitol-based iminosugars where the nitrogen is not directly connected to the quaternary spiro carbon have been described by the group of Miller [34]. The spirocyclic skeleton was efficiently constructed via the dialkylation of cyclopentadiene with *N*-protected bis-2-chloroethylamines **25** (Scheme 5). The spiranic dienes **26** were then engaged in a series of iminonitroso Diels-Alder reactions to yield the corresponding cycloadducts **27** in good yields. In addition to the stereocontrolled formation of two carbon-heteroatom bonds, one advantage of this cycloaddition process is that the resulting endocyclic alkenes are converted into the corresponding diols **28** with high diastereoselectivity under Upjohn reaction conditions (OsO_4, *N*-methylmorpholine-*N*-oxide, NMO). The one-step cleavage of the N–O bond and deprotection of the Cbz group provided the 4-amino-8-azaspiro[4.5]decanols **29** in high yields (racemic form).

3 Carbohydrate-Based Azaspiranes

It is not surprising that several synthetic carbohydrate-based spiroazacycles have been described in the literature. The preparation of such compounds takes indeed advantage of the use of carbohydrates as starting materials. Biologically relevant carbohydrate-based azaspiranes are also found in nature as shown by the herbicide hydantocidin, a nucleoside having a unique structure with a spiro-annelated hydantoin and ribose [35–37].

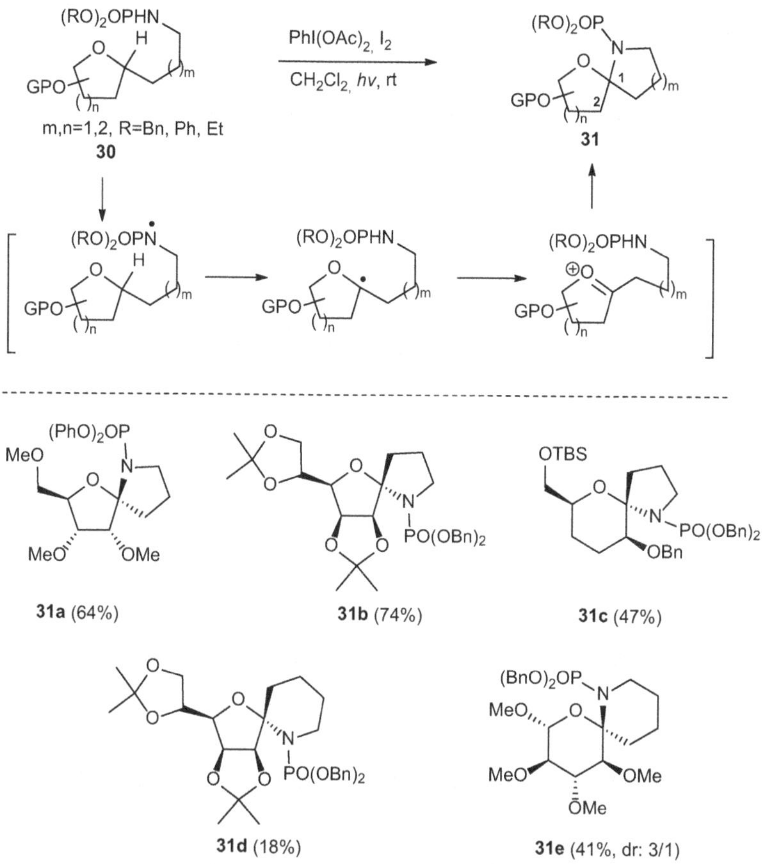

Scheme 6 Synthesis of 1-azaspirocycles **31** by intramolecular hydrogen atom transfer

3.1 *Systems with a Nitrogen Atom at α-Position to the Spirocyclic Center*

Suarez and co-workers reported the regio- and stereoselective formation of oxa-1-azaspirocycles by way of intramolecular hydrogen atom transfer (HAT) (Scheme 6) [38, 39]. This strategy is based on the formation of an electrophilic *N*-radical followed by subsequent abstraction of a hydrogen atom at the *pseudo*-anomeric position. The nucleophilic *C*-radical thus formed is oxidized to an oxonium ion which reacts with the amine group [40]. The authors synthesized several carbohydrate substrates **30** with different ring sizes and protecting groups. Irradiation of compounds **30** in the presence of molecular iodine and $PhI(OAc)_2$ afforded spiroaminals **31** [1, 38, 39]. Better yields were obtained for the formation of five-membered in comparison to six-membered rings. This difference of reactivity could be explained by a more favorable 1,5 HAT through a six-membered transition state.

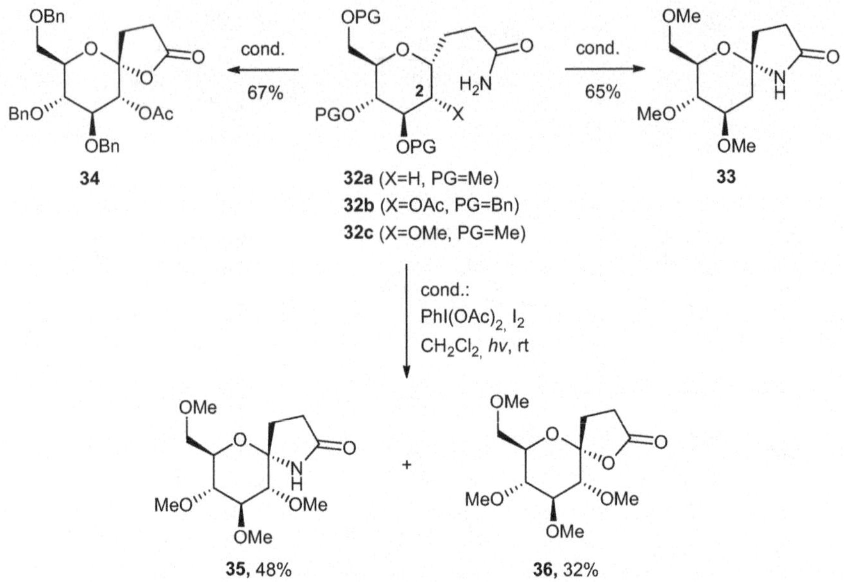

Scheme 7 Synthesis of carbohydrate-based spirolactams or spirolactones by intramolecular hydrogen abstraction

Furthermore, the methodology described was compatible with the most common protecting groups in carbohydrate chemistry. However, the introduction of an electron-withdrawing protecting group (acetate) at C-2 inhibits the radical abstraction at C-1, and the substrate was recovered unchanged.

The same group has applied the HAT strategy to the synthesis of spirolactams involving amidyl radicals [41] (Scheme 7). It is noteworthy that depending on the substitution at C-2, spirolactones or spirolactams are obtained. Indeed, 2-deoxysugar **32a** afforded lactam **33**, whereas 2-acetyl-D-glucose **32b** gave spirolactone **34** in good yield. This difference of reactivity is explained by the authors on the basis of the hard and soft acids and bases (HSAB) principle. In the case of **32b**, the oxocarbenium ion intermediate is a harder electrophile than **32a** due to the presence of the electron-withdrawing group at C-2. As the oxygen atom of the amide is considered to be a harder nucleophile than the nitrogen atom, it attacks the hard electrophilic oxocarbenium ion. In the case of **32c**, substituted at C-2 with an electron-donating group, a mixture of spirolactam **35** and spirolactone **36** was obtained. The highly stereoselective formation of the products with axial C–O (N) bonds is explained by a better stability of the bicyclic products **33–36** due to the anomeric effect.

Spirolactams have also been obtained from nitrosugars by way of Michael addition (Scheme 8) [42]. For example, reaction of nitrosugar **37** [43] with methyl acrylate in the presence of TBAF gave unstable adduct **38** which after column chromatography produced **39** in good yield ([42], for a review on carbanionic reactivity of the anomeric center in carbohydrates, see [44]). On the other hand, azidation of the crude

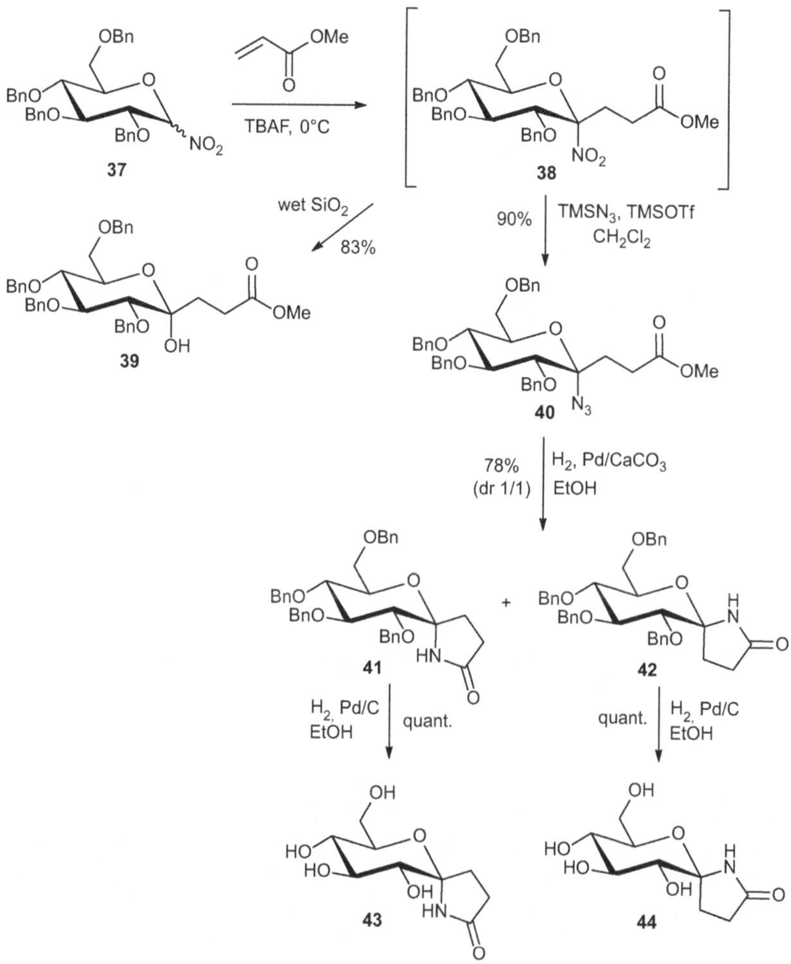

Scheme 8 Synthesis of carbohydrate-based spirolactams **43** and **44** from nitrosugar **37**

Michael adducts **38** with $TMSN_3$ in the presence of TMSOTf afforded 1-azido sugar **40** in 90% yield as a single diastereomer. This methodology has also been applied to other electrophiles such as acrylonitrile or ethyl propiolate and in other sugar series (D-manno). Amides **41** and **42** were then obtained after reduction of 1-azido sugar **40** and subsequent cyclization. Ring opening of the transient glucosylamine intermediate formed during the process followed by ring closing of the corresponding open imine led to the formation of a mixture of epimeric spirolactams **41** and **42** [42]. Deprotection of the benzyloxy groups afforded the final products **43** and **44** in 70% yields from nitrosugar **37**. Related spiro-carbamate and spiro-sulfamate glycosides were obtained by means of intramolecular rhodium-catalyzed amination of pseudo-anomeric C–H bond in 2-deoxy-*C*-glycosides [45].

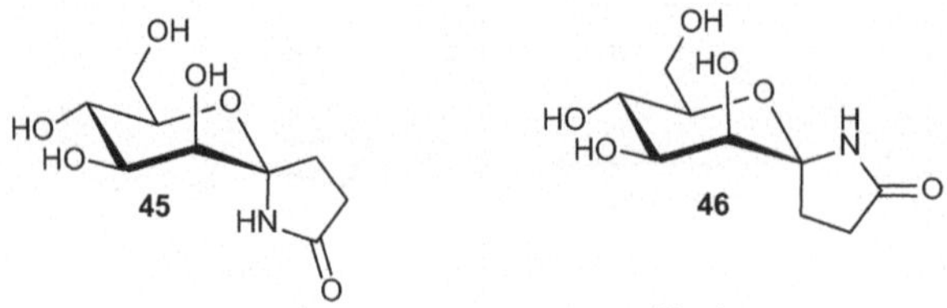

Enzyme	IC$_{50}$ (μM) 43	44	45	46
α-glucosidase (yeast)	NI	NI	NI	NI
β-glucosidase (almonds)	NI	200	NI	NI
α-galactosidase (coffee beans)	NI	77	NI	NI
β-galactosidase (bovine)	207	160	NI	400
α-mannosidase (Jack beans)	NI	NI	NI	NI

Fig. 3 Inhibition profile of spirolactams **43–46**

Spiroaminals **45** and **46** were obtained following a similar strategy [42]. The inhibitory activities of compounds **43–46** were evaluated on several commercially available glycosidases (Fig. 3). These compounds showed no inhibition against yeast α-glucosidase and Jack beans α-mannosidase [42]. Among the spirolactams tested, compound **43** with a glucose core gave the most interesting result as a highly selective inhibitor of bovine β-galactosidase.

Several authors prepared original spiroaziridines as valuable intermediates for the synthesis of carbohydrate derivatives of interest [24, 46–49]. In the course of their study on carbohydrate-based spiro 1,3-oxazolidine-2-thiones, Tatibouët et al. reported the synthesis of spiroaziridines **51** and **53** (Scheme 9) [47]. The strategy was based on the stereoselective formation of cyanohydrin derivatives **49** and **50**. The stereochemistry of the new stereogenic center could be controlled under kinetic or thermodynamic conditions from ketone **48**, obtained after oxidation of alcohol **47** [47–50]. Reduction of nitriles **49** and **50** afforded, after spontaneous intramolecular cyclization, aziridines **51** and **53**, respectively. However, amino alcohols **52** and **54** resulting from competing tosyl migration were also obtained in yields up to 33%.

3.2 *Systems with a Nitrogen Atom at β-Position to the Spirocyclic Center*

Several authors have reported the formation of carbohydrates spiro-linked with an azetidine [51–55]. The first examples were described by the groups of Fuentes and Mandal in 2006 [51, 52]. Fuentes and co-workers reported the formation of sulfoazetidine spiro-*C*-glycosides from ketose acetals [51]. For example, treatment of spiroacetal **55** obtained in four steps from D-fructose [56] with trimethylsilyl cyanide followed by reduction of the nitrile and dimesylation of the corresponding

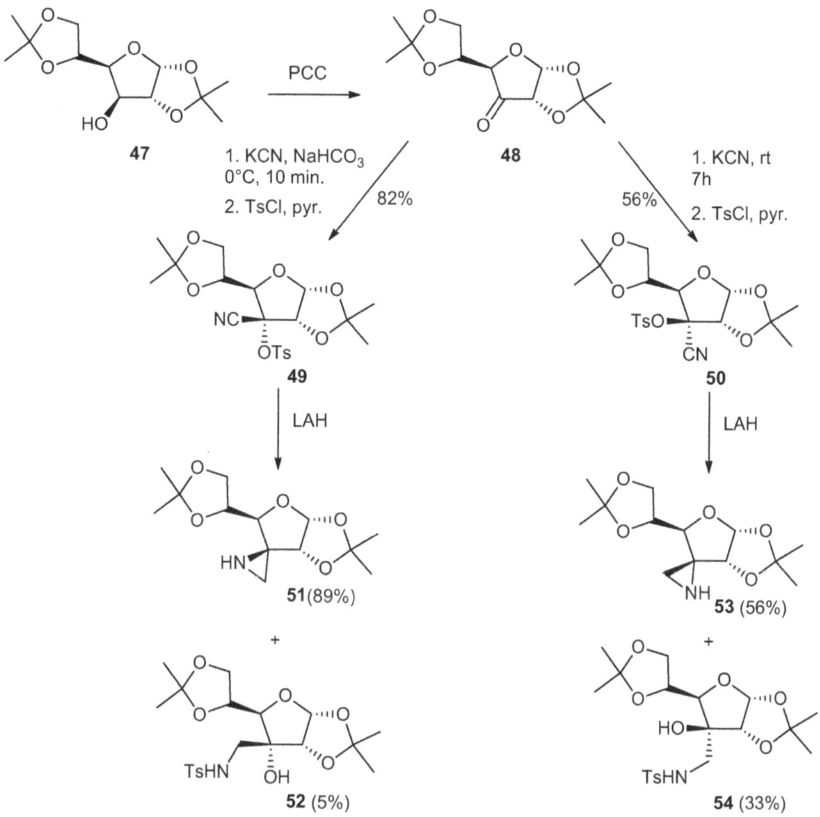

Scheme 9 Synthesis of spiroaziridines **51** and **53**

amino alcohol intermediate afforded compound **56** in 53% yield (three steps) [51, 57]. Base-mediated cyclization of **56** and deprotection of the resulting azetidine gave spiro iminosugar **58** in good yield (Scheme 10).

During their studies on the synthesis of spironucleosides [52], Mandal and co-workers reported the formation of nitrogen-containing spirosugars **61–63** through an aldol/Cannizzaro sequence (Scheme 11). Compound **59**, easily obtained from **47**, was treated by formaldehyde and sodium hydroxide to afford the corresponding diol which was then subjected to mesylation reaction with MsCl and Et_3N [52]. The resulting dimesylate **60** was reacted with benzylamine providing spiroazetidine **61**. After cleavage of the acetonide group and acetylation, a nucleobase could be introduced by treatment of compound **62** with 2,4-bis-(trimethylsilyloxy)pyrimidine. Deacetylation and debenzylation steps furnished spironucleoside **63**. Interestingly, during the debenzylation step using transfer hydrogenolysis with cyclohexene, introduction of an ethyl group on the amine was observed. The authors explained the formation of *N*-alkyl derivative **63** by the Pd-catalyzed oxidation of ethanol to acetaldehyde. Free amine may then condense with acetaldehyde to form an iminium

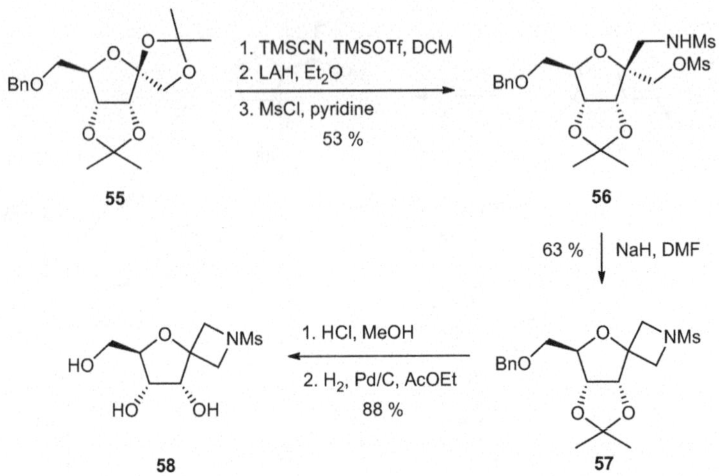

Scheme 10 Synthesis of spiroazetidine **58**

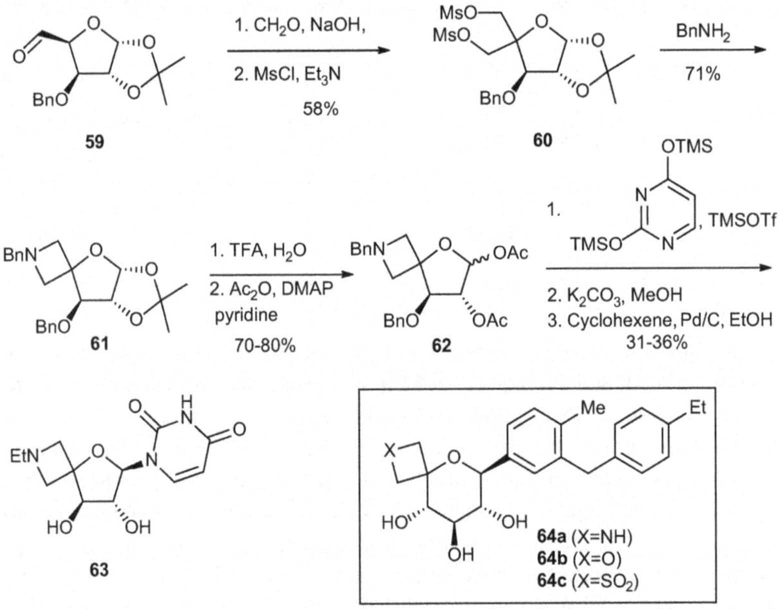

Scheme 11 Synthesis of spironucleosides **63** and **64** via Cannizzaro and cyclisation reactions

which could be reduced in situ [55]. Spirocyclic *C*-glycosides **64** have been obtained by a related synthetic strategy using an aldol/Cannizzaro sequence (Scheme 11) [53]. These compounds have been evaluated as SGLT2 (sodium glucose transporter of type 2) inhibitors. SGLT2 inhibitors are expected to display interesting therapeutic applications, in particular as antidiabetic agents [53]. Evaluation of compounds **64**

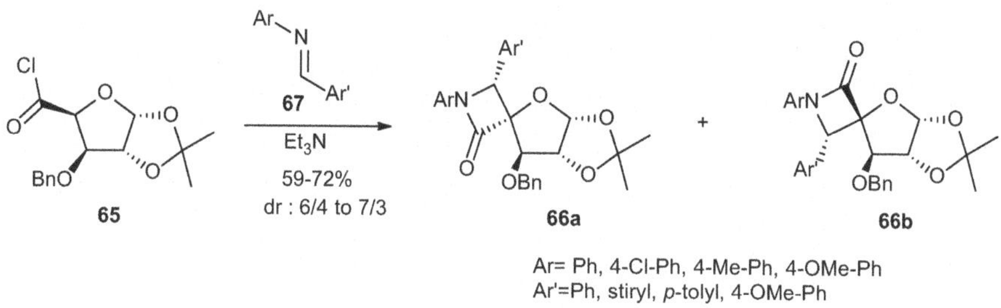

Scheme 12 Synthesis of lactams **66** via Staudinger cycloaddition

indicated that **64b** and **64c** were potent and selective SGLT2 inhibitors. Azaspironucleoside **64a** with IC_{50} values in the micromolar range was found to be up to 1,500-fold less potent as SGLT2 inhibitors than **64b** and **64c**.

Related carbohydrates spiro-linked with an azetidine may be also obtained by means of Staudinger cycloaddition (Scheme 12) [54]. Acid chloride **65** was prepared from D-glucose in five steps and was used as a ketene precursor which can react with imines **67** to provide lactams **66a** and **66b** (dr 6/4 to 7/3 in favor of **66a**). The [2+2] cycloaddition reaction proceeded with a good level of stereoselectivity; only two diastereomers were obtained among the four theoretically possible. The stereoselectivity observed could be rationalized by the torquoelectronic model.

Zhang and Schweizer reported the synthesis of carbohydrate-based azaspiranes with a piperidine or pyrrolidine moiety from 2,3,4,6-tetra-*O*-benzyl-D-gluconolactone (**68**) [58, 59]. The first key step of the synthesis involved the highly diastereoselective TMSOTf-mediated *C*-glycosylation of the exocyclic glucose-based epoxide **69** with allyltributylstannane (Scheme 13). The α-selectivity observed may be rationalized by preferential nucleophilic attack along axial trajectory on the most favored half-chair of the glycosyl cation intermediate **70** in which all the substituents are *pseudo*-equatorial. The resulting *C*-ketoside **71** was obtained in 80% yield. Conversion to the corresponding α-amino ester **72** was performed by way of reductive amination. The iodocyclization reaction in the presence of molecular iodine turned out to be not completely regioselective; iodo-compounds **73** and **74** were obtained in quantitative yield as an inseparable mixture of spirocyclic products of various ring sizes. Attack of the amino group at the most substituted site of the cyclic iodonium intermediate was found to be favored and occurred without stereocontrol. Treatment of **73** and **74** with silver acetate followed by saponification of the resulting acetates afforded alcohols **75–76** that could be separated by flash chromatography. Their structures and stereochemistry were established in part by NMR analysis and NOE experiments. *Pseudo*-iminosugars **77** were finally obtained after deprotection of spirocyclic pyrrolidines **75** in the presence of Pearlman's catalyst. The authors reported also the synthesis of analogues of **77** in which the primary alcohol of the D-glucose moiety has been replaced by an amine [59].

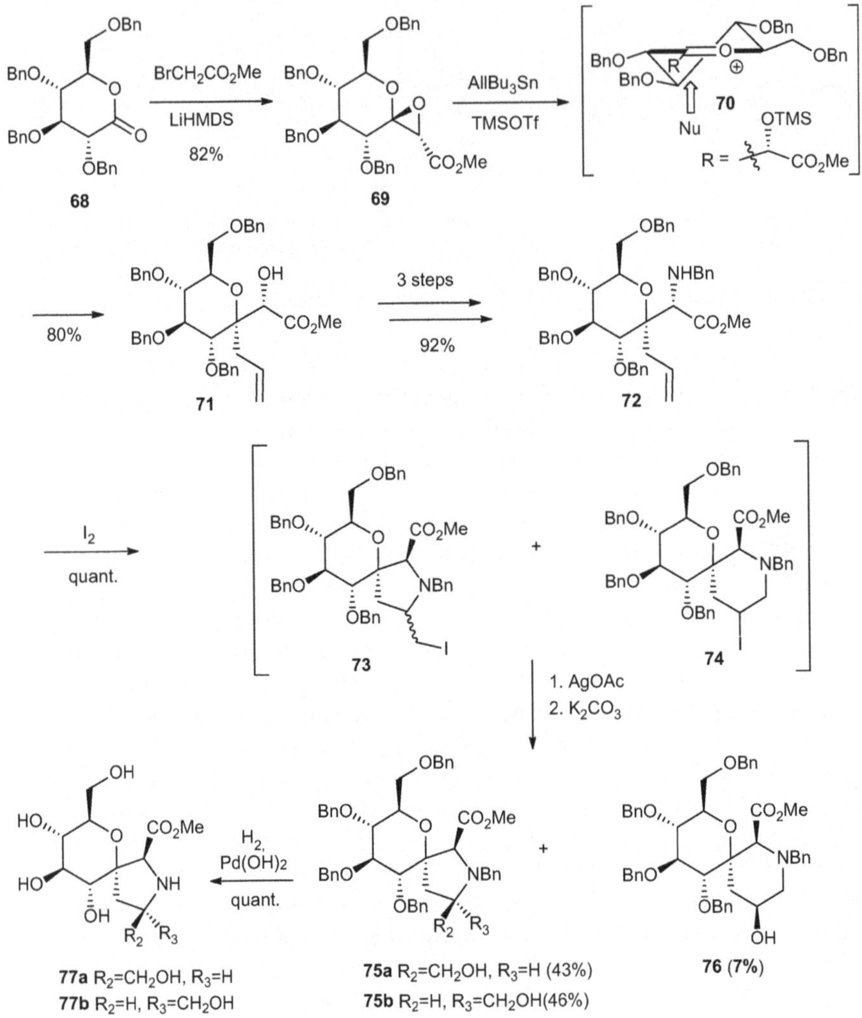

Scheme 13 Synthesis of spiro iminosugars via iodocyclization

Ring-closing metathesis (RCM) is one of the most powerful methods to access amine-containing heterocycles in particular for the formation of medium or large rings (for selected reviews, see [60–64]). The group of Vankar has used this methodology for the formation of oxa-aza spiro sugars containing a piperidine or azepane ring (Scheme 14) [65]. Intermediates **78** were prepared from gluconolactone **68** by addition of vinyl- or allylmagnesium bromide. *C*-glycosylation using trimethylsilyl cyanide in the presence of TMSOTf afforded nitriles **79** in good yields and high diastereoselectivity. Key RCM precursors **80** were then synthesized following a three-step sequence: reduction of nitrile group, protection of the resulting amine, and *N*-allylation. Ring-closing metathesis of dienes **80** gave the

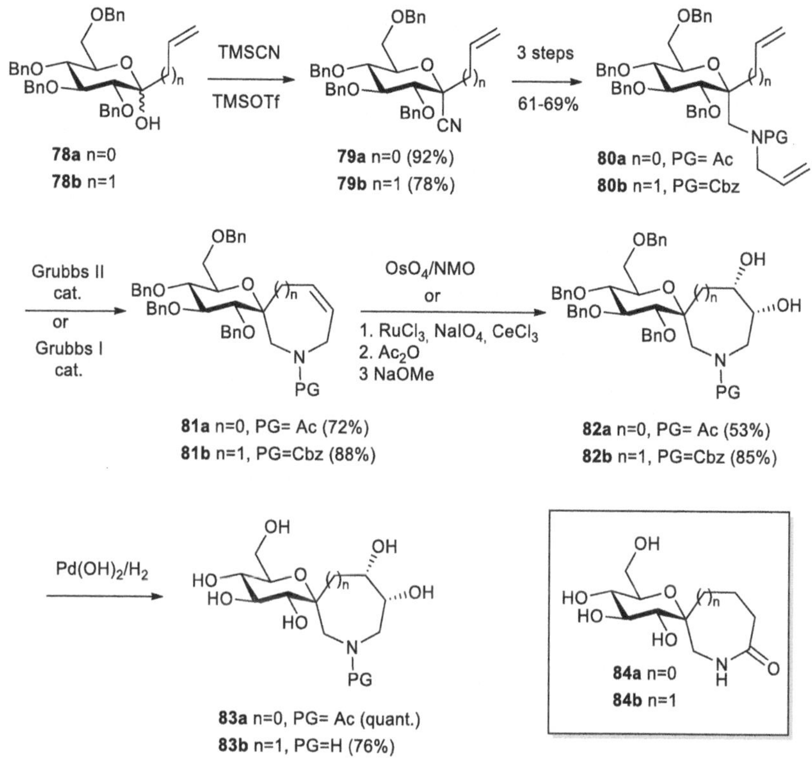

Scheme 14 Synthesis of spiro iminosugars via ring-closing metathesis

corresponding azacycles **81** in good yields. It is noteworthy that compound **81a** was not formed in the presence of Grubbs I catalyst. Substantial substrate reactivity differences were also observed for the dihydroxylation step. Whereas treatment of **81b** with osmium tetroxide in the presence of *N*-methylmorpholine-*N*-oxide (NMO) provided diol **82b** in good yield and as a single diastereomer, no conversion was observed when the same conditions were applied to piperidine **81a**. To overcome this lack of reactivity, bis-hydroxylation was performed with $RuCl_3$ in the presence of $NaIO_4$ and $CeCl_3$ to give the expected diol which was directly acetylated and then deprotected to afford **82a** in 53% yield for the three steps. Spiro iminosugars **83** were obtained after debenzylation. A related synthetic strategy was used for the formation of lactams **84**. Evaluation of spiro compounds **83–84** as inhibitors of a panel of five glycosidases indicated that **83** were weak but selective inhibitors of Jack beans α-mannosidase. Furthermore lactams **84** showed no inhibition or inhibition in the mM range.

4 Iminosugars Spiro-Linked with Carba-, Oxa-, and Azacycles

In addition to the structures presented before, in which an azacycle is attached to a polyhydroxylated scaffold, spiro compounds featuring a nitrogen-containing heterocyclic polyol have also attracted great attention. Aside from compounds **24** (Scheme 4) and **83** (Scheme 14) encompassing multiple OH on both the N and X (X=C,O) cycles, a number of additional spiro-iminoalditols have been built in an attempt to affect the peculiar biological properties of their monocyclic templates. Simple aminocyclitols are usually tailored to adapt the catalytic site of a given glycoenzyme, mimicking the corresponding glycoside to be processed [8]. Spiro-linked iminoalditols might be structurally related to fused bicyclic iminosugars (Fig. 1); however, the peculiar orthogonal orientation of the extra spirocycle offers scope for unprecedented interactions in the active site. Furthermore, synthetic issues to access the two kinds of adjoined cycles are rather distinct. Contrarily to their fused analogues [66–69], the construction of spirocyclic iminoalditols has been scarcely reviewed [70]. Main strategies are described below.

4.1 Iminosugar-Carbacycle Spiro Systems

L-Fucose and L-rhamnose are 6-deoxyhexoses commonly found in complex carbohydrates from all organisms, the trimming of which is performed by fucosidases and rhamnosidases. Simple iminosugars featuring both the methyl substituent and the adequate hydroxyl distribution are generally potent inhibitors of these enzymes. Analogues with a spiro carbacycle in place of the C-5 methyl group have been designed in order to evaluate the tolerance of the corresponding binding pocket to steric strain and to improve either potency or hydrophobicity. To access spirocyclopropyl iminosugars in L-fuco or L-rhamno series, the group of Behr applied a Kulinkovich-Szymoniak-Bertus cyclopropanation of designed glycononitriles. This reaction of very wide application [71–75] is based on the generation of a titanacyclopropane intermediate, a bis-anionic reactive species able to add twice to the –CN electrophilic partner to afford a spirocyclopropyl primary amine (Scheme 15). Hence, D-lyxose- or L-arabinose-derived nitriles **85** or **90** gave linear cyclopropylamines **86** (80%) and **91** (49%), the latter being formed after in situ Bz transfer [76, 77]. A game of protection/deprotection combined with activation of the primary hydroxyl to induce intramolecular nucleophilic displacement furnished the expected spirocyclopropyl piperidines **89** and **94**, analogues of L-rhamnose and L-fucose, respectively. Unsaturated pyrroline **95** was obtained from intermediate **87** after oxidative cleavage of the free vicinal diol and isopropylidene deprotection. The same synthetic sequence was applied for the preparation of **96**, the

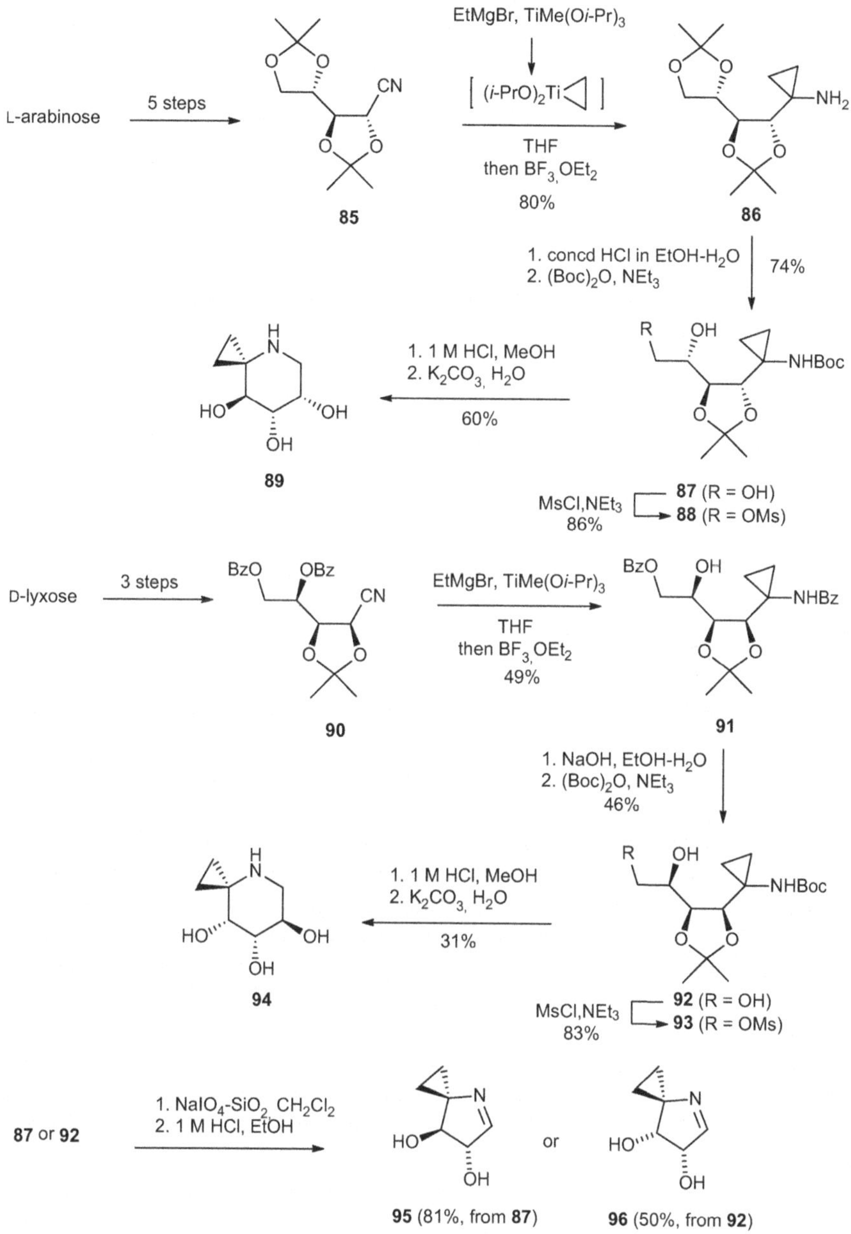

Scheme 15 Synthesis of spiro iminosugars from glycononitriles

C-3 epimer of **95**, from **92**. Spirocyclopropyl iminosugars showed some activity against fucosidase and rhamnosidase, best results being obtained with **94** ($K_i = 18$ μM against α-fucosidase from bovine kidney).

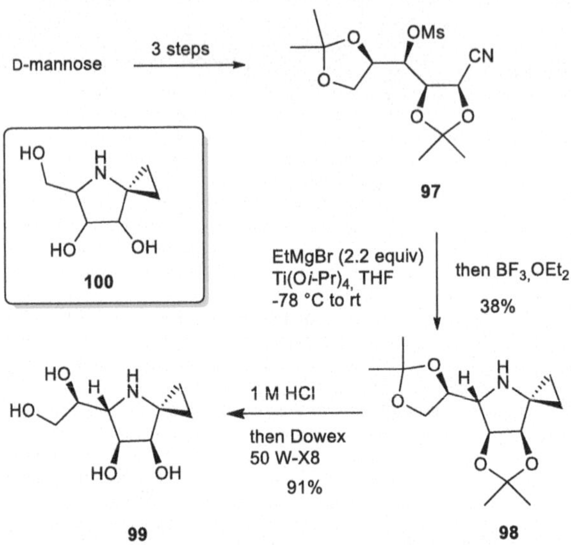

Scheme 16 Synthesis of spiro iminosugar **99**

To access spirocyclopropyl pyrrolidines, the same group adapted their strategy to 4-methanesulfonyl-glycononitriles such as **97** (Scheme 16), the cyclopropanation of which was accompanied by concomitant cyclization [78, 79]. By this very straightforward method, spirocyclopropyl iminocyclitol **99** was obtained in only five steps starting from D-mannose. Further stereoisomeric pyrrolidines of general structure **100** were prepared following the same strategy through variation of the starting sugar. Among these, only **99** showed inhibition potency in the micromolar range ($K_i = 1.6$ μM against α-fucosidase from bovine kidney). In general, biological assays with such compounds suggest that the replacement of a methyl substituent by a spirocyclopropyl group reduces the inhibitory potency toward the corresponding enzymes.

A spirocyclohexyl analogue of **99** was synthesized more recently in the racemic series from a hydroxyalkyldihydropyrrole precursor [80]. Elaboration of these latter substances involved the coupling of propargylamines with α-chloroaldehydes, followed by alkyne reduction and one-pot epoxide formation/ring-opening sequence (Scheme 17). Thus, the alkynyl chlorohydrin **102** was prepared first following addition of Boc-protected propargylamine **101** to 2-chlorocinnamaldehyde. Then, Lindlar reduction followed by acidic treatment of the crude product afforded the expected dihydropyrrole **103** in good overall yield (78%) after neutralization. It was then converted into spirocyclohexyl iminosugar **104** via reaction with phosgene and subsequent dihydroxylation. The relative stereochemistry of final compound **104** was assessed by 1D NOESY spectra.

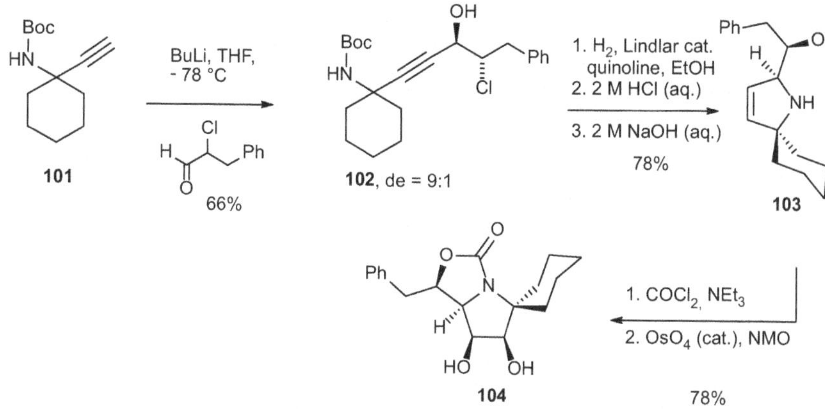

Scheme 17 Synthesis of spirocyclohexyl iminosugar **104** via dihydropyrrole intermediate

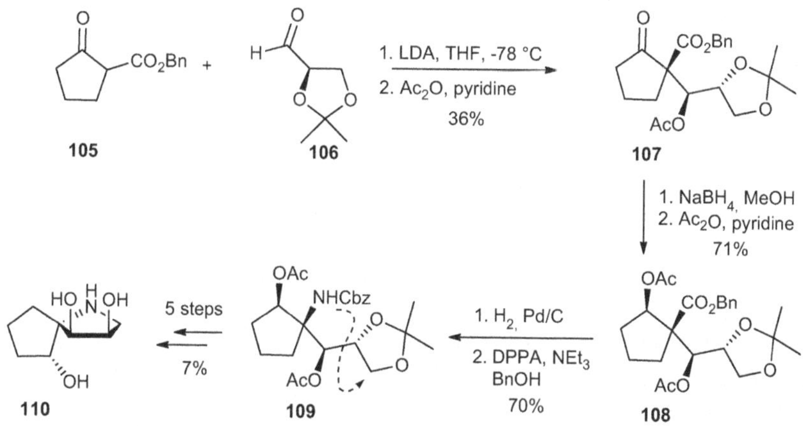

Scheme 18 Synthesis of swainsonine analogue **110** via Curtius rearrangement

In such azaspiroheterocycles, partial hydroxylation of the hydrocarbon portion could afford new binding contributions to improve inhibition potencies against a given enzyme. Therefore, Pinto and Chen prepared swainsonine analogue **110** (Scheme 18), which was expected to interact strongly with the hydrophobic pocket of Tyr727, Phe206, and Trp415 in *Drosophila* Golgi α-mannosidase II [81]. The key quaternary center was formed stereoselectively by aldol condensation of ketone **105** with (*R*)-isopropylidene glyceraldehyde **106** and subsequent acylation. From the four possible stereoisomers (ratio 30:6:5:1), the desired compound **107** was isolated by silica gel chromatography and reduced/acylated to obtain **108**. A Curtius rearrangement allowed the introduction of the amine-derived functional group at this stage, affording carbamate **109**. Additional standard procedures comprising ring closure by intramolecular nucleophilic displacement delivered target compound **110**.

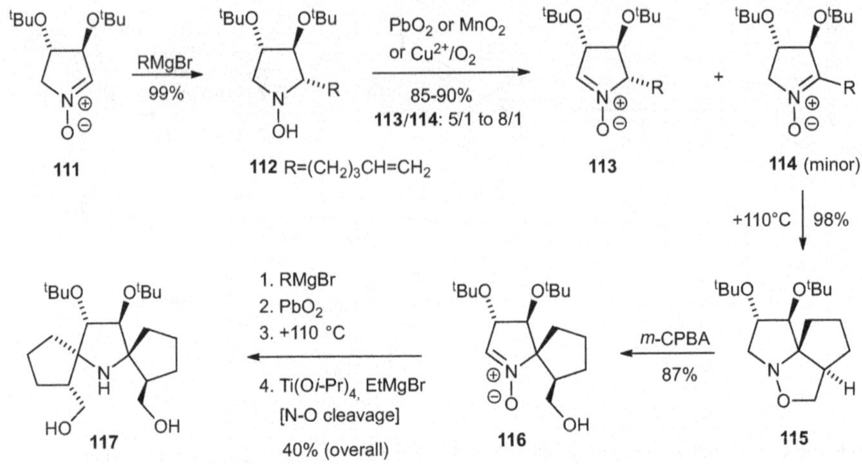

Scheme 19 Synthesis of bis-spiranic pyrrolidine **117** via iterative nitrone chemistry

Unfortunately, it did not show effective inhibition of human maltase glucoamylase or Golgi mannosidase II.

An analogue of **110**, the bis-spiranic C_2-symmetric pyrrolidine **117**, was successfully synthesized by Morozov et al. via iterative nitrone chemistry [82]. As depicted in Scheme 19, quaternization of the two α-carbon atoms resulted from successive completely regio- and stereoselective organometallic addition and intramolecular cycloaddition reactions on nitrones **111** and **116**. Transient oxidation of hydroxylamine **112** proved however moderately regioselective, affording the two isomeric nitrone intermediates **113** and **114**. Following the iterative synthetic strategy, *tert*-butoxy-protected dispiro-pyrrolidine **117** was eventually obtained in reasonable overall yield.

Jarosz and co-workers prepared a series of highly hydroxylated spiro iminosugars, i.e., azaspiro[4.5]decanes **121** and **123**, using a double addition of allyl Grignard to polyhydroxylated ω-bromonitrile **118** (Scheme 20) [83]. While monoallylated product was isolated in good yields when 1.3 equivalents of allylMgBr were used, the reaction with 5 equivalents of this same reagent afforded the bis-allylated piperidine **119** in 70% yield. The presence of DMPU or HMPA clearly favored the addition of a second allyl nucleophile to the putative imine intermediate [84]. Protection of the free NH proved somewhat tricky due to the presence of bulky substituents in the proximity of the nitrogen atom. However **119** was efficiently protected as a trifluoroacetate, which was subsequently subjected to RCM (Grubbs II catalyst) to afford azaspiro[4.5]decene **120**. Upcoming osmylation of **120** with osmium tetroxide and *N*-methylmorpholine-*N*-oxide (NMO) was completely stereoselective, and the configuration of the formed diol **121** was firmly assigned by X-ray analysis of a hexaacetate derivative. RCM was also applied to amine **119** by using transient in situ protection with a Brønsted acid. Interestingly, osmylation occurred from the opposite

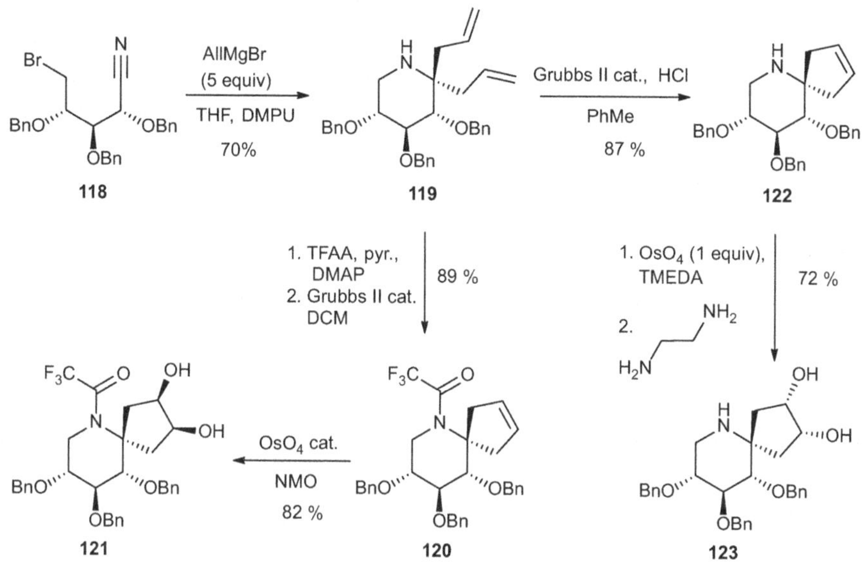

Scheme 20 Synthesis of azaspiro[4.5]decanes **121** and **123** via RCM

face with spirodecene **122** than it did in the case of olefin **120**, the free amine certainly acting as a directing group in such transformations.

Finally, spirocycles were also installed at the nitrogen atom of iminosugars to generate a permanent positive charge (as a quaternary ammonium salt) in order to favor ionic interactions at the catalytic site of glycosidases [85]. To this end, a double nucleophilic substitution reaction on tetrabenzylated DNJ **124** with either 1,4-dibromobutane or 1,5-dibromopentane followed by hydrogenolysis afforded *N*-spirofused iminosugars **126a,b** (Scheme 21). An analogous strategy gave 1-deoxynojirimycin derivatives **127** and **128** with isoindoline and morpholino extra rings, respectively. The glycosidase inhibition activities of the bromide salts of **126–128** were evaluated at 500 μM concentration on various commercially available glycosidases from different natural origins (α-mannosidase from Jack bean, α-glucosidase from *S. cerevisiae*, amyloglucosidase from *A. niger*). Best results were obtained with **126b** for which 30% residual activity was observed toward amyloglucosidase. Ammonium salts are also frequently used as antibacterial agents, disrupting cell membranes through ionic interactions. Therefore, along with their antiglycosidase potential, antibacterial effect of *N*-spirofused iminosugars was evaluated on the Gram-positive *S. aureus* and the Gram-negative *E. coli*. However, no activity against the two selected microorganisms could be detected up to 5 μM.

Compound **128** prepared by the same method differs from its congeners by the presence of an oxygen atom in the additional spirocycle. Other examples of such iminosugars with an adjoining oxacycle are given in the next section below.

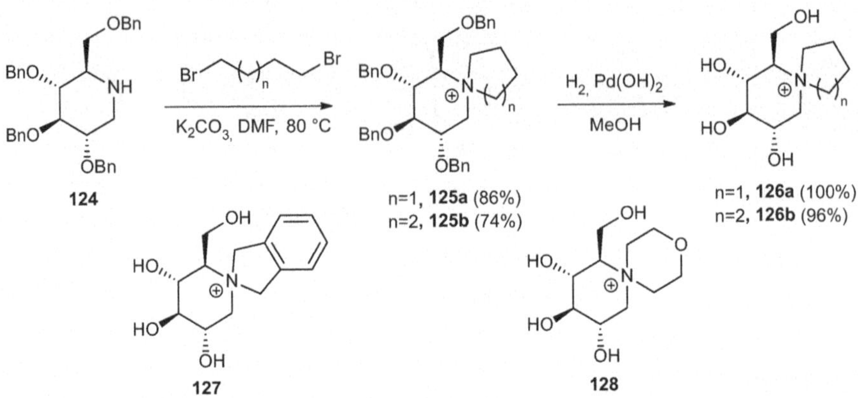

Scheme 21 Synthesis of *N*-spirofused derivatives of DNJ

4.2 Iminosugar-Oxacycle Spiro Systems

In medicinal chemistry, hybrid drugs in which two pharmacophores are present in one molecule have attracted a great deal of interest. The resulting molecule might show dual therapeutic mechanism with possible synergism in bioactivity. In this direction, some iminosugars tethered with additional pharmacophores have been prepared to improve or localize the delivery of an anticancer-active iminosugar [86–90]. Dhavale and co-workers designed a series of azepines and piperidines spiro-linked with morpholine-fused 1,2,3-triazole, which combine the glycosidase inhibitory potency of the iminosugar platform with the antifungal potential of the 1,2,3-triazole pharmacophore [91]. Quaternization of the C-3 carbon of D-glucose affords unstable 1,2;5,6-di-*O*-isopropylidene-3-*O*-propargyl-3-azido-D-glucofuranose **129**, prone to intramolecular azide-alkyne cycloaddition (AAC) at ambient temperature [92]. The triazole **130** thus obtained is a pivotal intermediate in the synthesis of the targeted compounds (Scheme 22). Access to the morpholino-piperidines required the installation of an azido function at C-5 of the D-gluco moiety yielding **131**, or **134** when azidation was coupled with periodate degradation. Amination of the anomeric carbon under reductive conditions afforded the expected spirofused piperidines **132a** and **135a**, which were *N*-alkylated further to generate the library of compounds **132b-g** and **135b-g**. Their glycosidase inhibitory and antifungal activities were evaluated. Compounds **132a** ($IC_{50} = 0.075$ μM) and **135a** ($IC_{50} = 0.036$ μM) showed potent α-glucosidase (rice) activity when compared to the standard miglitol ($IC_{50} = 0.100$ μM) and proved highly active against *Candida albicans* yeast cells with minimum inhibition concentration (MIC) = 0.85 μM and MIC = 0.025 μM, respectively.

Alternatively, other spiro iminosugar-oxacycle molecules were obtained, rather as reaction intermediates. Thus, during their study on metal-catalyzed reactions of α-diazo-ß-hydroxyamino esters such as **136**, Py and co-workers observed the unexpected formation of the spirotetrahydrofuran **137** in 36% yield in the presence of a

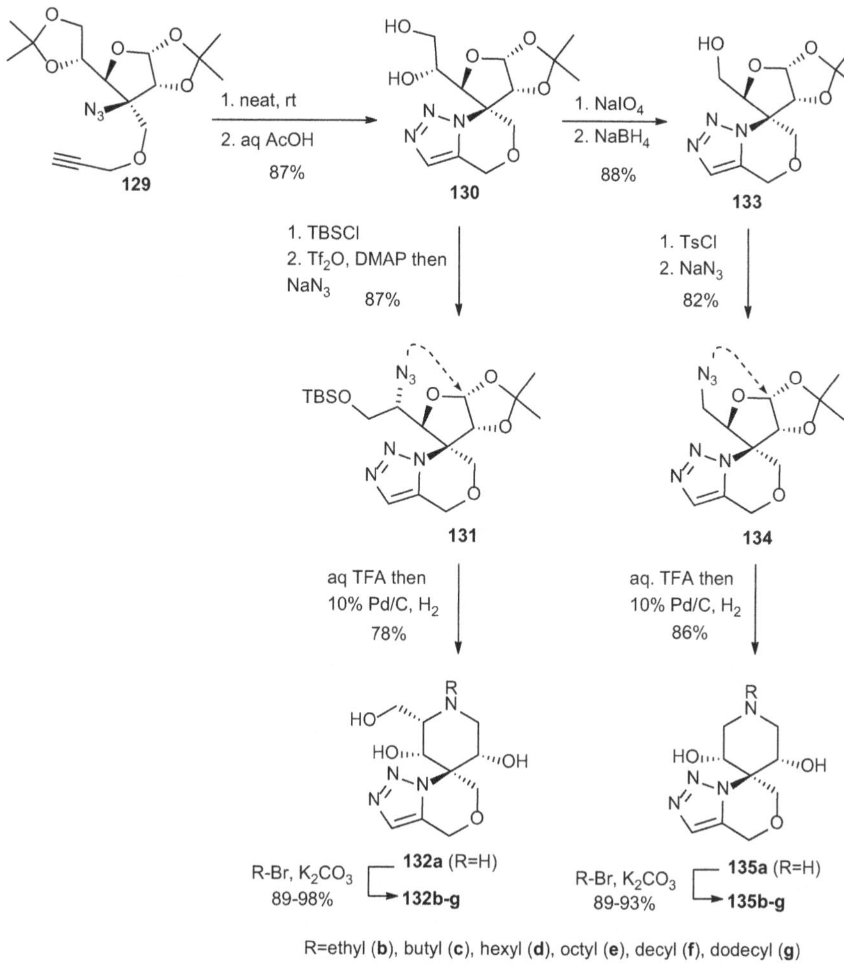

Scheme 22 Synthesis of morpholino-piperidines via intramolecular AAC

copper catalyst (Scheme 23) [93]. The compound arose from C–H insertion of the transient carbenoid into a benzylic C–H bond of the nearby OBn substituent.

Regioselective ring opening of functionalized aziridines has been exploited by Eum et al. for the synthesis of pyrrolizidines and indolizidines [94]. During the synthesis of a nonnatural analogue of castanospermine, the spiro-hemiaminal intermediate **140** could be isolated and characterized (Scheme 23). The treatment of aziridine **138** in strongly acidic media afforded aziridinium **139** through intramolecular hemiaminal/hemiacetal formation. A subsequent nucleophilic opening reaction with water from the less hindered *si*-face at the more substituted carbon gave spiro-piperidine **140** which was further transformed into the targeted indolizidine.

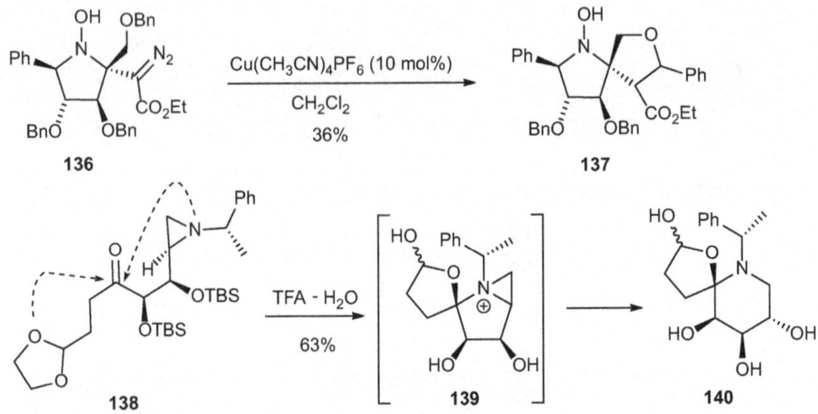

Scheme 23 Synthesis of spiro iminosugar-oxacycles **137** and **140**

4.3 Iminosugar-Azacycle Spiro Systems

The synthesis of a new class of diazaspiro-iminosugars has been reported by Dhavale and co-workers, the structures of which comprise a nitrogen atom in both rings [95]. Diazaspiro skeleton is an important structural motif in biologically active molecules as exemplified by spirocyclic pyrrolidone **141**, a HIV-1 protease inhibitor [96], or the NK1 receptor antagonist **142** [97]. Glucose-derived azido compound **143** with a geminal α,ß-unsaturated ester was used as building block to access the expected spirolactams **145–148** (Scheme 24). Intramolecular lactamization occurred spontaneously after azide reduction, to furnish the spiro-γ-lactam **144** after selective 5,6-O-acetonide hydrolysis. This key intermediate was then converted into the targeted iminosugars following strategies analogue to those described in Scheme 22. Spiro-*bis*lactams **146** and **148** were obtained by additional Schmidt-Boyer-Aube rearrangement that took place between azide and carbonyl functions in an intramolecular fashion. Glycosidase inhibitory activity of diazaspiro-iminosugars **145–148** was studied against α-mannosidase (Jack bean), α-galactosidase (green coffee bean), and α-glucosidase (yeast). Strong inhibition of the galactosidase occurred with **147** ($IC_{50} = 0.029$ μM), whereas **148** was a potent inhibitor of Jack bean α-mannosidase ($IC_{50} = 0.080$ μM).

5 Conclusion and Outlook

Nitrogen-containing analogues of carbohydrates are not only important in medicinal chemistry as therapeutic agents and as biological tools to study carbohydrate-processing enzymes, but they represent also fascinating chemical targets for synthetic organic chemists. A number of novel methodologies have found application in

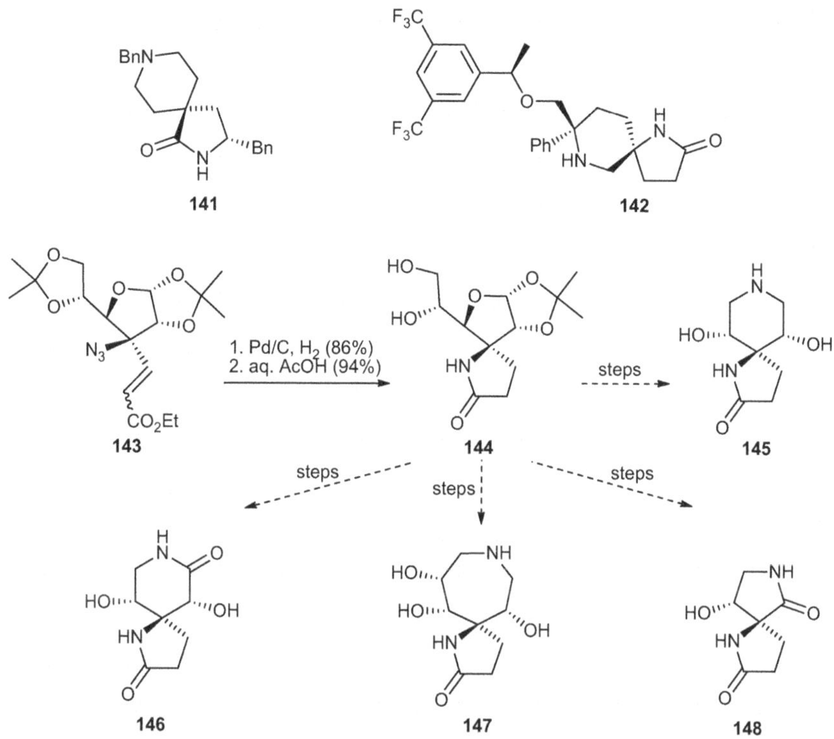

Scheme 24 Synthesis of diazaspiro-iminosugars **145–148**

the synthesis of unprecedented iminosugar-like structures. As a consequence, a growing number of nonnatural amino polyols have emerged in the literature, presenting a wide spectrum of chemical profiling. Among them, spiro iminosugars have the unusual particularity to merge substructures by means of a quaternary apical carbon in almost perpendicular planes. The design and synthesis of azaspirocycles have been largely guided by this distinctive structural feature enabling conformational diversity and unprecedented interactions with biological receptors. As exemplified in this chapter, the main issue to be addressed during the synthesis of spiro iminosugars is the installation of the pivotal quaternary spiro atom. It might rarely originate from the starting material, but it is mostly elaborated throughout the synthesis prior to or concomitantly with cyclization. On a biological point of view, in some cases the extra ring hinders recognition for a given target due to deleterious steric interactions. But, encouragingly, some azaspirocycles showed more potent or at least distinct biological activities than their non-spirocyclic homologues. This feature should stimulate research and innovation for the development of further structures of this kind to complement the repertoire of iminosugar analogues.

References

1. Sinibaldi ME, Canet I (2008) Eur J Org Chem 2008:4391–4399
2. Dake G (2006) Tetrahedron 62:3467–3492
3. Galliford CV, Scheidt KA (2007) Angew Chem Int Ed 46:8748–8758
4. Singh GS, Desta ZY (2012) Chem Rev 112:6104–6155
5. Fujita S, Nishikawa K, Iwata T, Tomiyama T, Ikenaga H, Mastumoto K, Shindo M (2018) Chem Eur J 24:1539–11543
6. Altman RA, Nilsson BL, Overman LE, Read de Alaniz J, Rohde JM, Taupin V (2010) J Org Chem 75:7519–7534
7. Burkhard JA, Wagner B, Fischer H, Schuler F, Müller K, Carreira EM (2010) Angew Chem Int Ed 49:3524–3527
8. Compain P, Martin OR (2007) Iminosugars from synthesis to therapeutic applications. Wiley, Weinheim
9. Winchester BG (2009) Tetrahedron Asymm 20:645–651
10. Nash RJ, Kato A, Yu CY, Fleet GWJ (2011) Future Med Chem 3:1513–1521
11. Horne G, Wilson FX (2011) Prog Med Chem 50:135–176
12. Orsato A, Barbagallo E, Costa B, Olivieri S, De Gioia L, Nicotra F, La Ferla B (2011) Eur J Org Chem 2011:5012–5019
13. Decroocq C, Stauffert F, Pamlard O, Oulaïdi F, Gallienne E, Martin OR, Guillou C, Compain P (2015) Bioorg Med Chem Lett 25:830–833
14. Santos C, Stauffert F, Ballereau S, Dehoux C, Rodriguez F, Bodlenner A, Compain P, Génisson Y (2017) Bioorg Med Chem 25:1984–1989
15. Markham A (2016) Drugs 11:1147–1152
16. Compain P, Decroocq C, Iehl J, Holler M, Hazelard D, Mena Barragán T, Ortiz Mellet C, Nierengarten JF (2010) Angew Chem Int Ed 49:5753–5756
17. Compain P, Bodlenner A (2014) ChemBioChem 15:1239–1251
18. Gouin S (2014) Chem Eur J 20:11616–11628
19. Lillelund VH, Jensen HH, Liang X, Bols M (2002) Chem Rev 102:515–553
20. Gloster TM, Davies GJ (2010) Org Biomol Chem 8:305–320
21. Zheng Y, Tice CM, Singh SB (2014) Bioorg Med Chem Lett 24:3673–3682
22. Maaliki C, Gauthier C, Massinon O, Sagar R, Vincent SP, Blériot Y (2014) In: Pilar Rauter A, Lindhorst T, Queneau Y (eds) Carbohydrate chemistry, vol 20. Royal Society of Chemistry, Cambridge, pp 418–444
23. Lahiri R, Ansari AA, Vankar YD (2013) Chem Soc Rev 42:5102–5118
24. Yanagisawa H, Kinoshita M, Umezawa S (1971) Bull Chem Soc Jap 44:3399–3405
25. Rios R (2012) Chem Soc Rev 41:1060–1074
26. Ding A, Meazza M, Guo H, Yang JW, Rios R (2018) Chem Soc Rev 47:5946–5996
27. Kapferer P, Birault V, Poisson JF, Vasella A (2003) Helv Chim Acta 86:2210–2227
28. Ahmadian M, Khare NK, Riordan JM, Klon AE, Borhani DW (2001) Tetrahedron 57:9899–9909
29. Brewster K, Harrison JM, Inch TD, Williams NJ (1987) J Chem Soc Perkin Trans 1:21–26
30. Nocquet PA, Hensienne R, Wencel-Delord J, Wimmer E, Hazelard D, Compain P (2015) Org Biomol Chem 13:9176–9180
31. Nocquet PA, Hensienne R, Wencel-Delord J, Laigre E, Sidelarbi K, Becq F, Norez C, Hazelard D, Compain P (2016) Org Biomol Chem 14:2780–2796
32. Nocquet PA, Hazelard D, Gruntz G, Compain P (2013) J Org Chem 78:6751–6757
33. Hazelard D, Compain P (2017) Org Biomol Chem 15:3806–3827
34. Lin W, Gupta A, Kim KH, Mendel D, Miller MJ (2009) Org Lett 11:449–452
35. Haruyama H, Takayama T, Kinoshita T, Kondo M, Nakajima M, Haneishi T (1991) J Chem Soc Perkin Trans 1:1637–1640
36. Nakajima M, Itoi K, Takamatsu Y, Kinoshita T, Okazaki T, Kawakubo K, Shindo M, Honma T, Tohjigamori M, Haneishi T (1991) J Antibiot 44:293–300

37. Soengas RG, da Silva G, Estévez JC (2017) Molecules 22:2028
38. Freire R, Martín A, Pérez-Martín I, Suárez E (2002) Tetrahedron Lett 43:5113–5116
39. Martín A, Pérez-Martín I, Suárez E (2009) Tetrahedron 65:6147–6155
40. Fransisco CG, Herrera AJ, Martín A, Pérez-Martín I, Suárez E (2007) Tetrahedron Lett 48:6384–6388
41. Martín A, Pérez-Martín I, Suárez E (2015) Org Lett 7:2027–2030
42. Pal APJ, Vankar YD (2010) Tetrahedron Lett 51:2519–2524
43. Aebisher B, Vasella A (1983) Helv Chem Acta 66:789–794
44. Somsák L (2001) Chem Rev 101:81–135
45. Toumieux S, Compain P, Martin OR (2005) Tetrahedron Lett 46:4731–4735
46. Alves RJ, Castillon S, Dessinges A, Herczegh P, Lopez JC, Lukacs G, Olesker A, Thang TT (1988) J Org Chem 53:4616–4618
47. Tardy S, Vicente JL, Tatibouët A, Dujardin G, Rollin P (2008) Synthesis 2008:3108–3120
48. Bourgeois JM (1976) Helv Chem Acta 59:2114–2124
49. Bourgeois JM (1974) Helv Chem Acta 57:2553–2557
50. Bourgeois JM (1975) Helv Chem Acta 58:363–372
51. Sayago FJ, Pradera MÁ, Gasch C, Fuentes J (2006) Tetrahedron 62:915–921
52. Roy A, Achari B, Mandal SB (2006) Tetrahedron Lett 47:3875–3879
53. Mascitti V, Robinson RP, Préville C, Thuma BA, Carr CL, Resse MR, Maguire RJ, Leininger MT, Lowe A, Steppan CM (2010) Tetrahedron Lett 51:1880–1883
54. Chincholkbar PM, Puranik VG, Deshmukh ARAS (2007) Tetrahedron 63:9179–9187
55. Maity JK, Ghosh R, Drew MGB, Achari B, Mandal SB (2008) J Org Chem 73:4305–4308
56. Mio S, Kumagawa Y, Sugai S (1991) Tetrahedron 47:2133–2144
57. Sano H, Mio S, Kitagawa J, Sugai S (1994) Tetrahedron Asymm 5:2233–2240
58. Zhang K, Schweizer F (2005) Synlett 20:3111–3115
59. Zhang K, Schweizer F (2009) Carbohydr Res 344:576–585
60. Hazelard D, Compain P (2017) Top Heterocycl Chem 47:111–154
61. Compain P (2007) Adv Synth Catal 349:1829–1846
62. Felpin FX, Lebreton J (2003) Eur J Org Chem 2003:3693–3712
63. Dragutan I, Dragutan V, Mitan C, Vosloo HCM, Delaude L, Demonceau A (2011) Belstein J Org Chem 2011:699–716
64. Deiters A, Martin SF (2004) Chem Rev 104:2199–2238
65. Pal APJ, Gupta P, Reddy YS, Vankar YD (2010) Eur J Org Chem 2010:6957–6966
66. Robertson J, Stevens K (2017) Nat Prod Rep 34:62–89
67. Davies SG, Fletcher AM, Roberts PM, Thomson JE (2017) Synlett 26:2697–2706
68. Pansare SV, Thorat RG (2013) Targets Heterocycl Syst 17:57–86
69. Kim IS, Jung YH (2011) Heterocycles 83:2489–2507
70. Rowicki T (2016) Targets Heterocycl Syst 20:409–447
71. Forcher G, Clousier N, Beauseigneur A, Setzer P, Boeda F, Pearson-Long MSM, Karoyan P, Szymoniak J, Bertus P (2015) Synthesis 47:992–1006
72. Szalata C, Szymoniak J, Fabis F, Butt-Gueule S, Rault S, Bertus P, Gerard S, Sapi J (2013) ChemMedChem 8:70–73
73. Wolan A, Six Y (2010) Tetrahedron 66:15–61
74. Bertus P, Szymoniak J, Jeanneau E, Docsa T, Gergely P, Praly JP, Vidal S (2008) Bioorg Med Chem Lett 18:4774–4778
75. Bertus P, Szymoniak J (2007) Synlett 2007:1346–1356
76. Pearson MSM, Plantier-Royon R, Szymoniak J, Bertus P, Behr JB (2007) Synthesis 2007:3589–3594
77. Pearson MSM, Floquet N, Bello C, Vogel P, Plantier-Royon R, Szymoniak J, Bertus P, Behr JB (2009) Bioorg Med Chem 17:8020–8026
78. Laroche C, Plantier-Royon R, Szymoniak J, Bertus P, Behr JB (2006) Synlett 2006:223–226
79. Laroche C, Behr JB, Szymoniak J, Bertus P, Schütz C, Vogel P, Plantier-Royon R (2006) Bioorg Med Chem 14:4047–4054

80. Dhand V, Draper JA, Moore J, Britton R (2013) Org Lett 15:1914–1917
81. Chen W, Pinto BM (2007) Carbohydr Res 342:2163–2172
82. Morozov DA, Kirilyuk IA, Komarov DA, Goti A, Bagryanskaya IY, Kuratieva NV, Grigor'ev IA (2012) J Org Chem 77:10688–10698
83. Malik M, Witkowski G, Ceborska M, Jarosz S (2013) Org Lett 15:6214–6217
84. Behr JB, Kalla A, Harakat D, Plantier-Royon R (2008) J Org Chem 73:3612–3615
85. D'Orazio G, Martorana AM, Filippi G, Polissi A, De Gioia L, La Ferla B (2016) ChemistrySelect 1:2444–2447
86. Hottin A, Scandolera A, Duca L, Wright DW, Davies GJ, Behr JB (2016) Bioorg Med Chem Lett 26:1546–1549
87. Hottin A, Wright DW, Dubar F, Steenackers A, Delannoy P, Biot C, Davies GJ, Behr JB (2013) Chem Eur J 19:9526–9533
88. Hottin A, Dubar F, Steenackers A, Delannoy P, Biot C, Behr JB (2012) Org Biomol Chem 10:5592–5597
89. Garcia-Moreno MI, Diaz-Perez P, Ortiz Mellet C, Garcia Fernandez JM (2002) Chem Commun 8:848–849
90. Zhou Y, Zhao Y, Boyle KMO, Murphy PV (2008) Bioorg Med Chem 18:954–958
91. Chavan SR, Gavale KS, Khan A, Joshi R, Kumbhar N, Chakravarty D, Dhavale DD (2017) ACS Omega 2:7203–7218
92. Chavan SR, Gavale KS, Kamble KM, Pingale SS, Dhavale DD (2017) Tetrahedron 73:365–372
93. Kui EL, Kanazawa A, Philouze C, Poisson JF, Py S (2017) Eur J Org Chem 2017:363–372
94. Eum H, Choi J, Cho CG, Ha HJ (2015) Asian J Org Chem 4:1399–1409
95. Parihar VS, Pawar NJ, Ghosh S, Chopade B, Kumbhar N, Dhavale DD (2015) RSC Adv 5:52907–52915
96. Kazmierski WM, Furfine E, Spaltenstein A, Wright LL (2002) Bioorg Med Chem Lett 12:3431–3433
97. Duffy RA, Morgan C, Naylor R, Higgins GA, Varty GB, Lachowicz JE, Parker EM (2012) Pharmacol Biochem Behav 102:95–100

Index

The manufacturer's authorised representative in the EU is Springer Nature Customer Service Centre GmbH, Europaplatz 3, 69115 Heidelberg, Germany. If you have any concerns regarding our products, please contact ProductSafety@springernature.com

Printed and bound by CPI Group (UK) Ltd, Croydon, CR0 4YY
25/09/2026
02222859-0018